WILD AND SOWN GRASSES

Profiles of a temperate species selection: ecology, biodiversity and use

Alain Peeters

Professor of Agronomy and Grassland Ecology
Laboratory of Grassland Ecology
University of Louvain
Louvain la Neuve
Belgium

with contributions from

Cécile Vanbellinghen

on grass diseases

John Frame

who provided detailed comments
on the manuscript during its preparation

Published by the
FOOD AND AGRICULTURE ORGANIZATION OF THE UNITED NATIONS
and
BLACKWELL PUBLISHING
Rome, 2004

A catalogue record for this book is available from the Library of Congress and the British Library

FAO ISBN 92-5-105159-3
Blackwell ISBN 1-4051-0529-1

Foreword

The Food and Agriculture Organization of the United Nations (FAO) has long been concerned with grassland species as part of its focus on pastoral development issues and sustainable production from grassland systems of the world. *Inter alia*, it has previously published a number of books on tropical grasses and tropical forage legumes.

In temperate areas, and in particular in Western Europe, grasslands occupy on average almost 40 percent of the agricultural area and almost 100 percent in some regions, thus playing an essential role in farming systems. Grasslands may be temporary or permanent and may be grazed, cut or managed by alternate cutting and grazing. In mountain areas, on moors and marshlands, and in places where crop cultivation is either not possible or land is kept under natural cover, grasslands dominate.

These grasslands provide the bulk of the forage eaten by ruminants during the grazing season and, in the winter period, hay and grass silage are usually a major part of the diet. Surprisingly, only a small number of grass species are regarded as significant in these grasslands. The trade of forage grass seeds is dominated by the ryegrasses, but timothy, cocksfoot, meadow fescue, tall fescue and some bromes and meadow grasses also play a part. These species are well known and various publications have described their characteristics and management in detail. However, there is much less information available on the many indigenous species that occur, particularly in permanent grasslands, such as bent, couch, foxtail, mat, oat, purple moor and sweet vernal grasses, although some undoubtedly have many desirable qualities and are well suited to specific ecological or management situations. In addition to forage production, the need to maintain biodiversity and the important role of various grasses in soil erosion control, carbon sequestration, watershed protection and in recreational areas are subjects that should receive increasing attention.

This book profiles 43 grass species, sown or naturally present in temperate grasslands. While comprehensive descriptions and information are provided as far as possible, many gaps in our knowledge are also revealed, especially for the lesser known species. In line with FAO's role of disseminating available information, it is hoped that this book will promote discussion and prompt further research.

The major contribution made by the author, Professor Alain Peeters, of the Laboratoire d'Ecologie des Prairies, Université catholique de Louvain, Belgium, is much appreciated by FAO. Thanks are particularly due to Caterina Batello and Stephen Reynolds of the Grassland and Pasture Crops Group for ensuring that the book was brought to publication.

Eric Kueneman
Chief
Crop and Grassland Service
Plant Production and Protection Division
Agriculture Department
FAO, Rome, Italy

Contents

PART I

CHAPTER 1

INTRODUCTION

CHAPTER 2

IMPORTANCE AND DIVERSITY OF GRASSES AND GRASS COMMUNITIES

CHAPTER 3

DIVERSITY OF GRASS USES

CHAPTER 4

MORPHOLOGY AND PHYSIOLOGY OF GRASSES

PART II

Acknowledgements

The author would like to thank Mrs C. Batello and Dr S. Reynolds of the Grassland and Pasture Crops Group (AGPC) of the Food and Agriculture Organization of the United Nations (FAO) for their support.

Dr John Frame, past President of the British Grassland Society and well known European grassland scientist, provided detailed comments on the manuscript during its preparation.

Mrs C. Van Bellinghen, Laboratory of Plant Diseases, University of Louvain (Belgium), contributed to section 2.10 in Chapter 2 and to the paragraphs on diseases in Chapter 5.

The author is grateful to Mrs A. Lannoye and Mrs B. Fettweiss for the grass drawings, to Mr S. Lambay for the computer management of pictures, to Mrs Bourguignon for the typing and to Mr F. Quertainmont for his project of the cover layout.

Mrs A.L. Jacquemart, Professor of Plant Ecology at the University of Louvain (Belgium) and Messrs J.M. Couvreur, researcher in the Laboratory of Grassland Ecology (UCL, Belgium), R. Delpech, Professor Emeritus of Plant Ecology at the National Agronomical Institute of Paris-Grignon (INA-PG, France), J. Lambert, Professor Emeritus of Forage Production at the University of Louvain (Belgium), J.F. Maljean, researcher in the Laboratory of Grassland Ecology (UCL, Belgium) and M. Vivier, former Director at the National Institute of Agronomical Research (INRA-Normandy, France) agreed to give their opinion on the text on species ecology.

Mr C. Decamps provided data on agronomical aspects.

The author is grateful to the following persons for the right to reproduce photographs: P. Allard; P. & W. Guinchart, BIOS; Ph. Levêque, BIOS; F. Quertainmont; R.E.N. Smith, IGER and C. Vanbellinghen, UCL.

The collaboration of Mrs Josie Severn and Mr Nigel Balmforth of Blackwell Science is particularly noted, as is the cooperation of the Publishing Management Service of FAO and of the editor, Ms Hazel Bedford.

Abbreviations

NUTRITIONAL TERMS

ADF
Acid detergent fibre

ADG
Average daily gain

CP
Crude protein

DDM
Digestibility of the DM

DM
Dry matter

DOM
Digestible organic matter

DOMD
Digestibility value = proportion of DOM in the DM = D-value =
% OM x % OMD

D-value
Digestibility value = proportion of DOM in the DM = DOMD =
% OM x % OMD

FM
Fresh matter

LW
Live weight

LWG
Live weight gain

NDF
Neutral detergent fibre

OM
Organic matter

OMD
Organic matter digestibility

COUNTRIES

B
Belgium

CAN
Canada

CH
Switzerland

CZ
Czech Republic, the

D
Germany

DK
Denmark

F
France

I
Italy

N
Norway

NL
Netherlands, the

NZ
New Zealand

PL
Poland

S
Sweden

UK
United Kingdom, the

USA
United States of America, the

DISEASES

BYDV
Barley Yellow Dwarf Virus

RMV
Ryegrass Mosaic Virus

PART I

Chapter 1

Introduction

The aim of this book is to increase awareness and knowledge about grasses from rangelands and grasslands[1] of temperate regions, and in particular about a selection of wild grasses, so that these species can be better known and, hopefully, more widely used. It is hoped that the book will stimulate research on little-known but widely distributed grasses. Many species have considerable production potential, and they sometimes also have characteristics complementary to those of the few species selected to date mainly for forage production. Many grass species can be used for purposes other than forage production. These other uses of grasses are also summarized.

The intensification of forage production and utilization, especially during the second half of the twentieth century, led to a drastic reduction in the number of grass species in sown mixtures, as well as in permanent swards. Only a small number of species, mainly the ryegrasses (*Lolium perenne* and *L. multiflorum*), timothy (*Phleum pratense*), cocksfoot (*Dactylis glomerata*) and two fescues (*Festuca arundinacea* and *F. pratensis*), are highly suited to intensive management. The two *Lolium* species became dominant in the European seed trade (Kley, 1995) and in heavily nitrogen-fertilized swards in many regions. The interest in other species by farmers and agronomists decreased dramatically. However, some of the other grass species, for example meadow grasses (*Poa* spp.), Yorkshire fog (*Holcus lanatus*), bents

(*Agrostis* spp.) and meadow foxtail (*Alopecurus pratensis*), remained extremely widespread in low-input grasslands and rangelands, and even in intensive permanent grasslands. Considering European permanent grasslands as a whole, many grass species are still much more abundant than the two *Lolium* species and their contribution to animal feeding is also more important. Moreover, from 1980 onwards, many forage systems were deintensified for several reasons and wild or little-selected species became more important in such systems.

Wild species are also a source of genes that could be used in plant breeding, whether by traditional breeding methods or by genetic manipulation. Genes for disease resistance, for drought or cold tolerance and for aluminium toxicity tolerance come to mind. Other characteristics, such as early spring growth, late heading or high digestibility could certainly be transferred between species. Conserving species biodiversity therefore permits the conservation of a genetic heritage that may be necessary for future ruminant production systems that are as yet unforeseen.

Grasses, and especially wild or little-selected forage grasses, are becoming increasingly important for environmental purposes. Currently, the seed trade for amenity and landscape improvement is already comparable to the forage seed trade, and is still developing (Kley, 1995). In this context, new species are being used for nature conservation (biodiversity restoration), amenity (gardens, parks, sports pitches, lawns), landscape enhancement (verges of railways, waterways and roads) and restoration of degraded sites

---

[1] For a definition of grassland and rangeland, see Chapter 2, section 2.4.

(abandoned industrial sites, polluted soils, bare soils after civil engineering works).

There is a huge difference in the information available on the different forage species. There is a wealth of published literature on perennial ryegrass (*Lolium perenne*) and other widely sown species, while there is very little or nothing on the forage potential of many less widely known or used species.

The book focuses on temperate Europe and thus excludes the Mediterranean part of the continent. However, several species described in the book are used in other temperate areas of the world: North America, some parts of Asia, New Zealand and southern Australia, and the south of South America. Readers in those parts of the world should thus find useful information for their conditions.

This text is mainly written for students of agriculture, lecturers, researchers, agricultural advisers and consultants. Much information has also been added that will serve the needs of ecologists, nature conservationists and others involved in the diverse uses of grass communities as a vegetative land cover.

The book is designed in three parts.

The first part comprises four chapters: an introduction (Chapter 1); a description of the importance and the diversity of grasses and grassland communities (Chapter 2); a discussion of the various uses of grasses (Chapter 3); and a description of their physiology and morphology (Chapter 4).

The second part (Chapter 5) contains the main body of the text, namely descriptive data for individual species, with wild, spontaneous species figuring prominently. Chapter 5 provides detailed information on 43 temperate species, indicating what is known, but also identifying the many gaps that exist in our knowledge.

The third and final part of the book comprises two chapters: the advantages of wild or secondary grasses and the role of multispecies and multivariety mixtures (Chapter 6); and some ideas for future research (Chapter 7).

The main references consulted are listed and two useful annexes are added, namely, an identification key of the grasses at a vegetative stage and lists of the names of the species in Latin, English, French and German.

Chapter 2

Importance and diversity of grasses and grass communities

2.1 IMPORTANCE

Natural vegetation dominated by grasses corresponds to different biomes of the biosphere: in the tropics there are savannahs, tropical steppes and some high altitude formations; in temperate regions, there are temperate steppes (prairies in North America, steppes of central Asia) and high altitude, alpine ecosystems. In addition, many biomes have been profoundly transformed by human beings and their livestock into grasslands and rangelands. This happened in the past for the temperate deciduous forest, for the taiga and for the mountain and subalpine belts of highlands. Currently, this is the case in the tropical rainforest, where grasslands have been sown on vast areas, especially in South America.

Temperate forage grasses are the main components of grasslands and rangelands. Demonstrating the importance of these species thus amounts to defining the many functions fulfilled by these two ecosystems. They occupy the land, providing a habitat and a source of food for domestic livestock, thus ensuring a supply of livestock products, which contributes to rural agricultural and economic development. The ecosystems also protect soil and water resources and they provide diverse habitats for many different forms of wildlife. They are an important element of landscapes and so are part of the heritage of regions, particularly where they cover large areas. Increasingly, green landscapes are part of the ecotourism catering for the demands of urban dwellers.

Grass vegetation of different types occupies vast areas of the world's land area. Shantz (1954) estimated such cover at about

46 000 000 km², that is, about 25 percent of the world's plant cover. Grass covers allow the utilization of marginal soils that are unsuitable for arable cropping. Soils that are steep, superficial, wet or very dry, sites at high altitudes or latitudes with a short growing season and steppe regions, are environments that cannot be cultivated but are nevertheless useful for agriculture because of their grass vegetation. In other regions, dominated by soils that are suited to arable farming, stock-rearing on associated marginal soils gives farmers the opportunity to increase their income. In Western Europe, grasslands occupy almost 40 percent of the agricultural area. In some countries or some regions, the share of grasslands is higher (Ireland 76 percent, Switzerland 72 percent, United Kingdom 65 percent, Austria 57 percent). However, in Central and Eastern Europe (the Russian Federation not included) in countries that previously had centrally planned economies, grasslands occupied only about 30 percent of the agricultural area (Romania 33 percent, Poland 22 percent, Hungary 19 percent) (FAO, 2002: data of 1999). Nevertheless, it is foreseen that this percentage will increase in the future, as the population's purchasing power for animal products increases and animal husbandry develops.

Domestic animals find space and conditions on the grasslands and rangelands close to that of the natural habitats of their wild ancestors, where they can feed, ruminate, rest, develop social relations with other members of the herd and reproduce. These conditions are not so well fulfilled when animals are housed all year round. So long as the stock-rearing system is well managed, grasslands and rangelands obviously increase the animals' well-being.

Grasses are a 'natural' food for ruminants. Their leaves are broad and supple enough to be easily gripped by the mouth organs of cattle, sheep and goats, and fine enough to be easily attacked by rumen micro-organisms. Rich in fibre, they provide the bulk necessary for good rumen function. Eaten at a sufficiently young stage, they are highly digestible and contain almost as much energy as cereal grains. At this stage, they also have high mineral and protein contents that cover a large part of animals' requirements (Demarquilly, 1981; Jarrige, 1988). Grass is above all a healthy food, usually free of residues or toxic substances. Recent crises over hormones, dioxin and BSE remind us that animal husbandry should be based on healthy principles and 'natural' foods, as close as possible to the diet of wild ruminants. This does not exclude the quest for zootechnical performance to ensure the profitability of agricultural enterprises. Nevertheless, healthy feeding of ruminants ought to be based on grass, without excluding other natural products like maize silage, cereals, oilseeds and protein-rich crops or roots. Grass is also a balanced diet for horses, even if their dietary needs are somewhat different from those of ruminants.

Grasslands and rangelands usually produce the majority of the forage ingested by ruminant animals during the grazing season. During winter housing, hay and grass silage are often major parts of the staple diets. On an annual basis, in Europe, it is common for 50 to 75 percent of cattle and 90 to 95 percent of sheep fodder requirements to be met by grasslands.

The final aim of exploiting grass in agriculture is the transformation of the herbage into livestock products. Thus meat and milk should not only meet society's needs in terms of quantity and feeding value, but should also be free of toxic substances,

have high organoleptic qualities and be produced according to ethical rules acceptable to society. Livestock products based on the utilization of grass are in an excellent position to meet these conditions. In this way, grasslands and rangelands contribute to economic development.

Grass cover protects the soil because the grasses and their root systems limit water and wind erosion. Water erosion is much less than that which occurs with annual fodders such as maize crops, in which the soil is bare for much of the year – from November to May in many regions (in the northern hemisphere) – unless intercropping is practised. Winter erosion can be serious but erosion is most spectacular in spring, in May to early June, when spring rainstorms wash away the newly worked soil just after the sowing of maize (Kwaad, 1994; Van Dijk, van der Zijp and Kwaad, 1996; Kwaad, van der Zijp and van Dijk, 1998). In North America and in Asia, ploughing up the steppes brought about catastrophic wind erosion in many places (Texas, Oklahoma, Colorado, Dakota, Kazakhstan, Uzbekistan) (Ramade, 1992). Such erosion was unknown before cropping, which demonstrates the effectiveness of grass cover for protecting the soil of that habitat.

A grass cover also allows good infiltration of water into the soil, much better than other vegetation like crops and dense coniferous forest. In coniferous forests, great quantities of water evaporate or are transpired before they reach the soil, even in winter. Some plantations of trees in wet grasslands, like poplars, are also known for drying the environment and are used for this end. Rangelands and cutting grasslands protect the quality of surface water and groundwater tables, because water that infiltrates the soil from such a cover contains little or no pesticides or nitrates, unlike cropland (Benoit, 1994; Thélier-Huché *et al.*, 1994). With a crop of maize, leaching of nitrogen reaches 100–200 kg $N-NO_3$/ha annually, while from a hay meadow, the leaching is maintained at about 5 kg $N-NO_3$/ha annually (Le Gall, Legarto and Pfimlin, 1997). These figures are representative of heavily fertilized crops or grasslands that produce about 15 tonnes DM/ha annually. Some pesticides are also carried by the water percolating below cultivated soils. Atrazine, for example, which is still often used in maize growing, can pollute the groundwater table (Di Muccio *et al.*, 1990; Demon *et al.*, 1994; Dousset, Mouvet and Schiavon, 1995; Bottoni, Keizer and Funari, 1996; Funari *et al.*, 1998). On grasslands and rangelands, very small amounts of pesticides are used and so the risk of pollution of groundwater by such substances is limited.

The diversity of environments and types of community in grasslands and rangelands is immense (see section 2.5.). Some extensive grass covers are extremely rich in plant and animal species, a feature that can be of great importance for the natural heritage. Other heavily fertilized grasslands are much poorer in plant species. Furthermore, all grass vegetation areas favour the development of organisms that decompose organic matter. Earthworm populations, for example, are much greater under grassland than in arable land or forests. An earthworm biomass of 1 000–5 000 kg/ha can be found in grassland, compared to less than 500 kg/ha under crop (Granval, Muys and Leconte, 2000). This quantity is greater in grazed than in mown grassland. Biomass values only reach about 500 kg/ha in forests with mull humus and about 50 kg/ha in forests with mor humus (Duvigneaud, 1974). Extensive mown meadows can be breeding habitats for birds of great heritage interest like the corncrake (*Crex crex*), the curlew (*Numenius arquata*), the quail (*Coturnix coturnix*), the skylark (*Alauda arvensis*), the yellow wagtail (*Motacilla flava*) and the whinchat (*Saxicola rubetra*). They can also be rich in plant species, especially rare or endangered ones like orchids (*Orchis* spp.), marsh orchids (*Dactylorhiza* spp.) and bee orchids (*Ophrys* spp.). The European steppes are also very rich in plant species adapted to very dry conditions. In Hungary, they are home to the severely endangered great bustard (*Otis tarda*). The intensively fertilized grasslands of maritime areas provide feeding and breeding sites for waders and provide winter feed for wild geese. Several butterfly species are intimately associated with grassland plants with which they have coevolved, and are directly dependent on these plants for the development of their caterpillars. Grass covers are also a strategic habitat for the development of many orthopteran species.

Grass covers are key components of the landscape and of human culture. In pastoral regions, pastures are major constituents of the landscape, notably in zones of farmland criss-crossed with hedges and trees (the 'bocage' of Normandy and in southern England), in the hills and in the highlands. The grazing livestock contribute to the attractiveness of the countryside. Grassland landscapes have deeply affected the cultures, the traditions and the lifestyles of the people who live within them. In the past, hay-making was a particularly important moment, which necessitated logical, coordinated work and mutual assistance between individuals. Numerous proverbs and expressions bear witness to the importance of grass and grasslands in the collective psyche. On the other hand, moorland has always been associated with desolate, savage, mysterious places where exceptional and extraordinary events could take place (sightings of the *bête du Gévaudan* and the Dartmoor beast, for example). Grass landscapes therefore fulfil a function of 'aesthetic information' that is all the more important when there is great heterogeneity and complexity of the landscape. Such landscapes also have a recreational function (tourism) that is reinforced by the region's cultural and aesthetic characteristics. Grasslands can also contain historical and archaeological elements or be associated with religion, e.g. churches, standing stones, dolmens. Grass-dominated ecosystems are also often used to create awareness of nature and environmental problems among the public.

✓ 2.2 GRASS SYSTEMATICS AND EVOLUTION

Grasses are monocotyledons (Angiosperma) and are grouped in the family of *Poaceae* (*Gramineae*) which comprises six subfamilies worldwide: *Bambusoideae, Centostecoideae, Arundinoideae, Chloridoideae, Panicoideae* and *Pooideae* (Table 2.1) (Watson and Dallwitz, 1994). In their turn, the six subfamilies contain more than 50 tribes, 650 genera and 10 000 species. European grasses belong to the subfamilies *Arundinoideae* and *Pooideae*.

The number of species is very high worldwide and in Europe (Table 2.1); nevertheless about a third of the world's species belong to only ten genera: *Agrostis, Aristida, Calamagrostis, Digitaria, Eragrostis, Festuca, Panicum, Paspalum, Poa* and *Stipa*. Each of these genera contains at least 200 species (Watson and Dallwitz, 1994). Among these genera, the following contain species that are widespread in temperate Europe: *Agrostis, Calamagrostis, Festuca, Poa* and *Stipa*.

The genera containing the most important fodder species in temperate zones are notably *Bromus, Dactylis, Festuca, Lolium, Phleum* and *Poa*. The number of species belonging to each of them is given in Table 2.1.

Europe is the centre of origin of many temperate forage grasses. Hartley and Williams (1956) placed the area of the highest diversification in Italy and the southern half of France. More than 20 species are indigenous to this area. Sixteen to 20 are native to the main part of Europe. North and central Asia, as well as a small part of Quebec in Canada, have only six to ten indigenous forage species. In contrast, the native species of the United States of America (USA), namely the spontaneous species of the prairie, have a low forage potential and most of them have a low nutritive value.

The *Poaceae* family is one of the most highly evolved families of the plant kingdom. Grasses probably appeared in the Cretaceous period about 70 million years ago, but the first traces of their existence are from the Palaeocene period about 65 million years ago, by way of pollen grains (Muller, 1981).

Most grasses coevolved with wild herbivores and are therefore adapted to frequent defoliations through grazing, since morphologically most of their vegetative

TABLE 2.1

Systematic diversity of grasses (*Poaceae*) in the world and in Europe

		World	Europe
Subfamilies		*Bambusoideae, Centostecoideae, Arundinoideae, Chloridoideae, Panicoideae, Pooideae*	*Arundinoideae, Pooideae*
Tribes of the subfamily *Pooideae*		Poeae, Aveneae, Agrostideae, Bromeae, Triticeae, Stipeae	
Number of genera		about 650	155
Total number of species		about 10 000	about 900
Number of species of genus:	*Bromus*	150	7
	Dactylis	1 to 5	1 to 5
	Festuca	360	170
	Lolium	8	5
	Phleum	15	11
	Poa	500	45

Source: After Watson and Dallwitz, 1994 and Tutin *et al.*, 1980

buds are close to the soil surface where they cannot be destroyed by grazing herbivores. Therefore grasses have not generally developed defence mechanisms against herbivores, in contrast to numerous species of dicotyledons, which have physical defences or synthesize secondary metabolites, which are bitter, toxic or which reduce digestibility (Chapman, 1998). On the contrary, the high digestibility of grass leaves and their high energy and protein contents encourage consumption by herbivores. This apparently suicidal evolutionary strategy actually gives grasses a great advantage compared to other species that are less well adapted to defoliation, as are most dicotyledons. It allows grasses to dominate herbaceous covers, thus leaving little space for dicotyledons in most swards that are regularly grazed by livestock or mown by human beings.

2.3 WILD AND BRED GRASSES

Forage grasses were domesticated later than crops such as wheat and barley. The first domesticated species of forage grass was Italian ryegrass (*Lolium multiflorum*) in the twelfth century, in Piedmont and Lombardy, northern Italy (Borrill, 1976).

Fallows were utilized for soil fertility restoration in Europe from the start of farming between 5 000 and 6 000 years ago (Ammerman and Cavalli-Sforza, 1971; Zohary, 1986). Starting in the sixteenth century, farmers began to improve fallows by sowing mixtures of grasses and legumes instead of letting wild plants colonize the soil after a cropping period. These two types of plants provided better forage than that from natural regeneration and the legumes improved soil fertility thanks to symbiotic fixation of nitrogen with bacteria of the genus *Rhizobium. Lolium perenne* was often used in these mixtures in northwestern Europe and notably in the United Kingdom (UK) where this technique developed strongly. Other species were used in regions less suited to ryegrass, namely *Festuca pratensis* and *Phleum*

pratense, and of course *Lolium multiflorum*. The rotation based on the succession of crops and forage plants was called 'ley farming' (Stapledon and Davies, 1948).

Dactylis glomerata, Festuca arundinacea, F. pratensis, Lolium multiflorum, L. perenne and *Phleum pratense* were introduced to North America in the mid-eighteenth century (Leafe, 1988). Timothy, the English name for *Phleum pratense*, comes from Timothy Hanson who introduced it to Maryland in 1720. Seeds of the species were thereafter exported to England in 1760 under the name of timothy (Leafe, 1988). Until the eighteenth century, most seeds were collected in the wild. In grazed pastures, the use of *Lolium perenne* spread in England around 1650, along with other grasses used to create swards of various life spans, but mainly to establish permanent pastures. *Lolium perenne* was not immediately recognized as the most valuable species. Seeds of meadow foxtail (*Alopecurus pratensis*), crested dog's tail (*Cynosurus cristatus*), sweet vernal-grass (*Anthoxanthum odoratum*) and even annual meadow grass (*Poa annua*) were used in sowings (Leafe, 1988). Smooth brome (*Bromus inermis*) was imported to North America from Hungary in 1880 (Borrill, 1976). Towards the end of the eighteenth century, production of *Lolium perenne* seed was widespread in England. Nevertheless, systematic grass breeding did not really begin until 1919, when R.G. Stapleton and T.J. Jenkin began to work in Great Britain with several species including *Lolium perenne* (Leafe, 1988). In the USA, the first species bred were *Festuca arundinacea, Dactylis glomerata* and *Bromus inermis* (Hanson and Carnahan, 1956).

The number of species that have been subject to selection is very small compared to the total number of species available. On a world scale about 100 to 150 forage species have undergone selection or at least been cultivated (Skerman and Riveros, 1990; Barnes, Miller and Nelson, 1995) out of a total of 10 000 species of grasses, that is, 1–1.5 percent. In Europe, selection has involved about 20 species out of a total of 900, that is, 2 percent. The main selection work has, however, concentrated on only six forage species: *Lolium perenne, L. multiflorum, Festuca pratensis, F. arundinacea, Phleum pratense* and *Dactylis glomerata,* or 0.7 percent of the European flora. Selection work was undertaken by public or private organizations, but all chose to work on species that had the greatest commercial potential, namely, species of good nutritive value with a high yield potential, species that were suitable for turf, and also species that could be used in a wide range of situations. Thus, the species bred also include those of highest value for intensive systems and for soils of good agronomic quality. In consequence, less productive, less digestible species, though suited to particular ecological conditions or more extensive farming systems, have not received a lot of attention in spite of their potential value in many situations and farming systems.

The number of bred varieties has reached large numbers for some species (Table 2.2). In Europe, *Lolium perenne* has the greatest number, with 565 varieties registered in the European Catalogue (European Commission, 1999). Then in decreasing order come *Lolium multiflorum, Festuca arundinacea, Dactylis glomerata, Phleum pratense,* hybrid ryegrass (*L. x hybridum*) and *F. pratensis.* These six species together with the hybrid *Lolium x hybridum* comprise 93 percent of the forage grass varieties sold in Europe. The number of varieties of red fescue (*Festuca rubra*) listed is 206 and that of smooth meadow grass (*Poa pratensis*) 110, but the varieties of these two species are mainly used for amenity purposes. In contrast, some widespread forage species have hardly been selected at all and the number of available varieties is small or even nil. No variety of *Holcus lanatus* is registered in the 1999 European catalogue, *Trisetum flavescens* has one, *Alopecurus pratensis* has two, and *Arrhenatherum elatius* has five. These numbers underline how much breeding effort is concentrated on only a few species. If these numbers are compared with

those of 1989, the dominance of *Lolium* spp. varieties is accentuated (Table 2.2). *Lolium* spp. varieties have risen from 44 percent to 54 percent of the total number of varieties in a decade. The number of varieties of *Lolium multiflorum* and *Festuca arundinacea* have doubled. The increase in *Lolium x hybridum* varieties is also noteworthy.

The quantity of seed sold is also indicative of the domination of the market by only a few species. Indeed, the trade of forage grass seeds used in the European Union is largely dominated by the *Lolium* spp. For forage production, *Lolium multiflorum* represents 42 percent of the quantities sown by farmers, *L. perenne* 41 percent, then comes *Festuca pratensis* with 6 percent, *Dactylis glomerata* with 4 percent, *Phleum pratense* with 4 percent, *F. arundinacea* with 2 percent and *Poa pratensis* with 1 percent (Kley, 1995).

The wild, non-bred species are sometimes called 'secondary' because their seeds are not sold in large amounts and indeed the seeds of many are unavailable (Frame, 1982 and 1991; Frame and Tiley, 1988) while the widely sown grass species can be termed 'primary species'.

Since the 1990s, there has been a general tendency in Europe for the public sector to reduce forage grass breeding markedly and this has left the field open to private companies. Obviously, these companies give more importance to criteria of profitability in their breeding activities and thus concentrate on producing varieties of the most widely sown species. Nevertheless, collaboration between the public and private sectors in order to start or reinforce breeding programmes for species that have been little worked on to date, should not be excluded in the future.

2.4 AGRICULTURAL TYPOLOGY OF COMMUNITIES

Grasslands may be permanent or temporary (regularly sown) but rangelands are normally permanent.

Permanent grasslands are usually more than 20 years old. Some were sown with fairly

TABLE 2.2
**Number of varieties registered in the European Catalogue
(European Commission, 1989 and 1999)**

	1989	1999
Lolium perenne	233	565
Lolium multiflorum	122	237
Festuca rubra	133	206
Poa pratensis	92	110
Festuca arundinacea	49	106
Dactylis glomerata	62	86
Phleum pratense	40	67
Lolium x hybridum	18	62
Festuca pratensis	39	45
Festuca ovina	19	33
Agrostis capillaris	20	27
Agrostis stolonifera	4	16
Bromus catharticus	7	10
Poa trivialis	6	7
Arrhenatherum elatius	4	5
Bromus sitchensis	1	4
Agrostis gigantea	4	3
Agrostis canina	0	2
Alopecurus pratensis	3	2
Trisetum flavescens	1	1
Total number of varieties	857	1 594
Number of varieties of *Lolium* spp.	373	864
% of *Lolium* spp. varieties	44	54

complex mixtures as early as the end of the nineteenth century. Other grasslands were never sown and often developed from grazed moorland or from the spontaneous colonization of previous arable land by wild grasses. Regardless of whether or not they were sown, the vegetation types of these grasslands evolved considerably under various farming practices, e.g. stocking rates, cutting frequency, fertilization, oversowing, use of herbicides, drainage. Agronomic classifications of permanent grasslands are often based on the percentage of *Lolium perenne* or on the percentage of 'good' or 'preferred' grasses in the sward, or on the 'pastoral value' (see section 2.9.).

Temporary grasslands are regularly resown and are usually kept for one to five years depending on the type of management used. Some temporary grasslands (leys) are associated with crop rotations. Other temporary grasslands are resown immediately and thus do not take part in a crop rotation. This is particularly true for mainly grazed grassland resown every four or five years with *Lolium perenne*. The agronomic classification of temporary grasslands often distinguishes between pure-sown grass swards and grass-legume mixtures.

Regardless of their longevity, grasslands can be grazed only, cut only or managed alternately by cutting and grazing (mixed-use grasslands).

Rangelands differ from grasslands in that the livestock graze vast unfenced areas, generally looked after by a herder. The vegetation of rangelands comprises herbaceous or mixed herbaceous and woody plants and grasses are usually well represented (Allen, 1991).

2.5 ECOLOGICAL TYPOLOGY OF COMMUNITIES

In the phytosociological system, plant communities are classified in a hierarchic system analogous with the classification of species. The basic typological unit is the association (Braun-Blanquet, 1964). The higher levels are the alliance, the order and the class. The levels inferior to the association are the subassociation, the facies and the variante. The names of these typological units are based on the names of one or two dominant or characteristic species. The suffix 'etosum' is added to these names to create the name of a subassociation, the suffix 'etum' for an association, 'ion' for an alliance, 'etalia' for an order and 'etea' for a class. The names of these species are given in the genitive but not used so as not to overload the nomenclature when the name of the genus does not lead to confusion. An example is given in Table 2.3.

This system is widely used in several regions of the world, notably continental Europe where it was first developed from the beginning of the twentieth century by Braun-Blanquet, who mainly studied communities in Switzerland, France and Germany (Braun-Blanquet, 1964). The methodology later reached the rest of continental Europe, and then many other regions of the world.

The grasslands of the European Union are classified in eight habitats according to the CORINE (1991) typology, which is inspired by the phytosociological classification. The ninth habitat is that of grasses used in ornamental and sports turfs (Table 2.4).

Within these various habitats, the main grass-dominated communities are given in Table 2.5.

The classification system for plant communities used in Great Britain is slightly different. It does not use species names in community nomenclature, but a combination of letters and numbers (Rodwell, 1992):

- MG6: *Lolium perenne-Cynosurus cristatus* mesotrophic grasslands,
- CG5: *Bromus erectus-Brachypodium pinnatum* calcareous grasslands,
- U5: *Nardus stricta-Galium saxatile* acid grasslands.

It is possible to establish a correspondence between the two systems, as shown in the following examples (Rodwell, 1992):

- MG6: *Cynosurion, Lolio-Cynosuretum cristati,*
- CG5: *Mesobromion, Cirsio-Brometum erecti,*
- U5: *Violo-Nardion, Nardo-Galietum saxatilis.*

It is not possible in this work to cite all associations dominated by grasses, but the long list of alliances in Table 2.5 suffices to show the diversity of communities in which grasses participate and in which they play an

TABLE 2.3

Example of the system of phytosociological classification for some types of grassland

Level of classification	Characteristic species	Nomenclature
CLASS of grasslands	*Arrhenatherum elatius*	*Arrhenatheretea elatioris*
ORDER of grazed and mown grasslands	*Arrhenatherum elatius*	*Arrhenatheretalia elatioris*
ALLIANCE of grazed pastures	*Cynosurus cristatus*	*Cynosurion cristati*
ASSOCIATION of grazed pastures established on fertile soils	*Lolium perenne* and *Cynosurus cristatus*	*Lolio-Cynosuretum cristati*
dry SUB-ASSOCIATION	*Ranunculus bulbosus*	*Ranunculetosum bulbosi*
FACIES of		*Lotus corniculatus*

TABLE 2.4
**List of the principal grass-dominated habitats in the CORINE
(1991) typology**

CORINE reference number and habitat description
15. Salt marshes
16. Dunes: 16.2. Grey dunes
34. Dry calcareous grasslands and steppes
35. Dry siliceous grasslands
36. Alpine and subalpine grasslands
37. Humid grasslands and tall herb communities
38. Mesophile grasslands
81. Improved grasslands
85. Urban parks and large gardens

essential role. The great majority of these communities are not characterized by selected species but by other species adapted to a great range of habitats. However, these habitats could not produce forage without this diversity of species and communities adapted to different situations and conditions.

2.6. EVOLUTION OF COMMUNITIES

Plant communities are not unchangeable since they evolve as species do, according to modifications of habitat, whether these be 'natural' or imposed by human action.

Before agriculture appeared, forage grasses were limited to smaller areas than at present. Many grass species developed after the forest clearings were created by physical disturbances (fire, high winds), by the natural cycle of evolution and the ageing of the forest cover, but also by the activity of large herbivores. The populations of aurochs (*Bos primigenius*), European bison (*Bison bonasus*), tarpan (*Equus caballus*), deer including elk (*Alces alces*) and red deer (*Cervus elaphus*), as well as wild boar (*Sus scrofa*) exerted such pressure on the woody vegetation and its regeneration that grassy clearings could form and maintain themselves on significant areas of forest zones. Furthermore, even if the forest is the final stage in the evolution of plant cover in most European habitats, there

have always been some habitats that are particularly hostile to tree life, even in the interglacial periods. These are mainly zones of high altitude and latitude, marshy areas, saline muds, valley bottoms liable to flooding and very dry habitats regularly subject to natural fires. Later, populations of hunter-gatherers also pushed back the forest cover by various means including fire.

With the appearance of pastoral practices and the beginning of farming in Europe between 5 000 and 6 000 years ago, forest cover has constantly retreated, even if some periods have seen temporary increases in forest area (Ammerman and Cavalli-Sforza, 1971; Zohary, 1986; Hancock, 1992; Smith, 1995). New habitats thus appeared that were generally richer in plant species, notably grasses, than the forest cover from which they emanated. It was in this way that pastoral activity created grassland 'bocage', i.e. farmland criss-crossed by hedges and trees, extensively cut meadows, moors and pastures of very low carrying capacity and limestone grasslands grazed by sheep. All these habitats are particularly rich in species, the greatest diversity being probably attained in traditional mown meadows cut in July with the aftermath grazed in September. Grazed swards on limestone are also very rich in species. Farming and stock-raising

TABLE 2.5
List of grass-dominated communities and characteristic species

CORINE reference number and habitat description		Name of the community (classes, orders, alliances)	Characteristic grass species
15.3.	Atlantic salt meadows	*Glauco-Puccinellietalia maritimae*	
15.32.	Saltmarsh grass communities	*Puccinellion maritimae*	*Puccinellia maritima*
15.33.	Upper shore communities	*Armerion maritimae*	*Agrostis stolonifera, Festuca rubra*
16.221.	Coastal sand dunes: Northern grey dunes	*Galio-Koelerion albescentis*	*Koeleria albescens*
34.3.	Dense perennial grasslands and middle European steppes	*Festuco-Brometea*	
34.31.	Subcontinental steppe grasslands	*Festucetalia valesiacae*	*Festuca* spp., *Stipa* spp., *Koeleria macrantha, Agrostis capillaris, Poa* spp., *Melica ciliata, Brachypodium pinnatum*
34.312.	Central European steppe grasslands	*Festucion valesiacae*	
34.32.	Sub-Atlantic semi-dry calcareous grasslands	*Mesobromion*	*Bromus erectus, Brachypodium pinnatum, Koeleria pyramidata, Festuca* spp., *Avenula pubescens, Sesleria albicans, Briza media*
34.33.	Sub-Atlantic very dry calcareous grasslands	*Xerobromion*	*Bromus erectus, Brachypodium pinnatum, Koeleria vallesiana, Sesleria albicans, Melica ciliata, Stipa* spp., *Phleum phleoides*
35.1.	Atlantic mat-grass swards and related communities	*Nardetalia: Violo-Nardion:* Mat-grass (*Nardus stricta*) swards, *Agrostis-Festuca* grasslands, *Deschampsia flexuosa* grasslands, namely	*Nardus stricta, Festuca filiformis, F. ovina, F. rubra, Agrostis capillaris, Sieglingia decumbens, Anthoxanthum odoratum, Deschampsia flexuosa, Poa angustifolia*
36.3.	Alpine and subalpine acidophilous grasslands	*Caricetea curvulae*	
36.31.	Mat-grass swards and related communities	*Nardion*	*Nardus stricta, Festuca eskia, F. rubra, Alopecurus gerardii, Bellardiochloa (Poa) violacea, Anthoxanthum odoratum*
36.33.	Subalpine thermophile siliceous grasslands	*Festucion spadiceae, Poion violaceae, Festucion eskiae, Festucion variae*	*Festuca paniculata (spadicea), Festuca varia* group, *Festuca eskia*
36.34.	Crooked-sedge swards and related communities	*Caricion curvulae, Festucion supinae*	*Festuca* spp.
36.4.	Alpine and subalpine calciphilous grasslands	*Elyno-Seslerietea*, numerous alliances	
36.5.	Alpine and subalpine fertilized grasslands	*Arrhenatheretalia elatioris*, several alliances	
36.51.	Subalpine yellow oat-grass hay meadows	*Polygono-Trisetion*	*Trisetum flavescens, Anthoxanthum odoratum, Festuca rubra*, namely

(Continued)

TABLE 2.5

List of grass-dominated communities and characteristic species (Cont.)

CORINE reference number and habitat description	Name of the community (classes, orders, alliances)	Characteristic grass species
36.52. Rough hawkbit pastures	*Poion alpinae*	*Agrostis alpina, Phleum alpinum, Poa alpina*
37.1. Meadowsweet stands and related communities	*Filipendulion ulmariae*	*Deschampsia caespitosa, Phalaris arundinacea*
37.2. Eutrophic humid grasslands	*Molinietalia*	
37.21. Atlantic and sub-Atlantic humid meadows	*Bromion racemosi (= Calthion palustris), Deschampsion caespitosae*	*Bromus racemosus, Deschampsia caespitosa, Holcus lanatus, Alopecurus pratensis, Festuca pratensis*
37.22. Sharp-flowered rush meadows	*Juncion acutiflori*	*Holcus lanatus, Deschampsia caespitosa*
37.24. Flood swards and related communities	*Agropyro-Rumicion crispi*	*Elymus repens, Agrostis stolonifera, Alopecurus geniculatus*
37.3. Oligotrophic humid grasslands	*Molinion caeruleae, Juncion squarrosi*	
37.31. Purple moor-grass meadows and related communities	*Molinietalia: Molinion caeruleae*	*Molinia caerulea, Deschampsia caespitosa*
37.32. Heath rush meadows and humid mat-grass swards	*Nardetalia: Juncion squarrosi*	*Nardus stricta, Festuca ovina*
38.1. Mesophile pastures	*Cynosurion*	*Lolium perenne, Cynosurus cristatus, Poa spp., Festuca spp.*
38.2. Lowland hay meadows	*Arrhenatherion, Brachypodio-Centaureion nemoralis*	*Arrhenatherum elatius, Trisetum flavescens*
38.3. Mountain hay meadows	*Polygono-Trisetion*	*Trisetum flavescens, Anthoxanthum odoratum, Festuca rubra, namely*
81.1. Dry or mesophile intensive grasslands		*Lolium perenne, Dactylis glomerata, Holcus lanatus, Poa pratensis, P. trivialis, Festuca rubra, Agrostis capillaris*
81.2. Humid improved grasslands		*Lolium perenne, Agrostis stolonifera, Holcus lanatus, Poa trivialis, Alopecurus pratensis, A. geniculatus*
85.1. Large parks		*Lolium perenne, Poa annua, P. pratensis, P. trivialis, Festuca arundinacea, F. rubra, F. ovina, Holcus lanatus, Agrostis capillaris*
85.12. Park lawns		
85.2. Small gardens and city squares		
85.3. Gardens		
85.32 Ornamental gardens		

thus allowed an expansion of the biological diversity of the land, at the level of genes as well as species, and especially at the level of communities.

Before the period of intensification, evolution had resulted in grasslands that were very rich in species. The evolutionary pathways from forest communities to extensive grasslands differed from one region of Europe to another and even between habitats within the same region. These evolutions differed markedly on different soil types, such as calcareous, clay or sandy soils. Nevertheless, from the 1960s onwards, once these grasslands were fertilized, cut more frequently or grazed at heavy stocking rates, the communities that they engendered evolved once more. But this time the number of species everywhere was reduced and all the different communities converged towards

a single model as their botanical composition simplified. The final stage of evolution of these communities corresponded more or less to swards dominated by *Lolium perenne* and a small number of other grasses including *Cynosurus cristatus, Dactylis glomerata, Holcus lanatus* and *Poa* spp., whether on calcareous, clay or sandy soil types. The agricultural pedo-climax (the culmination of the evolution of vegetation on a given soil type and under local agricultural management conditions) of grassland intensification on a vast proportion of the temperate European territory therefore consists of a sward of *Lolium perenne*, whatever the earlier stages of community evolution had been.

An example of this evolution of plant communities from species-rich grasslands to intensified swards was studied in the 1960s in Famenne, Belgium. The evolutionary

FIGURE 2.1

Evolution of plant communities and decrease of their number following the intensification of Belgian grasslands at Famenne

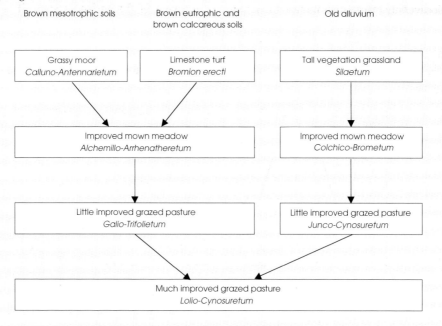

Note: Exclusive hay production on permanent grasslands was abandoned in intensive systems
Source: Sougnez and Limbourg, 1963

stages have been reconstituted as shown in Figure 2.1.

After the appearance of agriculture and animal husbandry, biodiversity therefore evolved according to two successive tendencies, which can be represented as two truncated triangles (Figure 2.2).

2.7. PRODUCTION LEVELS, FREQUENCY OF DISTURBANCE AND BIODIVERSITY

The biodiversity of grasses, as for any other plant group, can be considered at several levels: at the level of the **genetic diversity** within a species, of the number or proportion of **species** within a community, and of the diversity of **communities** within a landscape unit. These diversities can be seen in space as well as in time. For example, C4 grasses thrive in the hot season while C3 grasses appear in the cold season in the same place, or when forage mixtures are sown for short periods in crop rotations, or when the relative abundance of the different species changes during a growing season.

Besides this first distinction, two categories of biodiversity can be distinguished in forage species: the biodiversity planned by the farmer and that associated with the system (Maljean and Peeters, 2002).

Planned biodiversity (or **agricultural biodiversity**) is composed at the level of the genes, of the varieties or ecotypes of grasses used voluntarily by the farmer. At the species level, it is made up of the cultivated species, for example, *Lolium* spp., *Phleum pratense*, meadow fescue (*Festuca pratensis*), tall fescue (*F. arundinacea*), *Dactylis glomerata*. At the community level, it is defined by the kinds of cover (permanent or temporary grasslands, intensive or extensive grasslands, rangelands, annual forage crops, fallows or set-asides), by the size of the fields, by the network of herbaceous field margins and by the heterogeneity of the cropping pattern within the landscape.

Associated biodiversity comprises ecotypes, wild species and communities that appear spontaneously in production systems. Nevertheless, one part of that associated biodiversity, which is not planned by the farmer, plays a determining role in the working of the agro-ecosystem. It is called

FIGURE 2.2
Evolution of biodiversity in Europe over 6 000 years according to agricultural and pastoral activities

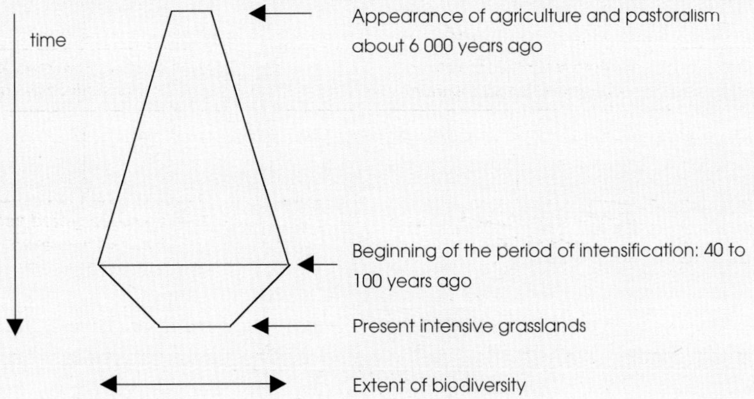

Source: after Kornas, 1983

functional biodiversity (or **para-agricultural biodiversity**). It concerns species that have a positive effect on forage production, for example, common bent (*Agrostis capillaris*), creeping bent (*A. stolonifera*), tall oat-grass (*Arrhenatherum elatius*), *Festuca rubra, Holcus lanatus,* rough meadow grass (*Poa trivialis*), yellow oat-grass (*Trisetum flavescens*). Other 'functional' species have at least a partially negative effect on agricultural production, such as grasses that livestock refuse to eat, e.g. tufted hair grass (*Deschampsia caespitosa*) or are toxic, e.g. reed canary-grass (*Phalaris arundinacea*). At community level, functional biodiversity is mainly made up of semi-natural permanent grasslands and extensive rangelands.

Among other wild ecotypes and species that are more or less closely associated with the system, but which play a less important role in it (**extra-agricultural biodiversity**), there are grasses that produce very little biomass so their contribution to overall production is negligible, e.g. velvet bent (*Agrostis canina*), *Poa annua*. Nevertheless, numerous species in this category have an important heritage interest, e.g. common quaking grass (*Briza media*), heath grass (*Sieglingia decumbens*). Among elements of community diversity of lesser importance from a functional point of view, ponds can be mentioned, with their stream-edge and floating grasses, as can abandoned land.

It is clear that, in some cases, the difference between para-agricultural and extra-agricultural biodiversity is slim. A species can be put in one or another category according to its abundance in the community.

Most fodder grasses are cross-pollinating, although they may retain a certain degree of self-pollination (Chapman, 1998). This allows the appearance of a diversity of genetically distinct individuals, within the same population. Individuals belonging to different types can often be considered as distinct ecotypes since they have different ecological adaptations. For example, in a field

alternately mown and grazed, ecotypes with erect foliage, suited to mowing, and prostrate ecotypes adapted to grazing can be seen side by side. If the management of the field is modified, the frequency of some ecotypes will decrease to the benefit of others. If, for instance, a mixed sward is only grazed, the proportion of ecotypes with erect foliage decreases while the proportion of prostrate ecotypes increases (Borrill, 1976). In the same way, synthetically bred varieties are the product of crosses between clones with different characteristics; all individuals are not strictly identical and there are families of genotypes within the variety. Nevertheless, if a variety is used to establish a long-term grassland, it is possible that the management of this grassland will modify the initial proportion of the lines of the variety. After 20 years of exploitation, for example, it is possible that some families will have disappeared from the sward.

Conservation of grass biodiversity can be envisaged *ex situ* in seed banks, in living collections (notably in botanical gardens, in breeding establishments, by nursery workers or amateur collectors), or *in situ* in nature reserves, national parks or agricultural systems.

Ex situ conservation can be effective for protecting species or a small number of threatened varieties or ecotypes. On the other hand, it is not very effective for maintaining the vast genetic diversity characteristic of species, and the same is true for conserving the great diversity of communities. In addition, *ex situ* conservation is very expensive, especially when live plants are maintained.

Conservation efforts should therefore concentrate on *in situ* protection of grass diversity. Ecological factors and agricultural practices have created a vast biodiversity that can only be conserved by protecting the habitats and using management methods close to those that created that diversity. Nature reserves can play a role in this sort of conservation but, since their area is only a small proportion of the territory, at least in Europe, they cannot be used to solve the overall problem. Other land policies, such as national parks, allow the protection of greater areas. A European ecological network is also in the process of being set up, namely the ambitious programme, Natura 2000. The bulk of the diversity of genes and communities should therefore be ensured in agricultural conditions where often extensive ancestral traditional practices are kept up. Maintaining this kind of farming often requires structural and financial assistance from states, or federations of states, at least when agriculture is being simultaneously intensified in these regions. Aid to less favoured regions and the 'agri-environmental' scheme of the European Union are examples of practicable policies.

The species diversity of a sward evolves as a function of the community's production (Al Mufti *et al.*, 1977; Grime, 1979) (Figure 2.3).

Swards of very low productivity (about 1–3 tonnes DM/ha annually), like those on very nutrient-poor and very acid soils for example, contain few species. These communities are essentially composed of oligotrophic species like *Agrostis canina, Briza media*, purple moor-grass (*Molinia caerulea*), mat-grass (*Nardus stricta*) and *Sieglingia decumbens*. Moderately productive swards (about 5 tonnes DM/ha annually) growing on acidic or alkaline soils moderately supplied with nutrients, can contain many species. These communities are composed of a majority of mesotrophic species like *Agrostis capillaris, Festuca rubra* and *Trisetum flavescens*, but also some oligotrophic and eutrophic species. Very productive swards (up to 10 to 15 tonnes DM/ha annually), like grasslands containing legumes or fertilized with nitrogen, growing on soils rich in nutrients and slightly acid or alkaline, contain few species. These are essentially eutrophic species like *Dactylis glomerata,*

FIGURE 2.3

Relation between the maximum standing biomass + litter and the density of species

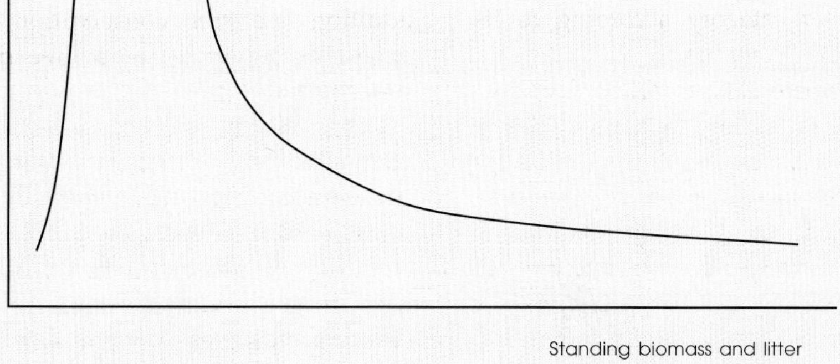

Source: after Al Mufti *et al.*, 1977

couch (*Elymus repens*), *Lolium perenne, Phleum pratense, Poa trivialis*, and some wide-ranging mesotrophic or meso-eutrophic grasses like *Arrhenatherum elatius* and *Holcus lanatus*.

It can also be seen in grass swards that an associated animal biodiversity often evolves in the same way, as a function of the primary biomass (Green, 1990) and this has been demonstrated for birds and insects. The maximum number of insect species is seen at levels of biomass production that are slightly higher than the levels that correspond to the maximum of plant species. Insofar as water birds are concerned (waders and geese), the maximum of species is seen in eutrophic habitats ('t Mannetje, 1994). Nitrogen fertilizing increases the production of plant biomass and thus also the amount of dead material produced. In turn, this stimulates decomposers in the soil including earthworms, which serve as food for waders. Nitrogen fertilization also encourages a better autumn regrowth of pasture, which provides more feed for grey (*Anser* spp.) and black and white geese (*Branta* spp.) in winter.

The species richness of grass communities is also influenced by the disturbance regime (Figure 2.4). Swards that have been undisturbed

FIGURE 2.4

Relation between the frequency of disturbance and specific diversity

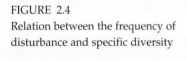

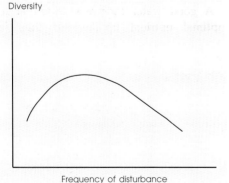

Source: after Grime, 1979 and Hutson, 1979

or little disturbed, like wasteland and pastures long abandoned by agriculture, are characterized by small numbers of species. In such swards, the following species can be found in the following conditions: tor grass (*Brachypodium pinnatum*) and upright brome (*Bromus erectus*) (oligotrophic habitat, calcareous, dry soil); *Molinia caerulea* (oligotrophic habitat, very acid, humid soil); *Arrhenatherum elatius, Dactylis glomerata* and *Elymus repens* (mesotrophic habitat to eutrophic, normal water supply to dry); *Deschampsia caespitosa* (mesotrophic habitat, damp soil with a fluctuating water table) or *Glyceria* spp. (mesotrophic to eutrophic habitat, very wet soil). Moderately disturbed swards, like those cut twice yearly, or cut in June-July and grazed in September, contain a great number of species, mainly *Agrostis capillaris, Anthoxanthum odoratum, Arrhenatherum elatius, Briza media, Dactylis glomerata, Festuca rubra, Holcus lanatus, Poa pratensis* and *Trisetum flavescens*. Swards regularly and often disturbed, like grazed pastures and turf defoliated between five and 20 times yearly contain few species. Among these are *Agrostis capillaris, Cynosurus cristatus, Festuca rubra, Lolium perenne, Poa pratensis, P. trivialis* and *P. annua*.

Some plant communities are richer than others in grass species, but not all grasses are equally competitive towards other plant species, whether or not they are grasses. Some species are especially aggressive and leave little room for others. For example, *Alopecurus pratensis, Arrhenatherum elatius, Dactylis glomerata, Lolium perenne* and *L. multiflorum* have this characteristic. Their aggressiveness increases as the availability of soil nutrients rises. Other species can be aggressive even on poor soils when their communities are not disturbed by regular defoliation; *Brachypodium pinnatum, Bromus erectus* and *Molinia caerulea* belong to this category. All these grasses easily exclude a great number of other species, notably those which grow slowly or have a small size, and so they are often part of species-poor communities. In constrast, *Agrostis capillaris, Anthoxanthum odoratum, Festuca rubra* and *Trisetum flavescens* form 'more welcoming' swards for other species. These grasses can therefore be part of species-rich swards.

2.8 ECOLOGY OF GRASSES

Despite the fact that many grasses exhibit a great similarity of growth habit, they have very different ecological requirements for many factors, such as soil moisture, soil texture, nutrient availability, soil pH, light and frequency of defoliation. They can also be more or less tolerant of limiting factors, such as drought, flooding, trampling, winter frost, snow periods and late frosts in spring. However, all ecological factors do not have the same importance. Four main factors influence sward composition:
- type of management,
- nutrient availability,
- soil moisture, and
- soil pH.

Grassland can be managed through exclusive grazing, exclusive cutting or by mixed use of cutting and grazing. The effect of cutting is very different from that of grazing for several reasons. When a sward is cut, it is usually less frequently defoliated than when it is grazed. There can be two to three cuts per year for hay production and three to four cuts for silage production, while an intensive rotational grazing implies six to seven defoliations. Moreover, with grazing, the sward is not defoliated homogeneously as it is with cutting. Areas rejected by the animals appear where plants can produce seeds. Nutrient exports are important in cutting systems, while with grazing, the majority of the nutrients in the grass ingested by the animals is recycled back to the soil via the dung and urine. Finally, the trampling by animals can have a considerable adverse effect on species that are sensitive to this factor. All these aspects are influenced by the frequency of defoliation, the stocking rate and the grazing system (rotational or continuous). Table 2.6 shows some species ranked according to their sensitivity to grazing.

Most of the nutrients required by plants come from the soil, e.g. by mineralization, from the colloid fraction, and from organic or mineral fertilizers spread by the farmer, while nutrient nitrogen can also come from nitrogen fixation by legumes. High nutrient availability favours the fast-growing species to the detriment of others. In a heavily fertilized sward, a small number of grasses (eutrophic grasses) stifle other species. When the nutrient availability is moderate, there are a great number of species, mainly mesotrophic types. In some cases, the number of plant species per unit area can be very high and reach 55 per 100 m². On a soil that is very poor in nutrients, only a handful of oligotrophic species can survive. Table 2.7 ranks some widespread species according to their requirements for nutrients.

A good water supply is essential for optimal nutrient uptake and thus for

TABLE 2.6
Species sensitivity to grazing

Lolium perenne	tolerant to grazing
Poa pratensis	
Festuca rubra, F. pratensis, Holcus lanatus	
Alopecurus pratensis	
Phleum pratense	
Elymus repens, Arrhenatherum elatius	sensitive to grazing

productive grasses to express their potential. Some species (*Agrostis canina, A. stolonifera, Alopecurus geniculatus, A. pratensis, Deschampsia caespitosa, Glyceria fluitans, G. maxima, Lolium multiflorum, L. perenne, Phleum alpinum, P. pratense* and *Poa trivialis*, for instance) are particularly sensitive to drought. They quickly show a reduction in growth when water becomes limited and, if the drought is prolonged, they suffer from a high mortality rate. Some other species (*Arrhenatherum elatius, Bromus carthaticus, B. erectus, B. inermis, Dactylis glomerata, Lolium multiflorum* and *L. perenne*, for instance) are sensitive to flooding. They do not persist if the flooding period, which usually occurs in winter, is prolonged. Several species that are relatively resistant to summer drought, like *Festuca arundinacea* and *F. pratensis*, are also resistant to winter flooding. However, several drought-tolerant species, like *Arrhenatherum*

elatius, Bromus erectus, B. inermis and *Dactylis glomerata*, are not well adapted to flooding, and many species adapted to very wet conditions, like *Agrostis canina, A. stolonifera, Alopecurus geniculatus, A. pratensis, Deschampsia caespitosa, Glyceria fluitans* and *G. maxima*, are not drought-tolerant. Table 2.8 ranks some productive grasses according to their drought tolerance, while Table 2.9 shows the requirements of a wider range of species for soil moisture.

Extreme values of soil pH reduce nutrient availability, especially phosphorus availability. In very acid soils, phosphorus is strongly adsorbed by iron and aluminium oxides and, in alkaline soils, phosphorus is blocked as calcium phosphate, which also makes it unavailable for plants. Moreover, soil organic matter mineralization is reduced at low and high values of pH and consequently reduces the nitrogen and phosphorus supply by the

TABLE 2.7
Species requirements for nutrient availability

Lolium multiflorum	high requirements
Alopecurus geniculatus, A. pratensis, Elymus repens, Festuca arundinacea, Lolium perenne, Phleum pratense, Poa trivialis	
Agrostis gigantea, A. stolonifera, Arrhenatherum elatius, Bromus hordeaceus, Dactylis glomerata, Festuca pratensis, Holcus lanatus, Poa pratensis	
Bromus inermis, Cynosurus cristatus, Deschampsia caespitosa, Festuca rubra, Hordeum secalinum, Phleum alpinum, Poa alpina	
Trisetum flavescens	
Agrostis capillaris	
Anthoxanthum odoratum	
Agrostis canina, Avenula pratensis, A. pubescens, Briza media, Bromus erectus, Deschampsia flexuosa, Festuca ovina, Koeleria pyramidata, Holcus mollis, Molinia caerulea, Nardus stricta	low requirements

TABLE 2.8
Species tolerance of drought

Festuca arundinacea	high drought tolerance
Dactylis glomerata, Arrhenatherum elatius, Poa pratensis	
Festuca pratensis	
Lolium perenne, L. multiflorum	low drought tolerance

TABLE 2.9
Species requirements for soil moisture

Glyceria maxima	high requirements
Glyceria fluitans, Phalaris arundinacea	
Agrostis canina, Alopecurus geniculatus	
Deschampsia caespitosa	
Alopecurus pratensis, Agrostis stolonifera	
Festuca arundinacea, F. pratensis, Holcus lanatus, Poa trivialis	
Phleum alpinum, P. pratense	
Hordeum secalinum, Poa annua	
Briza media, Bromus hordeaceus, Cynosurus cristatus, Elymus repens, Lolium multiflorum, L. perenne, Poa alpina	
Agrostis capillaris, Anthoxanthum odoratum, Arrhenatherum elatius, Avenula pubescens, Festuca rubra, Poa pratensis, Trisetum flavescens	
Briza media, Dactylis glomerata	
Bromus inermis	
Avenula pratensis, Bromus erectus, Festuca ovina, Koeleria pyramidata	low requirements

soil. Aluminium can affect root development in very acid soils because this toxic element becomes more soluble at very low pH values. Some species, like *Agrostis canina, A. capillaris, Anthoxanthum odoratum, Deschampsia flexuosa, Holcus mollis, Molinia caerulea* ssp. *caerulea* and *Nardus stricta*, are particularly tolerant of low pH values, but others, like *Alopecurus pratensis, Lolium multiflorum* and *L. perenne*, can only survive in conditions close to neutrality or on slightly acid soils. Some other species, like *Arrhenatherum elatius, Avenula pratensis, A. pubescens, Briza media* (calcicolous ecotype), *Bromus erectus, B. inermis, Dactylis glomerata, Festuca arundinacea, F. ovina, Koeleria pyramidata, Molinia caerulea* ssp. *arundinacea, Poa pratensis* and *Trisetum flavescens* for instance, are particularly tolerant of the toxic effect of an excess of calcium in the soil. In some species, different ecotypes or subspecies show contrasting behaviours for pH. One taxon is adapted to very acid soils, while the other thrives on alkaline soils; *Briza media* and *Molinia caerulea* are two species that show this behaviour. Table 2.10 ranks grass species according to their requirements for soil pH.

Other factors, like light intensity and temperature, are also important for plant growth and competition, but to a lesser extent than the four previous factors. Most forage grasses are 'full light' plants, but several are shade-tolerant to some extent. *Dactylis glomerata, Deschampsia caespitosa, D. flexuosa, Holcus mollis, Molinia caerulea* and *Poa trivialis* are the most shade-tolerant species, while *Arrhenatherum elatius, Avenula pratensis, A. pubescens, Bromus erectus, Festuca ovina* and *Koeleria pyramidata* have high requirements for light. These species also have high requirements for temperature, and often grow well on south-facing slopes. However, *Dactylis glomerata*, which tolerates shade, also appreciates warm microclimates. Most temperate species can tolerate winter frost very well but some are sensitive to strong winter frosts, e.g. *Bromus catharticus, Holcus lanatus, Lolium multiflorum* and *L. perenne*, and some are particularly sensitive to late frosts in spring, e.g. *Arrhenatherum elatius, Lolium multiflorum* and *L. perenne*.

Altitude is not an ecological factor in itself, but this parameter influences several ecological factors. For each 100 m of elevation, temperature decreases by about 0.4 °C in

TABLE 2.10
Species requirements for soil pH

Avenula pratensis, A. pubescens, Bromus erectus, Koeleria pyramidata	high pH requirements
Arrhenatherum elatius, Bromus inermis, Hordeum secalinum, Trisetum flavescens	
Agrostis gigantea, A. stolonifera, Alopecurus geniculatus, A. pratensis, Bromus hordeaceus, Cynosurus cristatus, Dactylis glomerata, Elymus repens, Festuca arundinacea, F. pratensis, Glyceria fluitans, G. maxima, Lolium multiflorum, L. perenne, Phalaris arundinacea, Poa trivialis, Phleum alpinum, P. pratense, Poa alpina, P. annua, P. pratensis	
Deschampsia caespitosa, Festuca rubra, Holcus lanatus	
Agrostis capillaris, Anthoxanthum odoratum	
Agrostis canina, Deschampsia flexuosa, Holcus mollis, Nardus stricta	low pH requirements

autumn and winter and by about 0.7 °C in spring and summer. In the Alps, the number of freezing days ranges from 150 days at 1 500 m to 190 days at 2 000 m and 310 days at 3 100 m. The isotherm of 0 °C lies between 2 600 m and 3 000 m (Rameau *et al.*, 1993). Light intensity is higher at high altitudes compared to the lowlands and the albedo of the soil and the vegetation is also much higher, which implies a faster decrease of temperature during the night and a higher risk of frost, even in summer. In the Alps, annual rainfall increases with altitude by about 100 mm per 100 m of elevation, but the maximum rainfall, especially in summer, is recorded at about 1 700–1 800 m (Rameau *et al.*, 1993). The absolute air humidity decreases above 1 800 m, which implies a higher evapotranspiration of plants, which is increased further by faster wind speeds at altitude. The duration of snow cover also increases with altitude; it ranges from three months at 600 m to ten months at 2 500 m in the central Alps (Rameau *et al.*, 1993).

Because altitude has such a strong influence on all these ecological factors and especially on the temperatures, the duration of the growing period decreases with altitude by about six to seven days per 100 m of elevation in the Alps (Favarger, 1962).

Altitude also influences the vegetation composition and classically five belts have been defined, according to dominant or characteristic species of the climax vegetation (Favarger, 1963 and 1966; Ozenda, 1985 and 1994; Rameau *et al.*, 1993) (Figure 2.5).

The **hill belt** corresponds to the presence of oaks (*Quercus* spp.). These trees reach their upper limit at the higher altitudes of this belt. Coniferous trees are normally absent, except for some species like Scots pine (*Pinus sylvestris*) in special situations. In the Alps, the average annual temperature in the hill belt ranges from 8 to 12 °C and the duration of the growing period is over 200 days.

The **mountain belt** is dominated by beech (*Fagus sylvatica*) often mixed with European silver-fir (*Abies alba*). *Pinus sylvestris* can be very abundant on south-facing slopes and in other dry conditions. Norway spruce (*Picea excelsa*) and European larch (*Larix decidua*) are sometimes noted in special situations. In the Alps, the average annual temperature in the mountain belt ranges from 4 to 8 °C and the duration of the growing period is still higher than 200 days.

The **subalpine belt** is characterized by the abundance of coniferous trees (*Abies alba, Larix decidua, Picea abies*, arolla pine [*Pinus cembra*] and mountain pine [*P. uncinata*]). However, all of these species are not always present; for instance *Larix decidua* and *Picea abies* are not naturally present in the Pyrenees and, in low-elevation mountains, *Abies alba*,

FIGURE 2.5
Vegetation belts and altitude limits for north- and south-facing slopes in the central Alps

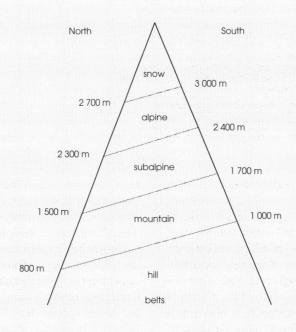

Fagus sylvatica and *Picea excelsa* in pure stand or in mixture can be important. In the upper part of this belt, *Ericaceae* shrubs, especially alpenrose (*Rhododendron* spp.) can be abundant in the understorey of trees or can constitute moors. On south-facing slopes, dwarf juniper (*Juniperus nana*) and purgative broom (*Cytisus purgans*) can also be present. In the Alps, the average annual temperature in the subalpine belt ranges from -2 to +4 °C and the duration of the growing period ranges from 100 to 200 days.

The **alpine belt** develops above the timberline and the moors of *Ericaceae*. The growth of trees is no longer possible because of extreme ecological conditions. This belt is characterized by a grassy vegetation dominated by several *Carex* species, namely *Carex sempervirens*, and several grass species including some *Festuca*. In the Alps, the average annual temperature in the alpine belt is lower than 0 °C and the

duration of the growing period is less than 100 days.

The **permanent snow belt** does not allow the development of higher plants except in some small locations with a warmer microclimate. Only algae and lichens can still survive.

Figure 2.5 indicates the approximate altitude limits of these vegetation belts in the central Alps.

The limits of these vegetation belts vary not only according to the altitude, but also according to the slope orientation, the degree of continentality, the latitude and the position of the site in the mountain range.

The requirements of a species for an ecological factor can be synthesized by an abundance curve (Figure 2.6). Several areas can be distinguished on this curve: two areas (A and E) where the species is absent at the extreme left and the extreme right, two areas (B and D) where the species is present but with a low proportion in the sward, and finally an area (C) where the species is particularly abundant. The areas B, C and D as a whole are the ecological range of the species for that factor, while area C is the ecological optimum. The curve can have different shapes (Figure 2.7), and can be symmetrical or asymmetrical, narrow (steno-oecic species) or wide (eury-oecic species), and there can be one or two optima (examples of this kind of figures are given in Grime, Hodgson and Hunt, 1988 for many grassland species in the Sheffield region, UK).

Some authors (Kruijne and De Vries, 1963; Ellenberg *et al.*, 1991) have characterized the requirements of a species for a factor by a single value on a semi-quantitative scale (range of 0 to 10 or -100 to +100 for instance). This value then corresponds to the ecological optimum. This method is more synthetic but

it does not give information on the shape of the curve and on the ecological range of the species.

In the wild, each species is of course under the influence of more than one factor; each species develops in a multidimensional ecological universe. It is possible to take this partially into account by combining the curves of a species for two factors on a single figure (Figure 2.8). This figure defines a two-dimensional ecological niche for that species. It is possible to draw three-dimensional niches, but from three dimensions and above it is more and more difficult to understand the figure. It is thus preferable to draw several two-dimensional figures (Rameau *et al.*, 1989 and 1993). This is the method adopted in this book for the species profiles in Chapter 5. In each profile, the first figure combines the requirement of a species for soil pH (R) with soil humidity (H), and the second combines

FIGURE 2.6
Example of an abundance curve of a species for an ecological factor

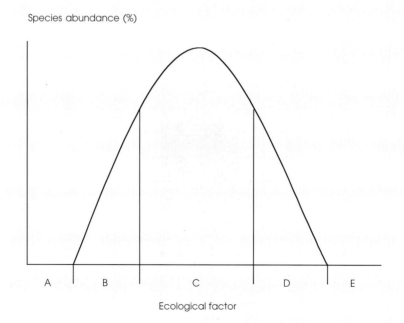

FIGURE 2.7

Examples of different shapes of abundance curves of species for an ecological factor

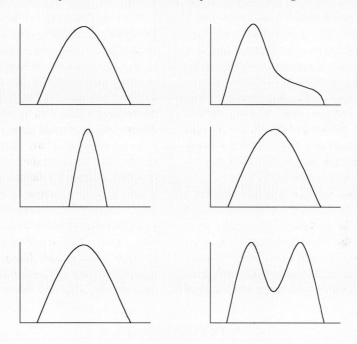

FIGURE 2.8

Theoretical example of an ecological optimum and an ecological range of a species for two factors

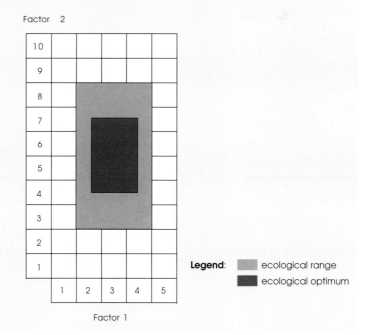

Factor 2

Factor 1

Legend: ecological range

ecological optimum

the requirement for nutrient availability (N) with the management regime (M).

The soil reaction index has 5 levels (Table 2.11). The highest levels include pH values above 7, while the lowest level comprises values below 4. The soil characteristics and the plant types are defined in the two right-hand columns of the table.

The soil humidity index has 10 levels because it is possible and useful to define more than 5 characteristics of grass habitats for that factor (Table 2.12). The plant types corresponding to the index values are given on the right of the table.

The nutrient availability index also has 5 levels (Table 2.13). The main nutrients taken into account in this scale are nitrogen (N) and phosphorus (P). P is mainly important at the lower values of the index where a small increase of availability can profoundly affect the competition between species. N is increasingly important when the value of the index rises. It is often the dominant factor above an index value of 2. The availability index takes the soil supply into account, together with organic and mineral fertilizations. For N, the soil supply is defined here as the sum of the atmospheric N deposition, the N fixation and the organic N mineralization. The plant types corresponding to the index values are given on the right of the table.

The management index is defined on a 5-level scale according to the number of defoliations, either by grazing or by cutting (Table 2.14). A sward can be defoliated more than 5 times per year, especially in intensive rotational or continuous grazing, but then the value of the index is still 5. Examples of possible managements are given on the right of the table.

TABLE 2.11
Soil reaction (pH) index: R

Index	pH	Soil characteristics	Plant types
5	> 7	alkaline, calcareous	basiphilous, calcicolous
4	6–7	neutral or slightly acid	neutrophilous
3	5–6	acid	acidoclinous
2	4–5	very acid	acidophilous
1	< 4	extremely acid	hyper-acidophilous

TABLE 2.12
Soil humidity index: H

Index	Soil characteristic	Plant type
10	permanently flooded	hygrophilous, helophyte
9	very wet, sometimes flooded	hygrophilous
8	permanently wet	meso-hygrophilous
7	rather wet, summer drought possible	hygroclinous
6	slightly wet, summer drought possible	meso-hygroclinous
5	average humidity or drainage	mesophilous
4	fast drainage (warm soil)	thermophilous
3	dry, in summer namely	meso-xerophilous
2	permanently very dry	xerophilous
1	extremely dry	hyper-xerophilous

TABLE 2.13

Nutrient availability index: N

Index	Nutrient availability (fertilization + soil supply)	Plant type
5	very high	eutrophic
4	high	meso-eutrophic
3	moderate	mesotrophic
2	low	meso-oligotrophic
1	very low	oligotrophic

TABLE 2.14

Management index: M

Index	Number of defoliations/year	Examples of management
5	5 or more	exclusive intensive grazing
4	4	4 silage cuts or 2–3 silage cuts + grazing or grazing + 1 silage cut
3	3	3 silage cuts or low stocking rate
2	2	2 hay cuts or 1 hay cut + 1 grazing
1	1	1 hay cut or one occasional grazing

In Chapter 5 each figure presents, for two factors, a black area corresponding to the ecological optimum of the species and a shaded area defining the rest of the ecological range (Figure 2.8). It is then very easy to compare the ecological requirements of different species; for instance in N-M figures, species adapted to intensive systems have their optimum in the upper right of the figure, while species adapted to extensive systems have their optimum in the lower left of the figure. If the niches of two species do not overlap, it is thus impossible to find them growing spontaneously in the same community and it is not appropriate to sow them in a mixture. On the other hand, species that present comparable niches are highly likely to be compatible and are to be found together in the same community.

The ecological optimum and range for a factor can be used to define the indicator value of species for that factor. The best indicators are species with a narrow optimum and range. A species with a narrow range is called an indicator of presence; its presence, even in a small amount, indicates a precise characteristic of the habitat. A species with a wider range but a narrow optimum is called an indicator of abundance; the presence in a small amount of this species has little indicator value but, when this species is abundant, it gives reliable information on the characteristic of the habitat. Indicator values are given in the species profiles of Chapter 5 whenever possible.

Many species have rather broad optimum and range and so their presence and abundance must be interpreted carefully. However, a set of species gives much more information on the habitat than a species taken individually, especially if several species have the same type of requirement for a factor. These species constitute a socio-ecological group, i.e. species that usually grow together and that have similar environmental

TABLE 2.15
Socio-ecological groups for different management regimes

GROUP OF ONLY-GRAZED GRASSLANDS

Eutrophic species
Agrostis stolonifera
Dactylis glomerata
Holcus lanatus
Lolium perenne
Poa annua
Poa pratensis
Poa trivialis

Mesotrophic species
Agrostis capillaris
Cynosurus cristatus
Festuca rubra

Oligotrophic species
Nardus stricta

GROUP OF MIXED (CUT AND GRAZED) GRASSLANDS

Eutrophic species
Agrostis stolonifera
Alopecurus pratensis
Dactylis glomerata
Festuca pratensis
Holcus lanatus
Poa pratensis
Poa trivialis

Mesotrophic species
Agrostis capillaris
Festuca rubra

GROUP OF ONLY-CUT GRASSLANDS

Agrostis capillaris
Arrhenatherum elatius
Avenula pratensis
Avenula pubescens
Bromus hordeaceus
Dactylis glomerata
Elymus repens
Festuca pratensis
Festuca rubra
Holcus lanatus
Phalaris arundinacea (very wet soil)
Phleum pratense
Poa trivialis
Trisetum flavescens

TABLE 2.16
Socio-ecological groups for the soil nutrient availability

GROUP OF RICH TO VERY RICH SOILS
Agrostis stolonifera
Alopecurus geniculatus
Alopecurus pratensis
Arrhenatherum elatius
Bromus hordeaceus
Dactylis glomerata
Elymus repens
Festuca pratensis
Holcus lanatus
Lolium perenne
Phleum pratense
Poa annua
Poa pratensis
Poa trivialis

GROUP OF MODERATELY RICH SOILS
Agrostis capillaris
Agrostis stolonifera
Arrhenatherum elatius
Bromus hordeaceus
Cynosurus cristatus
Dactylis glomerata
Festuca pratensis
Festuca rubra
Holcus lanatus
Poa pratensis
Trisetum flavescens

GROUP OF POOR SOILS
Usually acid soils
Agrostis canina
Agrostis capillaris
Anthoxanthum odoratum
Briza media
Deschampsia flexuosa
Festuca ovina
Holcus mollis
Molinia caerulea
Nardus stricta
Mainly chalk soils
Avenula pubescens
Avenula pratensis
Briza media
Bromus erectus
Festuca ovina
Koeleria pyramidata

requirements. Their use thus allows improvement of the evaluation of a habitat. Socio-ecological groups are identified for the management regime, nutrient availability, soil humidity and soil pH in Tables 2.15 to 2.18. A species can belong to several ecological groups if it has a large range for the considered factor.

Strength of competitiveness is an important characteristic of species, especially when the choice of a mixture is concerned. In intensively managed grasslands, fast-growing species are more competitive than slow-growing species. In cutting regimes, tall species, e.g. *Alopecurus pratensis*, *Arrhenatherum elatius*, *Phleum pratense*, can

TABLE 2.17
Socio-ecological groups for soil humidity

GROUP OF VERY WET GRASSLANDS AND BOGGY PLACES
Glyceria fluitans
Glyceria maxima
Phalaris arundinacea

GROUP OF WET GRASSLANDS
Soils moderately rich to rich in nutrients
Agrostis stolonifera
Alopecurus geniculatus
Alopecurus pratensis
Deschampsia caespitosa
Festuca arundinacea
Holcus lanatus
Poa trivialis
Poor soils
Agrostis canina
Molinia caerulea

GROUP OF COOL SOILS
Alopecurus pratensis
Festuca pratensis
Holcus lanatus
Poa trivialis

GROUP OF NORMALLY DRAINED SOILS
Eutrophic species
Agrostis gigantea
Dactylis glomerata
Elymus repens
Lolium perenne
Phleum pratense
Poa annua
Poa pratensis
Mesotrophic species
Agrostis capillaris
Bromus hordeaceus
Cynosurus cristatus
Festuca rubra
Trisetum flavescens
Oligotrophic species
Briza media
Deschampsia flexuosa
Holcus mollis
Nardus stricta

GROUP OF WELL DRAINED, WARM SOILS (THERMOPHILOUS SPECIES)
Arrhenatherum elatius
Dactylis glomerata
Poa pratensis
Trisetum flavescens

GROUP OF VERY DRY SOILS
Avenula pratensis
Avenula pubescens
Bromus erectus
Festuca ovina
Koeleria pyramidata

TABLE 2.18
Socio-ecological groups for soil pH

GROUP OF ALKALINE SOILS
Avenula pratensis
Avenula pubescens
Briza media
Bromus erectus
Festuca ovina
Koeleria pyramidata

GROUP OF NEUTRAL SOILS
Grazed grasslands
Agrostis stolonifera
Cynosurus cristatus
Dactylis glomerata
Festuca rubra
Holcus lanatus
Lolium perenne
Poa annua
Poa pratensis
Poa trivialis
Cut and grazed grasslands
Agrostis stolonifera
Alopecurus pratensis
Dactylis glomerata
Holcus lanatus
Festuca pratensis
Festuca rubra
Poa pratensis
Poa trivialis
Cut grasslands
Agrostis stolonifera
Arrhenatherum elatius
Bromus hordeaceus
Dactylis glomerata
Elymus repens
Festuca rubra
Holcus lanatus
Phleum pratense
Poa pratensis
Poa trivialis
Trisetum flavescens

GROUP OF ACID SOILS
Agrostis capillaris
Anthoxanthum odoratum
Cynosurus cristatus
Dactylis glomerata
Festuca ovina
Festuca pratensis
Festuca rubra
Holcus lanatus
Poa pratensis
Trisetum flavescens

GROUP OF VERY ACID SOILS
Agrostis canina
Briza media
Deschampsia flexuosa
Festuca ovina
Holcus mollis
Molinia caerulea
Nardus stricta

shade shorter species, e.g. *Festuca rubra*, *Lolium perenne*, especially under infrequent cutting regimes (Table 2.19). Under grazing conditions, the maximum height of a species is less important than its tillering density (Table 2.19). For instance, *Lolium perenne* and *Poa pratensis* can produce a high number of tillers per surface unit. Stolons are also effective means of species development in an intensively grazed sward, e.g. *Agrostis stolonifera*. Rhizomes can be efficient either in grazing, e.g. *Festuca rubra*, *Poa pratensis*, or in cutting, e.g. *Alopecurus pratensis*, *Elymus repens*.

The species profiles of Chapter 5 also give information on the occurrence of each species in plant communities. The alliances where the species is mostly frequent are listed. Only the perennial herbaceous vegetation is considered. The nomenclature of plant communities corresponds to the CORINE system summarized in section 2.5 of this chapter for the main grassland and rangeland communities. Sometimes, other communities are mentioned, such as weed alliances when they have a forage value. Other references consulted are Oberdorfer (1977, 1978, 1983 and 2001), Rodwell (1992)

and Bournérias, Arnal and Bock (2001).

2.9 FORAGE VALUE OF GRASSES AND GRASS SWARDS

The forage value of a species is difficult to define because it depends on several general parameters, such as production potential and quality. The quality itself is a rather vague term, which includes specific parameters like digestibility, energy content, crude protein content, mineral content, absence of toxic constituents, persistency, disease resistance and heading date.

Several authors have proposed scales of classification. Table 2.20 includes the three best known scales. The scale of Kruijne and De Vries ranges from 0 to 10, that of Delpech from 0 to 5 and that of Klapp from -1 to +8. Kruijne and De Vries have also classified the grasses in three categories: the good grasses, the medium grasses and the bad grasses. The scale of Delpech is based on the yield potential, while Kruijne and De Vries' scale combines yield and quality. These two scales differ mainly for *Festuca arundinacea* and *Dactylis glomerata*, which receive very high scores in the scale of Delpech. Kruijne and De Vries give them lower values because

TABLE 2.19
**Competitiveness of some widespread species
in heavily fertilized grasslands**

Mainly cut temporary grasslands	
Lolium multiflorum (short-term competition only)	very competitive
Dactylis glomerata	
Alopecurus pratensis, Arrhenatherum elatius	
Elymus repens, Festuca rubra, Holcus lanatus	
Lolium perenne, Phleum pratense	
Festuca pratensis	less competitive
Mainly grazed permanent grasslands	
Lolium perenne, Poa pratensis	very competitive
Dactylis glomerata, Holcus lanatus	
Festuca rubra	
Alopecurus pratensis, Festuca pratensis	
Arrhenatherum elatius, Phleum pratense	
Elymus repens	less competitive

TABLE 2.20
Indices of agronomic value of grasses

	Kruijne and De Vries (1976)	Delpech (1960)	Klapp (1965)
Good grasses			
Lolium perenne	10	5	8
Festuca pratensis	9	5	8
Phleum pratense	9	5	8
Poa pratensis	9	4	8
Arrhenatherum elatius	8	4	7
Avenula pubescens	8	-	4
Lolium multiflorum	8	4	-
Poa trivialis	8	3	7
Medium grasses			
Alopecurus pratensis	7	3	7
Cynosurus cristatus	7	1	6
Hordeum secalinum	7	1	-
Trisetum flavescens	7	3	7
Dactylis glomerata	6	5	7
Phalaris arundinacea	6	-	5
Agrostis gigantea	-	3	-
Agrostis stolonifera	5	3	7
Elymus repens	5	2	6
Holcus lanatus	5	2	4
Bad grasses			
Agrostis capillaris	4	2	5
Anthoxanthum odoratum	4	1	3
Briza media	4	-	5
Bromus erectus	4	-	5
Bromus inermis	4	-	-
Festuca arundinacea	4	4	4
Festuca rubra	4	2	4/5
Glyceria fluitans	4	-	4
Glyceria maxima	4	-	4
Poa annua	4	0	5
Poa nemoralis	4	-	-
Poa palustris	4	-	7
Alopecurus geniculatus	3	-	4
Bromus hordeaceus	3	0	3
Holcus mollis	3	-	3
Agrostis canina	2	-	3
Molinia caerulea	2	-	2
Festuca ovina	1	-	3
Deschampsia caespitosa	0	-	3
Deschampsia flexuosa	-	-	3
Koeleria pyramidata	-	-	3
Nardus stricta	0	-	2
Avenula pratensis	-	-	2

Legend:

Kruijne and De Vries:	0:	species of no forage value, 10: very good species
Delpech:	0:	species of no forage value, 5: very good species
Klapp:	-1:	toxic species, 0: species of no forage value, +8: very good species

these two species are often not well accepted by the animals. Klapp attributes a negative value to toxic species.

The indices of agronomic value can be used for calculating the pastoral value of a sward by multiplying the index of each species by its proportion (%) and by summing these products for all species:

$$\text{Pastoral value (PV)} = \sum_{i=1}^{n} \frac{AV_i \times P_i}{10}$$

with AV = agronomic value of each species
P = proportion (%) of each species in the sward
i = species
Maximum value of PV = 100

These relatively old classification scales can be improved by using the more recently acquired knowledge summarized in Chapter 5 in the species profiles. These data were used to rank the species for each parameter. The performances of each species were compared two by two and the rankings progressively built.

Production under cutting conditions and digestibility are combined in Table 2.21 for classifying the species on the basis of the data of Chapter 5 into 16 groups (one group does not include species). The most valuable

species are located in the upper and the right boxes of the table, especially in boxes: 4-4, 3-4, 4-3, 3-3, 4-2 and 3-2.

Table 2.22 also classifies the species in decreasing order of digestibility, but it is split into two parts to take into consideration the fact that some species in grazed swards are either not ingested at all or hardly ingested, or do not tolerate grazing.

A synthesis of Tables 2.21 and 2.22 has been attempted in Table 2.23 for grazing and conservation. This exercise is of course difficult, being partly subjective. However, the subjectivity is reduced to a minimum by the two-step reasoning referred to above.

In considering not only the yield potential and the digestibility, but also the other qualitative criteria mentioned above, e.g. persistency, disease resistance and heading date, the list is finally modified to provide a ranking of the species based on a global value (Table 2.24).

With regard to grazing, this classification is different from that of Kruijne and De Vries

TABLE 2.21
**Classification of grass species according to a production
and a digestibility scale**

Digestibility scale	Production scale			
	1	2	3	4
4	Poa annua	Poa trivialis Phleum alpinum	Lolium perenne Phleum pratense Festuca pratensis	Lolium multiflorum
3	Agrostis canina	Cynosurus cristatus Agrostis stolonifera Agrostis capillaris Anthoxanthum odoratum Poa alpina	Holcus lanatus Elymus repens Agrostis gigantea Poa pratensis Trisetum flavescens Glyceria fluitans	Dactylis glomerata Bromus catharticus Arrhenatherum elatius Bromus inermis
2	Briza media Bromus erectus Hordeum secalinum Molinia caerulea Deschampsia flexuosa Avenula spp.	Holcus mollis Bromus hordeaceus	Festuca rubra Alopecurus pratensis	Festuca arundinacea Phalaris arundinacea
1	Koeleria pyramidata Alopecurus geniculatus Festuca ovina Nardus stricta	Deschampsia caespitosa	Glyceria maxima	

Key: 1 = very low, 4 = very high

TABLE 2.22

Some widespread species in order of quality

Grazing	Conservation	
Lolium perenne	*Lolium perenne*	best quality
Phleum pratense	*Phleum pratense, Festuca pratensis*	
Festuca pratensis	*Lolium multiflorum*	
Poa trivialis	*Poa trivialis, Elymus repens*	
Holcus lanatus	*Holcus lanatus*	
Agrostis stolonifera	*Agrostis stolonifera*	
Dactylis glomerata	*Dactylis glomerata*	
Agrostis capillaris	*Agrostis capillaris*	
Poa pratensis	*Arrhenatherum elatius, Poa pratensis*	
Festuca rubra	*Festuca rubra*	
Festuca arundinacea	*Alopecurus pratensis, Festuca arundinacea*	
Festuca ovina		poor quality

TABLE 2.23

Some widespread species in order of production and quality

Grazing	Conservation	
Lolium perenne, Phleum pratense	*Lolium perenne, L. multiflorum, Phleum pratense*	high performance
Dactylis glomerata, Festuca pratensis, Holcus lanatus	*Dactylis glomerata, Festuca pratensis, Holcus lanatus*	
Poa trivialis	*Poa trivialis*	
Poa pratensis	*Poa pratensis, Arrhenatherum elatius, Elymus repens*	
Agrostis stolonifera, A. capillaris,	*Agrostis stolonifera, A. capillaris,*	
Festuca rubra	*Festuca rubra*	
Festuca arundinacea	*Festuca arundinacea*	
Festuca ovina	*Alopecurus pratensis*	low performance

TABLE 2.24

Some widespread species in order of global value

Grazing	Conservation	
Lolium perenne	*Lolium perenne, L. multiflorum, Phleum pratense*	high value
Phleum pratense	*Dactylis glomerata*	
Poa pratensis, Dactylis glomerata, Holcus lanatus	*Poa trivialis, Festuca pratensis*	
Festuca pratensis, Poa trivialis	*Poa pratensis, Arrhenatherum elatius, Holcus lanatus, Elymus repens, Festuca arundinacea*	
Agrostis stolonifera, A. capillaris, Festuca rubra	*Agrostis stolonifera, A. capillaris, Festuca rubra*	
Festuca arundinacea	*Alopecurus pratensis*	
Festuca ovina		low value

TABLE 2.25

Classification of grasses according to the intensification level

Grazing	Conservation
Species suited to intensive systems	
Dactylis glomerata, Festuca arundinacea, Holcus lanatus, Lolium perenne, Poa pratensis, P. trivialis	Bromus catharticus, Dactylis glomerata, Elymus repens, Festuca arundinacea, F. pratensis, Lolium multiflorum, L. perenne, Phleum pratense, Poa pratensis, P. trivialis
Species suited to semi-intensive systems	
Agrostis capillaris, A. stolonifera, Cynosurus cristatus, Dactylis glomerata, Festuca arundinacea, F. pratensis, F. rubra, Holcus lanatus, Lolium perenne, Poa pratensis, P. trivialis, Trisetum flavescens	Agrostis capillaris, A. gigantea, A. stolonifera, Alopecurus pratensis, Arrhenatherum elatius, Bromus hordeaceus, B. inermis, Dactylis glomerata, Elymus repens, Festuca pratensis, F. rubra, Holcus lanatus, Phalaris arundinacea, Phleum pratense, Poa pratensis, P. trivialis, Trisetum flavescens
Species suited to extensive systems	
Agrostis capillaris, Anthoxanthum odoratum, Briza media, Dactylis glomerata, Deschampsia flexuosa, Festuca ovina, F. rubra, Holcus mollis, Molinia caerulea, Nardus stricta, Phleum alpinum, Poa alpina, P. pratensis, Trisetum flavescens	Agrostis canina, A. capillaris, Anthoxanthum odoratum, Arrhenatherum elatius, Avenula pratensis, A. pubescens, Briza media, Bromus erectus, Dactylis glomerata, Deschampsia flexuosa, Festuca rubra, Holcus mollis, Molinia caerulea, Phleum alpinum, Poa alpina, P. pratensis, Trisetum flavescens
Species of no or low value in any system	
Alopecurus geniculatus, Deschampsia caespitosa, Hordeum secalinum, Koeleria pyramidata, Poa annua	

mainly because *Dactylis glomerata* and *Holcus lanatus* have since been better assessed. The classification is consistent with the opinion of Delpech on *Dactylis glomerata* with regard to cutting, in which the value of *D. glomerata* is emphasized. *Holcus lanatus, Elymus repens* and *Festuca arundinacea* rank higher in this classification, compared with the scale of Kruijne and De Vries, but *Alopecurus pratensis* ranks lower. Some of these modifications are justified by the existence of new varieties resistant to diseases.

The classification of species in good, medium and bad grasses is rather partial because this judgement of grasses is only based on the needs of intensive systems. However, a species can for example be adapted to an extensive system and not to an intensive one. Moreover, a species can be adapted to several systems and some species have low values in any system. It is thus more

justified to classify the grasses according to their value for different intensification levels. Table 2.25 presents this new and more positive type of classification.

2.10. DISEASES

A large range of diseases caused by fungi, bacteria or viruses affect forage grasses and these organisms have diverse effects. Some micro-organisms decrease forage yield and some particularly aggressive parasites can even lead to the death of the plants that they attack. This is the case for *Xanthomonas translucens* and snow mould (group of fungi including *Microdochium* [= *Fusarium*] *nivale* and *Typhula incarnata*). The diseases also influence forage quality by modifying the plant's chemical composition (protein, water-soluble carbohydrate and cellulose contents) and its digestibility. Toxins detrimental to cattle are produced by some fungi (Lancashire and Latch, 1966; Skipp and Hampton, 1996). In grazing, the development of fungi that attack the foliage can considerably reduce animal intake.

Table 2.26 presents the main diseases that develop on the most common forage grasses. The diseases that are particularly detrimental are highlighted. This table is a synthesis of the data in Chapter 5 on the grass species profiles.

2.10.1 Fungi

The diseases caused by fungi are the most widespread. They develop on the surface or in the tissues of the plant. Many are present

TABLE 2.26

Main diseases of the most common forage grasses

	ALPR	AREL	BRCA	DAGL	ELRE	FEAR	FEPR	HOLA	LOMU	LOPE	PHAR	PHPR	POPR	TRFL
FUNGI														
Rusts (*Puccinia* spp.)														
Black rust (*P. graminis*)	x	x	x	x	x	x	x	x		x		x	⊗	x
Yellow rust (*P. striiformis*)			x	⊗					x	x		x	⊗	
Brown rust (*P. recondita*)	x				x				x	x				x
Crown rust (*P. coronata*)	x	x		x	x	⊗	⊗	⊗	⊗	⊗	x		⊗	
Others		x		x	x				⊗	x	x	x	⊗	
Leaf blotches (*Rhynchosporium* spp.)														
R. secalis	x		x	x	x	x	x			x	x			
R. orthosporum				⊗		x			⊗	⊗				
Helminthosporium diseases (*Drechslera* spp.)														
D. dictyoides	x		⊗	x		x	x		⊗	⊗		x		
D. siccans				x		x	x		⊗	⊗		x		
Others	x	⊗	x	x	x	⊗	x	x	x	x		x	⊗	
Mastigosporium leaf flecks														
M. album	⊗											x		
M. rubricosum				⊗										
Others												x		
Timothy eyespot (*Cladosporium phlei*)													⊗	
Brown stripe (*Cercosporidium graminis*)	x			⊗			x		x	x		x		
Snow mould (*Microdochium nivale, Typhula incarnata, ...*)				⊗	x	x	x	x	⊗	x		x	x	
Smuts														
Ustilago spp.		x	x		x			x	x	x	x	x		x
Urocystis spp.							x		x	x				
Others									x	x	x			
Powdery mildew (*Blumeria graminis*)			x	x	x	x	x	x	x	x		x	x	
Choke (*Epichloë typhina*)			x	x				⊗				x		
Ergot (*Claviceps purpurea*)	x		x	x	x	x	x	⊗	x	x		x		
Spermospora leaf spot (*S. lolii*)						x	x		x	x			x	
Ramularia leaf spot (*R. holci-lanati*)				x				x						
Ascochyta leaf blight (*A.* spp.)					x	x				x		x	x	
Neotyphodium spp.						x				x				
BACTERIA														
Bacterial wilt (*Xanthomonas translucens*)		x	x			x	x		⊗	⊗		⊗		
VIRUS														
BYDV virus									x	x				
RMV virus									x	x				

Key: ALPR: *Alopecurus pratensis*   HOLA: *Holcus lanatus*   AREL: *Arrhenatherum elatius*   LOMU: *Lolium multiflorum*
   BRCA: *Bromus catharticus*   LOPE: *Lolium perenne*   DAGL: *Dactylis glomerata*   PHAR: *Phalaris arundinacea*
   ELRE: *Elymus repens*   PHPR: *Phleum pratense*   FEAR: *Festuca arundinacea*   POPR: *Poa pratensis*
   FEPR: *Festuca pratensis*   TRFL: *Trisetum flavescens*

x: disease present on the host plant   ⊗: disease particularly detrimental to the host plant

on the leaves where they often develop into spots of variable shapes and colours which help to identify them, e.g. *Puccinia* spp., *Rhynchosporium* spp., *Drechslera* spp., *Mastigosporium* spp. and *Cladosporium* spp.

Rusts, especially *Puccinia coronata*, seriously affect many species, particularly in summer and at the beginning of autumn. They are characterized by the presence of many yellow, orange or brownish powdery blisters (Plate 2.1). In warm and dry weather the spores are easily disseminated, thus favouring epidemics. These rusts lead rapidly to the senescence of the foliage and strongly reduce intake in grazing. On *Lolium perenne*, *Puccinia coronata* reduces yield, quality and competitiveness of the plants (Lancashire and Latch, 1966; Potter, 1987). However, varietal susceptibility to rusts differs significantly (Vanbellinghen, Moreau and Maraite, 1991). It is justifiable to take this varietal susceptibility into account, particularly for the choice of *Lolium* spp., *Poa pratensis* and *Dactylis*

PLATE 2.1
Puccinia coronata and *P. graminis* on *Lolium perenne*

PLATE 2.2
Varieties of *Lolium perenne* resistant and susceptible to *Puccinia coronata*

C. VANBELLINGHEN

PLATE 2.3
Varieties of *Dactylis glomerata* resistant and susceptible to *Puccinia striiformis*

C. VANBELLINGHEN

PLATE 2.4
Cladosporium phlei on *Phleum pratense*

glomerata varieties, when establishing grassland. When varieties of these species are harvested side by side in trial plots, the colour differences between plots are often spectacular at the end of summer. Yellow or orange-yellow plots indicate severe attack by the disease, in contrast to the dark green of more resistant varieties (Plates 2.2 and 2.3).

Phleum pratense is not attacked by *Puccinia coronata*, but it is very susceptible to *Cladosporium phlei*, which is specific to it. This fungus causes elliptical brown to purple spots with a pale centre when fructifying (Plate 2.4). It can be associated with yield losses (Roberts *et al.*, 1955) especially in fields for seed production (Smith, 1970).

Rhynchosporium spp. and particularly *Rhynchosporium orthosporum* can bring about severe attacks, mainly on *Lolium* spp. in spring and in summer, and on *Dactylis glomerata*. The symptoms are pale brown to greyish spots usually surrounded by a darker edging (Plate 2.5). Yield and quality losses were observed for *Lolium multiflorum* (Davies

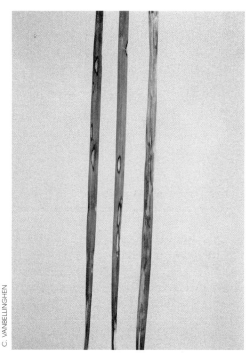

C. VANBELLINGHEN

PLATE 2.5
Rhynchosporium spp. on *Lolium perenne*

C. VANBELLINGHEN

PLATE 2.6
Drechslera siccans on *Lolium perenne*

and Williams, 1970; Lam, 1985). The agent of barley leaf blotch, *R. secalis*, can develop on many forage grasses. These fungi produce strains that vary in their host specificity and their aggressiveness (O'Rourke, 1976; Kastirr, 1998).

Drechslera (previously *Helminthosporium*) spp. affect mainly *Lolium* spp., *Poa pratensis*, *Festuca arundinacea, F. pratensis* and *Phleum pratense*. These very widespread diseases can appear at any moment of the year. They are characterized by dark foliar spots of different morphologies according to the host plant and the pathogenic species involved (Plate 2.6). Yield losses have been reported for *Lolium* spp. (Cook, 1975). Senescence of attacked leaves is accelerated while forage quality, especially digestibility, is reduced (Lam, 1985). *Drechslera* spp. are transmissible to seeds and can cause seedling blight or even an inhibition of germination (Raynal *et al.*, 1989; Ellis and Ellis, 1997).

Mastigosporium spp. are a particular case. They only massively attack *Alopecurus pratensis* and *Dactylis glomerata*; *A. pratensis* is invaded by *Mastigosporium album* while *D. glomerata* is attacked by *M. rubricosum*. In both cases, the foliage is often quickly covered by elliptical, dark brown spots whose centres become clearer during fructification. Necrosis brings these spots together thus reducing the area remaining for photosynthesis and so strongly reducing growth. Grazing animals are very sensitive to the development of these diseases and so the intake of diseased foliage is much reduced compared with that from healthy foliage. The choice of a *Dactylis glomerata* variety resistant to *Mastigosporium* spp. is essential. The new French varieties are much less susceptible to this disease than old varieties.

Snow mould is a common disease in regions where a heavy snow cover remains during winter. This disease, which is caused by a group of fungi including *Microdochium* (= *Fusarium*) *nivale* and *Typhula incarnata*, develops gaps in the sward by destroying

TABLE 2.27
Susceptibility to snow mould of some widespread grass species

Phleum pratense	low susceptibility
Dactylis glomerata, Poa pratensis	
Festuca pratensis	
Festuca arundinacea	
Lolium perenne	
Lolium multiflorum	high susceptibility

plants. Therefore, it decreases sward productivity and longevity and induces the need for early costly resowing. The susceptibility to this disease varies considerably from one species to another and so in regions prone to the disease, a resistant species must be sown. *Phleum pratense* is the most resistant species and *Lolium multiflorum* the most susceptible. Between these two extremes, the species can be classified according to susceptibility as shown in Table 2.27 (Raynal *et al.*, 1989; Ellis and Ellis, 1997).

Blumeria (= *Erysiphe*) *graminis* is a very widespread disease that can develop on most of the forage grasses. This fungus forms a whitish down on the upper part of the leaves and feeds in the leaves by haustoria (Plate 2.7). Yield and quality of forage can be decreased (Thomas, 1991). A dense tall sward creates a moist, shaded microclimate favourable to the development of this disease. Plants of the same variety can differ in degree of susceptibility because the parasite specialization is very pronounced.

Some fungi develop on stems and young inflorescences, such as the spectacular *Epichloë typhina*. This can result in important losses in seed production of *Dactylis glomerata* (Raynal, 1988). Other fungi, such as some *Ustilago* spp. or *Claviceps purpurea*, which can develop on many forage grasses, affect the inflorescences. *Claviceps purpurea* is particularly dangerous for animals since ingestion of ergot sclerots can cause abortions and deaths by poisoning.

Endophyte fungi such as *Neotyphodium* spp. colonize the leaves of grasses, though the leaves do not develop visible symptoms. In some cases, these infections are beneficial for the plant by increasing persistence and resistance to insects and diseases. Alternatively, these fungi can produce alkaloids which are toxic to animals (Fribourg, Hoveland and Codron, 1991).

PLATE 2.7
Blumeria graminis
on *Lolium perenne*

C. VANBELLINGHEN

PLATE 2.8
Xanthomonas translucens
damage on
Lolium perenne

2.10.2 Bacteria

Bacteria are less frequent but when they develop, they are often particularly detrimental. In warm and dry weather, plants attacked by *Xanthomonas translucens* develop bacterial wilt and disappear in groups after a few days (Sletten, 1989) (Plate 2.8). *Phleum pratense* and *Lolium* spp. are particularly susceptible to this disease, which is caused by several pathovars (*Xanthomonas translucens* pv. *phlei* and *X. translucens* pv. *graminis*). Cutting equipment can propagate the disease in the sward and from one sward to another. A frequent cutting regime associated with wet conditions encourages the spreading of these bacteria. Cutting at too young a stage in spring must be avoided otherwise the bacteria take advantage of the injuries caused to the plants and invade them.

2.10.3 Viruses

Many viruses have been noted on forage grasses, but their economic importance is not well known. Barley Yellow Dwarf Virus (BYDV), transmitted by several aphid species, and Ryegrass Mosaic Virus (RMV), transmitted by a dust mite (*Abaracus hystrix*), can cause severe yield losses in *Lolium* spp. (Catherall, 1987).

2.10.4 Forage plant protection methods

Curative treatments against diseases are not used in grasslands for economic reasons and so preventive means must be foreseen and utilized where possible in order to avoid the onset and development of the diseases. Choosing resistant species and varieties is essential when sowing a sward and, arguably, mixtures of species and varieties are preferable to pure sowings. A balanced fertilization associated with a suitable defoliation management system will help to maintain a healthy sward. Foliage fungi are favoured by nutrition stress in particular. However, the effect of fertilization on the development of diseases is a complex issue. It is obvious that the sward is especially attacked when the growing period is long. Since nitrogen fertilization speeds up the growth of grass and reduces the interval between successive defoliations, in cutting as well as in grazing, nitrogen application helps to curb foliar diseases. When a massive incidence of diseases occurs, for instance a rust attack at the end of summer, the best option is to cut and ensile the forage. The regrowth is very likely to be unharmed by the disease and can thus again be acceptable for grazing.

Chapter 3

Diversity of grass uses

3.1 INTRODUCTION

The use of grasses is not limited to the commercial production of milk, meat, leather or wool. It can take on a great diversity of forms, both commercial and non-commercial. In recent years there has been a marked tendency to use grasses in new or different ways and to develop novel products. A tentative list of some of the 'non-forage' uses of grasses is given below and these uses are briefly described. The fodder production aspects and the use of grasses as companion crops for forage legumes are omitted.

3.2 AGRICULTURAL USES

3.2.1 Seed production

The sowing of grasslands obviously necessitates the production of seeds. Seed production, whether for forage or amenity purposes, can be a valuable crop when diversifying from other specialized agricultural enterprises provided the conditions are satisfactory. In the European Union, the main seed production regions are mostly situated in Denmark, northern Germany and the Netherlands (Kley, 1995).

3.2.2 Source of residual nitrogen for crops in crop-grassland rotations

Temporary grasslands, composed of pure-sown grass fertilized with nitrogen or mixtures of grasses and legumes, allow the accumulation of organic nitrogen in the soil during their lifetime. When they are ploughed up, part of that nitrogen becomes available to the following crop. It is preferable to sow quickly, within a few days after ploughing to avoid leaching of nitrates. Estimates indicate that, after grassland of two or three years,

there can be up to about 200 to 300 kg of available nitrogen per ha for the following crop, but the amount of nitrogen available will decline progressively over several years (Le Gall, Legarto and Pflimlin, 1997; Viaux, Bodet and Le Gall, 1999).

3.2.3 Weed control in crop-grassland rotations

Temporary grasslands can be integrated into crop rotations in mixed farms or in livestock farms that grow annual forages (silage maize, cereals, fodder beet). Such grasslands generally last one to three years. Most arable weeds cannot develop during a period of temporary grassland; so they cannot add to or renew their seed bank in the soil. As a proportion of these seeds die each year and temporary grasslands occupy the land for several years, these grasslands have the effect of reducing the weed population in the following arable crops.

3.2.4 Green manures

The *Lolium* spp. can be sown as green manures in August (Pousset, 2000), with vegetation often being ploughed under in November. This contribution of green organic matter improves soil structure and provides humus and soil nutrients (Ninane *et al.*, 1995).

3.2.5 Soil cover in orchards, vineyards and nurseries

The spaces between the rows of trees in orchards, between vines or between plant beds in nurseries are very often sown with grasses to facilitate maintenance and transport and to increase the organic matter in the soil. The mixtures could contain *Lolium perenne* or *Festuca arundinacea*, or a mixture of

F. rubra, sheep's fescue (*F. ovina*), *Agrostis capillaris* and *Poa pratensis* (Parente, 1996). Sometimes a dwarf timothy (*Phleum bertolonii*) is also used.

3.2.6 Constituents of set-aside
Grasses are valuable plants for covering the soil and controlling weeds within the framework of 'set-asides' (Clarke, 1993; Straëbler, 1994; Clotuche *et al.*, 1997) and are mainly used to establish covers of over six months' duration. They can be mixed with clovers, especially red clover (*Trifolium pratense*) (Clotuche, Peeters and Van Bol, 1998; Clotuche, 1998). These mixtures leave a considerable residue of soil nitrogen available for the following crops (Clotuche and Peeters, 2000). *Lolium perenne* is the main companion grass sown for set-asides.

Some set-asides are sown and managed to favour game such as partridge (*Perdix perdix, Alectoris* spp.), pheasant (*Phasianus colchicus, Chrysolophus* spp.) and hare (*Lepus capensis*). *Dactylis glomerata* and *Arrhenatherum elatius*, for instance, sown at low density, allow the development of tall, but not too dense swards where game can hide, circulate and nest. Other mixtures based on *Festuca rubra* and white clover (*Trifolium repens*) form short swards that can be grazed by rabbits (*Oryctolagus cuniculus*) or even deer (Peeters and Decamps, 1998).

3.2.7 Production of turf rolls
Another niche connected with the multiplication of grasses consists of the production of rolls of turf that are used to create amenity swards, particularly lawns (Société française des gazons, 1990).

3.2.8 Use as bedding
Some grass swards have a low feeding value, but their biomass can be used as bedding, a practice that is traditional in the Alps, for example (Jeangros, Troxler and Schmid, 1991). In the lowlands, *Molinia caerulea* and *Phalaris arundinacea* communities are particularly well adapted to this practice. It can also be applied to material harvested on field margins which has to be cut late within the framework of the European Union's agri-environmental scheme. The bedding can then be composted as farmyard manure or added to raw excreta to give it body.

3.2.9 Production of edible seeds for humans and animals
The grasses of temperate regions generally have seeds of small dimensions, but there are some exceptions (see Chapter 6, section 6.6.1.). Species of the genus *Glyceria* and one species of fescue, *Festuca paniculata (F. spadicea)*, produce rather large seeds. Traditionally, the first are eaten by human populations living in humid zones of Poland. The seeds of the *Festuca* are eaten in the Pyrenees for example, by sheep that graze over the wild *Festuca* communities rejecting the leaves since they are not very digestible.

3.2.10 Meat quality
The prime factor in the organoleptic quality of meat is the age at slaughter (Young and Kauffman, 1978; Keane and Allen, 1998; Moloney, 1999). In intensive systems, the animals are fed forage of high nutritive value. They therefore grow quickly and can be fattened and slaughtered very young, but their meat has little taste (Reagan *et al.*, 1977; Schroeder *et al.*, 1980). In extensive systems, the growth of animals is slower, the fattening phase starts later, and so the meat of animals that are slaughtered older has more taste. There is therefore a relationship between grass communities and meat quality, and for example, the oligotrophic and mesotrophic communities only permit relatively low or moderate animal performance. The animals that have grazed these communities thus enter the fattening stage later and their meat has the likelihood of a stronger flavour. Detailed studies have begun recently on

these important aspects for the promotion of quality meat products.

3.2.11 Starch production

Grass rhizomes contain plant food reserves that are mobilized by the plant to facilitate regrowth. One of the constituents is starch that is usable in animal or human nutrition. The rhizomes of *Elymus repens* are particularly rich in starch (Steen and Larsson, 1986). These rhizomes have long been harvested during periods of famine to make a flour that is consumed in the form of cooked pancakes (Couplan, 1989). Rhizome flour was mixed with wheat flour until the early 1900s in Germany and Poland (Maurizio, 1933).

3.2.12 Production of protein cakes

Grasses at a young growth stage are rich in protein. The content can vary from 15–20 percent crude protein (CP) at the leafy stage and reach even 30–35 percent in young regrowth (Jarrige, 1981). This protein can be extracted to make very concentrated cakes that can replace soya in the feeding of swine and ruminants, and even in human nutrition. However, up to now the amount of fossil energy needed to produce them has been a brake on their production.

3.2.13 Production of alkaloids and aromatic substances

The grasses of temperate regions ordinarily produce few aromatic substances, but there are at least two species that contain important quantities of the alkaloid, coumarin. These are *Anthoxanthum odoratum* (Faliu, 1981) and holy grass (*Hierochloe odorata*). The former is not yet used for its alkaloid content but the latter has long been used in Poland to flavour the 'Zubrowka' or bison vodka. *Phalaris arundinacea* also contains high levels of alkaloids, namely, tryptamine-carboline, gramine and other indol-alkaloids (Sheaffer and Marten, 1995). These alkaloids give the plant hallucinogenic properties (Schultes, Hofmann and Rätsch, 2000).

3.2.14 Production of medicinal substances

The rhizomes of *Elymus repens* contain mucilage and other substances that are used in herbal medicine for their diuretic (Girre, 2001), emollient, vermifuge, febrifuge and depurative properties (Brüschweiler, 1999).

3.3 NON-AGRICULTURAL USES

3.3.1 Production of fibres

The fibres of some grasses can be used to produce pulp for paper-making. This is the case in North Africa with a Mediterranean grass, the 'alfa' (*Stipa tenacissima*). It could theoretically be extended to species from temperate regions such as common reed (*Phragmites australis*), if a market need was identified.

The rhizomes of *Elymus repens* are used to make hard brushes which are appreciated for their resistance to wear and tear. They were also used in the Alps for filtering milk (Brüschweiler, 1999).

3.3.2 Production of biomass for the production of energy

Grasses, pure or mixed with legumes, can reach very heavy yields of biomass in temperate regions – annual yields of 15 to 20 tonnes DM/ha are not exceptional (Peeters and Kopec, 1996). Biomass harvested once or twice annually has a high dry matter content, of the order of 30 percent, and could probably be used for the production of energy by anaerobic digestion and gasification (ethanol or biogas) of the biomass.

3.3.3 Production of building material

Grass stems can make an insulating material. For example, *Phragmites australis* stems are traditionally used to cover roofs in marshy regions and in a chipboard form for building construction. *Festuca rubra* and wavy hairgrass (*Deschampsia flexuosa*) are sometimes used as living covers on wood roofs in Scandinavia.

3.4 PROTECTION AND IMPROVEMENT OF THE ENVIRONMENT

3.4.1 Production of humus and improvement of the soil structure in grassland–crop rotations

Grasslands produce great quantities of dead material during their lifetime. This litter, which returns to the soil, is derived from the senescence of roots, stem bases and leaves. These constitute a large production of carbon, which is used by micro-organisms and other decomposing organisms in the soil to produce humus. This production of humus can be estimated at several tonnes per hectare per year (Voroney, van Veen and Paul, 1981; van Veen and Paul, 1981; Buchkina and Balashov, 2001).

The incorporation of fresh green matter when a grassland is ploughed up, and the consequent production of humus, contribute to improving soil structure, and therefore also improve the circulation of air and water in the soil. This makes working the soil easier and improves root penetration. The humus also augments the soil water and nutrient retention capacity, which in turn improves the supply of water and nutrients to crops during dry periods.

3.4.2 Carbon sinks

The humus produced in grassland is a carbon sink that contributes to reducing the CO_2 content of the atmosphere. Net fixation of carbon is especially great when arable land of low soil carbon content is transformed into permanent grassland (Garwood, Tyson and Clement, 1977; Curtin *et al.*, 1994). According to the model of Scholefield *et al.* (1991) for nitrogen and a C:N ratio of 10:1 in stable humus, the net accumulation of C can be evaluated at about 1–2 tonnes/ha/yr.

3.4.3 Nitrate catch crops

Lolium perenne and *L. multiflorum* can constitute very effective catch crops when sown after another crop has been harvested

in summer. These two species establish and develop rapidly in autumn and are capable of absorbing nitrogen residues in the soil profile up to a quantity of 50–130 kg $N-NO_3$/ha between August and November for a biomass of 3–6 tonnes DM/ha (Laurent *et al.*, 1995; Ninane *et al.*, 1995; Thomsen and Christensen, 1999). The grasses are then ploughed in between November and March, but much of the mineral nitrogen has meanwhile been transformed into organic nitrogen and brought to the top of the soil profile, a process that drastically reduces the risk of nitrate leaching.

Lolium spp. and other grass species can also be used under a maize crop. They are sown once the maize is at the four-leaf stage and though they develop little while the crop occupies the soil, they can grow rapidly under favourable conditions after maize harvest in October. They begin thereafter to absorb the excess nitrogen of the profile which reduces nitrate leaching (Simon and Le Corre, 1988; Vertès and Decau, 1992; Borin, Giupponi and Morari, 1997; Castillon, 2000).

3.4.4 Buffer zones and anti-erosive strips in arable land

Thanks to their great ability for absorbing nutrients and their density of soil cover, grasses can constitute buffer zones alongside watercourses. These zones stop surface runoff of nitrogen, phosphorus and pesticides, thus preventing them from reaching and polluting watercourses (Haycock and Pinay, 1993; Gilliam, 1994; Vought *et al.*, 1995; Borin and Bigon, 2002). Narrow strips, several metres wide, are usually sufficient (Dillaha *et al.*, 1989; Simmons *et al.*, 1992; Haycock and Pinay, 1993; Castelle, Johnson and Connolly, 1994; Vought *et al.*, 1995; Haycock *et al.*, 1997; Borin and Bigon, 2002).

It is also well known that contour strips can be sown along slopes on fragile cultivated soils to limit erosion.

3.4.5 Erosion control and soil stabilization

The capacity of grasses to cover the soil rapidly after sowing can be used to fix soils after earthworks, for example after the construction of ski pistes, the formation of roadside banks or motorways and profiling the banks of watercourses. On the poorest soils, grass seeds can be inoculated with mycorrhizae to improve plant nutrition in water and phosphorus.

The principal species that suit these uses are:
- Ski pistes: *Agrostis capillaris, Dactylis glomerata, Deschampsia flexuosa, Festuca rubra, Nardus stricta,* alpine cat's tail (*Phleum alpinum*), *Poa* spp., blue moor grass (*Sesleria caerulea*);
- Banks of roads and motorways: *Agrostis capillaris, Arrhenatherum elatius, Dactylis glomerata, Festuca rubra, Lolium perenne, Poa pratensis;*
- Banks of watercourses: *Agrostis stolonifera, Holcus lanatus, Lolium perenne, Phalaris arundinacea.*

Marram (*Ammophila arenaria*) and lyme grass (*Elymus arenarius*), are regularly used to fix moving sand dunes on beaches.

3.4.6 Rehabilitation of industrial or polluted sites

Thanks to their capacity to cover the soil and to their hardiness, the following grass species perform very well in the rehabilitation of abandoned industrial sites (quarries, slag heaps, tips, mines and industrial ruins): *Agrostis capillaris, Arrhenatherum elatius, Dactylis glomerata, Festuca rubra* and *Poa pratensis* (Muncy, 1985). *Agrostis capillaris, Deschampsia flexuosa* and *Festuca rubra* are grasses that can also develop on sites polluted by heavy metals.

3.4.7 Fodder production for deer in forests

Dense plantations of beech (*Fagus sylvatica*) or spruce (*Picea excelsa*) trees contain few fodder resources for deer. The animals often damage the trees for want of other possibilities of nourishment. Grassy covers are consequently often created in forest clearings to increase the grazing resource, though the soils of these sites are often acid and poor in nutrients. Grasses such as *Agrostis* spp., *Festuca rubra, Holcus lanatus* and *Poa pratensis*, that are adapted to low fertility, can be used in this context.

3.4.8 Biodiversity

Limestone grasslands, hay meadows on the plains or in the mountains, damp meadows and steppes contain the highest numbers of plant species of temperate region ecosystems. Acid grasslands generally contain fewer grass species, but also often contain dicotyledon species that have become rare and present great heritage interest. Some examples include: mountain tobacco (*Arnica montana*), *Dactylorhiza* spp., heathers (*Erica* spp.), trailing azalea (*Loiseleuria procumbens*), rampions (*Phyteuma* spp.), devil's-bit scabious (*Succisa pratensis*), bilberries (*Vaccinium* spp.) The grasses are the 'weft' of all these plant communities. The fauna associated with these swards is also very rich.

3.4.9 Components of field margins

Grasses can be used in field margins as part of the European Union agri-environmental scheme, that is, grassy strips several metres wide sown along the edges of cultivated fields, alongside watercourses, on the edges of woodlands or along roadsides. These field margins can have several functions. They can stop the run-off of nutrients and pesticides and limit water erosion (see also buffer zones above). They also constitute a habitat for wild flora and fauna, notably nesting sites for birds, and they have a landscape function if they contain dicot species with spectacular flowers.

3.4.10 Grasses, grass communities and landscapes

Grasses do not produce spectacular flowers, but their inflorescences can be attractive when they develop in 'carpets'. The colours of brown, yellow or silvery glumes and violet or whitish anthers are then visible from afar.

Grass stems which undulate in the wind also give a dynamic aspect to vast open landscapes. Close up, some inflorescences are particularly attractive, like those of *Briza media*, the spikelets of which are heart-shaped.

Communities dominated by *Alopecurus pratensis*, meadow buttercup (*Ranunculus acris*) and sorrel (*Rumex acetosa*) colour the landscape with the violet anthers of the grass, the yellow flowers of the buttercup and the brownish inflorescences of the sorrel. They give a characteristic touch of colour to alluvial valleys.

The tall yellow stems of the mature plants of *Arrhenatherum elatius* and *Bromus erectus*, which often form vast populations, are recognizable from afar and are very attractive when they form undulating 'waves' in the wind. The long and narrow white inflorescences of *Stipa* decorate the steppe and dry rocky slopes with their long white feathers. Great areas of *Agrostis* in flower give an impression of lightness and delicateness. This is also true for *Deschampsia flexuosa*, the spikelets of which give silvery reflections.

The robust stems of the plants of reed beds are very spectacular because they can be more than three metres tall. *Phalaris arundinacea* and *Phragmites australis* are the two most typical reeds. In addition to its tallness, reed sweet-grass (*Glyceria maxima*) has a particularly magnificent panicle the lightness of which contrasts with the robustness of the stems.

Grasses play an essential decorative role in the landscaping of leisure parks. They produce the same effect in forest parks created by pastoral activities (e.g. in Jura, Switzerland and in the Tatra mountains in Poland and Slovakia), in moorland, marshy meadows, alpages, bocages, tall orchards and rocky or limestone grasslands. All these landscapes, which are open or have a low tree density, are appreciated by walkers.

3.4.11 Turf for leisure areas

The market for turf seed is very important in the European Union and the amounts sold are comparable to the quantities of fodder seeds (Kley, 1995). The selling price is nevertheless markedly higher. The most used species are *Agrostis capillaris, Festuca arundinacea, F. ovina, F. rubra, Lolium perenne* and *Poa pratensis* (Kley, 1995; Société française des gazons, 1990). These species are used for turf in parks, gardens and sports grounds.

3.4.12 Decorative elements in parks and gardens

Grasses from temperate areas can be used in parks and gardens to contribute different decorative elements. Often horticultural varieties are used but sometimes wild types are also chosen. These grasses are incorporated in the following elements (Hansen and Stahl, 1992; Zinkernagel, 1995; Brinkforth, 1997; Greenlee, 2000; Grounds, 2002):

- In 'mixed borders': *Arrhenatherum elatius, Briza media, Dactylis glomerata, Deschampsia caespitosa, Molinia caerulea*;
- In rockeries: *Festuca ovina, F. scoparia, F. valesiaca*, hair grass (*Koeleria glauca*), melick (*Melica ciliata*), glaucous meadow grass (*Poa glauca*), *Sesleria caerulea*;
- Beside water features: *Glyceria maxima, Molinia caerulea, Phalaris arundinacea*;
- In understoreys: *Deschampsia caespitosa, D. flexuosa, Molinia caerulea*, broad-leaved meadow grass (*Poa chaixii*);
- As a basis for 'wild flower meadows': *Agrostis capillaris, Anthoxanthum odoratum, Briza media, Festuca rubra, Poa pratensis, Trisetum flavescens*.

Many other ornamental species, not native to Europe, are also regularly used in parks and gardens of temperate areas (Hansen and Stahl, 1992; Zinkernagel, 1995; Brinkforth, 1997; Greenlee, 2000; Grounds, 2002).

Very sober gardens, without flowers and composed of grasses, are more and more appreciated. These 'new wave' (Grounds, 2002) or 'zen' gardens are notably in fashion around office complexes, but also private homes.

Grasses can be very pretty in the growing

season thanks to their foliage or their inflorescences. They can also be very attractive in winter when these organs are covered in hoar frost.

3.4.13 Decorative elements in bouquets

Several species of grass are used in dry bouquets, namely, *Briza media, Deschampsia caespitosa, Glyceria maxima, Molinia caerulea* and *Phalaris arundinacea*. The inflorescences are used in particular, but sometimes the leaves are also used. This market has developed considerably in recent years.

3.5 ROLE OF GRASSES IN ENVIRONMENT AND LAND-USE PLANNING POLICIES

It is important to integrate biodiversity management of grass communities more effectively into land-use planning and to utilize research results in planning at local and regional levels. This may be done by auditing and diagnosing the biological quality of areas, by producing maps, or by writing local guidebooks. Proposals for developments and local initiatives should be discussed with all the interested parties of the areas concerned.

Within the framework of an environment and land-use planning policy, grasses can be used as indicators of environmental quality, particularly for monitoring the state of the environment. Grasses constitute an ideal group of bio-indicators because there are numerous species, each with very different ecological requirements, and because they are spread through a vast range of habitats. In addition, they are quite easy to record at all seasons of the year.

Chapter 4

Morphology and physiology of grasses

4.1 MORPHOLOGICAL STRUCTURE

Following seed germination and seedling growth, a grass plant is made up of roots and tillers. These roots and tillers begin at the 'tillering plateau' just below the ground surface. The roots of grasses are fasciculate and can be very dense at the base of the tillers. Roots can also appear in some conditions on the nodes of the tillers, e.g. when tillers are creeping on moist soil.

The meristems are concentrated at the base of the stems, at the tillering plateau. This part of the plant is really the 'heart' of the grass. A tiller is produced at the base of a leaf, from a bud or node. At the end of the winter, the nodes of each tiller are, in effect, piled up and so the tillers are then very short.

The stem of a grass includes leaves that originate at nodes, and an inflorescence. The space located between two nodes is called the internode and these internodes are very short at the beginning of spring, when it is almost impossible to see them with the naked eye. However, when the stem elongates and the inflorescence appears, the internodes are highly visible.

The stems of some species lie on the soil; such stems are considered as creeping, and they can root at the nodes, thus becoming stolons. The species that produce stolons are called stoloniferous, e.g. *Agrostis stolonifera*. In other species, the stems are first creeping but thereafter they rapidly become erect. They form an angle at a node and are called 'bent' or 'ascendent', e.g. marsh foxtail (*Alopecurus geniculatus*). In some species, the first internodes are swollen like a bulb, e.g. *Phleum pratense*.

Some species have underground stems (rhizomes) that grow horizontally below the soil surface. They are usually whitish or brownish, can be more vigorous than aerial stems, and the leaves on them are reduced to short scales. These rhizomes can produce roots and aerial stems at each node, e.g. *Alopecurus pratensis, Elymus repens, Poa pratensis*.

Grass species can differ dramatically in their growth habit. Some species form dense tufts because their tillering is very important, e.g. *Dactylis* spp., *Lolium* spp. Other species produce less dense tufts, with less tillering, e.g. *Phleum pratense*. The aerial stems are even more far apart in rhizomatous species, e.g. *Elymus repens, Alopecurus pratensis,* and the stems are located at more or less regular intervals on the rhizomes creeping below the soil surface.

The leaf of a grass (Figures 4.1A and 4.1B) has no distinctive petiole as is the case for most dicotyledons. It includes a sheath that embraces the stem and a blade or lamina that forms a variable angle with this stem. At the intersection of these two parts of the leaf, the ligule, and subsequently the auricles develop. The size of the ligule varies. A well developed ligule is an obstacle for water that trickles along the blade, thus preventing the water from soaking the space between the sheath and the stem. This can be an advantage for the plant since an excess of water can cause the stem to rot. In some species, the ligule is very short or reduced to a few hairs, this being mainly the case for some grasses in dry areas, but it is also the case for species elsewhere. The presence of auricles contributes to a better contact between the sheath and the stem. This function is also ensured when the sheath is

FIGURE 4.1A
Parts of a grass leaf: blade, ligule,
auricles, sheath

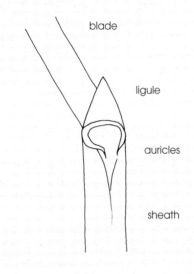

FIGURE 4.1B
The two types of young leaves: folded
and rolled

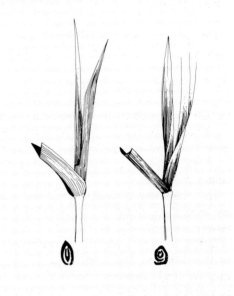

tubular for a part of its length.

The leaves of grasses all have the same general structure but they can be different in many characteristics (Annex 1). Some major differences are:

- Length and width of the blade;
- Colour of the blade: bluish green, yellowish green, pale green, or dark green;
- Hairiness: on the two parts (upper and lower) of the blade or on only one part, on the sheath only, on the auricles, or the leaf can be totally hairless. The hairs of the sheath can be perpendicular to the stem (*Elymus repens*) or bottom-oriented (*Trisetum flavescens*). The hairiness of the sheath can decrease from the node to the ligule (creeping soft-grass [*Holcus mollis*]);
- Surface of the blade: some blades are flat and only the central nerve can be seen; some have many veins, the alternance of veins embedded in the blade and of prominent ribs between these veins is easily seen; these blades look ridged lengthwise; in *Poa* species, the central nerve is flanked by two lines,

where the blade is translucent, called 'ski tracks' or 'tramlines';
- Top of the blade: it can be acute (*Lolium* spp., *Poa trivialis*) or rounded and hooded (*Dactylis glomerata, Poa annua, P. pratensis*);
- General aspect of the blade: wide and flat (*Dactylis glomerata, Phleum pratense*), rolled on itself in needle shape (*Deschampsia flexuosa, Festuca ovina, F. rubra, Nardus stricta*), or shiny in the lower part (*Festuca arundinacea, F. pratensis, Lolium* spp.);
- Colour of the sheath of the lower leaves: brown (*Cynosurus cristatus*), purple-red (*Festuca arundinacea, F. pratensis, Lolium* spp.), or ridged by vertical red lines (*Holcus lanatus*);
- Characteristics of the auricles: large (*Festuca arundinacea, F. pratensis, Lolium multiflorum*), long and narrow (*Elymus repens*), or hairy (*Festuca arundinacea*);
- Characteristics of the ligule: very short (*Elymus repens, Festuca rubra*), large and whitish (*Dactylis glomerata, Phleum pratense, Poa annua*), off-centre (*Agrostis capillaris*),

long and pointed (*Agrostis canina, A. stolonifera, Poa trivialis*), truncate (*Poa pratensis*), membranous or greenish (*Lolium multiflorum, L. perenne*), or reduced to some hairs (*Molinia caerulea*).

The inflorescence can be panicle-like or spike-like (Figure 4.2). A panicle is branched with the spikelets located on the branches of the inflorescence, e.g. *Dactylis* spp., *Festuca* spp., *Poa* spp. In a spike, the spikelets are located on the axis of the inflorescence itself, e.g. *Elymus* spp., *Lolium* spp. It is thus a spike of spikelets. Some panicles have very short branches, being contracted and looking like spikes (*Phleum* spp.).

The inflorescence includes spikelets that are themselves constituted by one or several flowers (Figure 4.3). The structure of the spikelets is very particular. It includes a pedicel, a lower glume, an upper glume, a rachilla and flowers. Each flower is constituted by a lower scale or lemma, an upper scale or palea, two lodicules located at the base of the ovary and the stamens. The stigmas are located at the top of the ovary. In some species, one or several elements can be lacking, like the upper glume for example. Some flowers are sterile since they do not include the ovary or stamens. Others are only males, e.g. *Arrhenatherum* spp., *Holcus* spp. The lemma can be extended by an awn, sometimes rather long and possibly bent.

Grasses can be annual, e.g. *Poa annua*, biennial, e.g. *Bromus hordeaceus* or perennial, e.g. *Lolium perenne*.

4.2 PHYSIOLOGICAL STAGES

When a grass seed germinates, the roots emerge first, protected by a membrane, the coleorhiza. Then comes the young stem or plumule, also protected by a membrane, the coleoptile.

A first leaf emerges through this coleoptile, ensuring the start of photosynthesis and the progressive freeing of the young plant from the seed nutrient store. Then, a second leaf appears, promptly followed by a third one. At that time, the 'tube' constituted by the

FIGURE 4.2

Two types of grass inflorescence: spike of spikelets (*Lolium perenne*) on the left, and panicle of spikelets (*Agrostis capillaris*) on the right

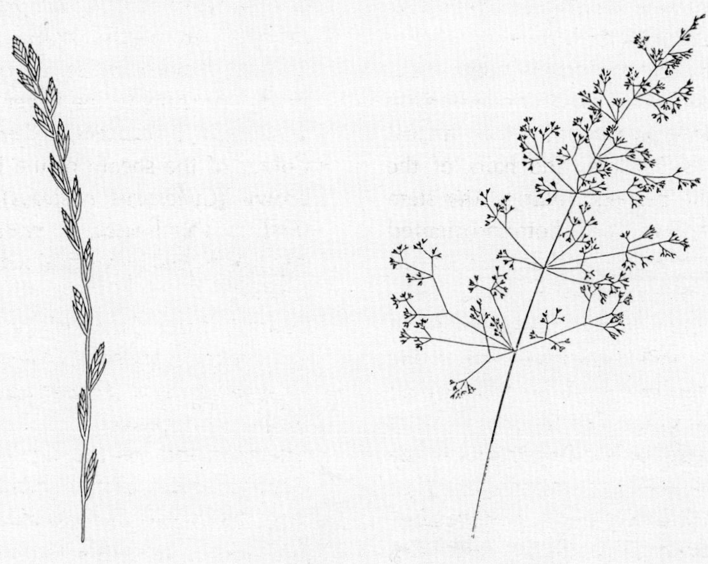

FIGURE 4.3
Grass spikelet

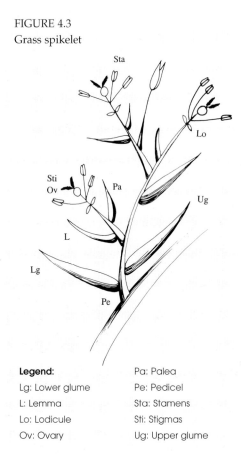

Legend:

Lg: Lower glume   Pa: Palea
L: Lemma   Pe: Pedicel
Lo: Lodicule   Sta: Stamens
Ov: Ovary   Sti: Stigmas
                    Ug: Upper glume

sheaths of these three leaves increases in size just below the soil surface and new roots, the secondary roots, appear. This thick part is still related to the seed by a rhizome. Some time later, the first roots, the primary roots, cease all activity and die, the seed is emptied of its food reserves and the rhizome also disappears. A bud then develops at the base of the first leaf. This bud, growing point, or meristem, will produce a stem and secondary roots. The same process will occur at a later stage at the base of the second leaf, of the third leaf and of the following leaves. These tillers will bear leaves and other tillers appear at the base of these leaves. The number of leaves and tillers increases exponentially provided conditions are met for their growth. This tillering stage begins with the appearance of the third leaf.

After that stage, the meristems, still located at the base of the stems and just below the ground surface, begin to change. At high magnification, it is possible to see that leaf primordia are replaced by spikelet primordia, a step that marks the beginning of the reproductive stage. In some grasses, this differentiation occurs under the effect of cold, so-called '**vernalization**'. It is a necessary step for the reproductive stage and these species are called 'non-alternative,' since they stay leafy without the effect of cold. In Western Europe, *Lolium perenne* reaches the full differentiation of its meristems at about the end of February. Other species like tropical grasses, but also *Phleum pratense* in temperate regions, do not need the action of cold to move on to the reproductive stage. They are called 'alternative' species.

At the time of the start of growth – about the beginning of April in temperate regions of the northern hemisphere – the stems, which are very short until this time, a couple of millimetres at most, start to elongate and the tillers, which were laxly spread out on the ground, start to become erect. The meristems then rise up in the tube formed by the sheaths of the leaves, a process called '**elongation**'. The inflorescence develops during this time and its volume increases causing the sheath to expand. The inflorescence continues to grow until the appearance of the last leaf at the top of the sheath. The '**heading**' stage (ear emergence) is reached when the inflorescence appears. The inflorescence finally emerges completely and separates from the upper level of the foliage. The stamens then appear from the spikelet: the '**flowering**' stage. The ovaries are soon fertilized and the seeds formed: the '**fructification**' stage. The ripened seeds then fall on the soil: the '**dispersal**' stage that ends the reproductive period.

All tillers that have produced seeds then die and so their maximum lifetime is one year. If the species is annual, the plant dies completely; if it is perennial, new vegetative tillers soon appear at the base of the stems. Subsequently, they develop in the same way as those deriving directly from a seed.

4.3 EARLINESS

Grass earliness can be defined in different ways: the earliness of spring growth, heading, flowering or seed ripening.

Early growth in spring allows earlier grazing and so reduces the work of the farmer (less winter feeding and housing maintenance). Early flowering is of small practical importance, but presents an interesting physiological reference point. The earliness of seed ripening mainly concerns breeders and seed producers. In contrast, the date of heading or ear emergence is an essential parameter for characterizing species and varieties. Moreover, the word 'earliness' relates mostly to the earliness of heading when it is used in a forage production context. The first silage cut is usually taken at the beginning of heading. Thus, the date of heading has a very concrete significance for the farmer.

The heading date depends on two factors: day length and temperature. Day length evolves similarly each year at a specific location and remains very stable over vast areas; therefore temperature fluctuations mainly influence the time of heading of a grass in a region. Obviously, heading dates vary in time and space, the heading of a variety at a specific location being earlier when winter and the beginning of spring are mild. The heading date is also later at high altitudes and latitudes and in continental climates than in lowlands, low latitudes and oceanic climates. However, the ranking of species and varieties according to their earliness is almost identical whatever the site or year of record.

At high latitudes and altitudes, the differences in earliness of heading among species and varieties are considerably reduced, the biggest differences being recorded in mild oceanic climates. The differences in earliness among species and varieties are also reduced with advancing plant maturity and so are smaller at the flowering stage, and especially at the time of seed ripening, than at the heading stage. The time interval between the heading dates of two taxa is always bigger than the time interval between the flowering dates of these two taxa.

Some species show a narrow heading date range, while other species have a very broad range. Figure 4.4 ranks a large number of grass species by order of earliness of heading and also shows the range of earliness. In *Lolium perenne*, the range is very wide, the heading of the earliest variety occurring more than one month before the latest variety.

The earliness of heading is usually linked to the earliness of spring growth. For instance, *Alopecurus pratensis* quickly produces biomass in spring and it heads very early. However, there are exceptions to this rule; for example, *Phleum pratense* starts its growth early, virtually at the end of winter, but heads very late.

Within a species, varieties are often classified according to their heading date, namely, early, intermediate or late, and these classes are specific for each species. For example, early *Phleum pratense* heads much later – by about three weeks – than intermediate varieties of *Lolium perenne*. In general, early varieties are better suited to cutting while late varieties are better suited to grazing. This is particularly true for *Lolium perenne*. The management of late varieties is more flexible than early varieties since their stem production is delayed; their leaf:stem ratio decreases more slowly and their digestibility declines less rapidly. Thus, a delay in management has less unfavourable consequences on forage quality. This is one advantage of *Phleum pratense*, a late heading species, compared with early and intermediate *Lolium perenne* and *Festuca pratensis*. Conversely, *Festuca arundinacea* and *Lolium multiflorum*, which head early, have a rapidly

FIGURE 4.4
Heading date range of 15 grass species

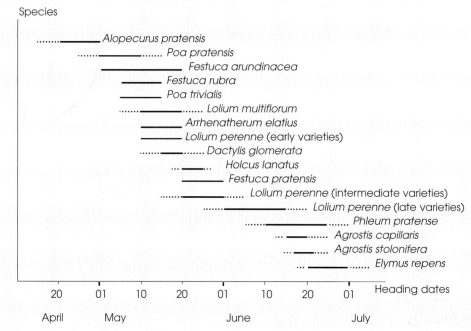

Data recorded at low altitude in Belgium; synthesis carried out by C. Decamps and A. Peeters

Legend: ———— most frequent heading dates

          --------- possible but less frequent heading dates

declining nutritive value compared with *L. perenne*.

On the basis of Figure 4.4, the following heading groups can be distinguished:
- Very early species:
Alopecurus pratensis;
- Early species:
Poa pratensis, Festuca arundinacea, F. rubra, P. trivialis, Lolium multiflorum, Arrhenatherum elatius, L. perenne (early varieties), *Dactylis glomerata;*
- Intermediate species:

Holcus lanatus, Festuca pratensis, Lolium perenne (intermediate varieties);
- Late species:
Lolium perenne (late varieties);
- Very late species:
Phleum pratense, Agrostis capillaris, A. stolonifera, Elymus repens.

In Western Europe, *Alopecurus pratensis* often heads as early as 15–20 April, while *Elymus repens* heads only at the beginning of July. The variation of heading between species is thus about 2.5 months.

PART II

Chapter 5

Profiles of individual grass species

Chapter 5

Profiles of individual grass species

This chapter is a database on 43 grass species classified in alphabetical order. The text for each species includes at least a morphological description, a definition of ecological requirements, and data on the agronomic characteristics of the species as far as the information is available. Each description is structured as follows:

SCIENTIFIC NAME

SYNONYM

ETYMOLOGY

COMMON NAMES
English
French
German

DISTRIBUTION

DESCRIPTION
• Morphological description
• Physiological peculiarities
• 1000-seed weight
• Chromosome number
• Biological flora

ECOLOGICAL REQUIREMENTS
• Altitude – vegetation belts
• Soil moisture
• Soil fertility
• Soil type and topography
• Climate requirements
• Tolerance of cutting and grazing
• Sociability – competitiveness – plant communities
• Compatibility with other forage grasses and legumes
• Indicator value

AGRONOMIC CHARACTERISTICS
• Sowing rates
• Varieties
• Fertilization
• Dry matter yields and seasonality of production
• Nutritive value
• Acceptability
• Toxicity
• Animal performance
• Diseases
• Use for purposes other than forage production
• Main qualities
• Main shortcomings

AGROSTIS CANINA L.

ETYMOLOGY

LATIN: *agrostis* or GREEK: αγρωστις
Name of an undetermined grass but also
GREEK: αγρωσ = *agros* = field; species that
ornaments fields.
LATIN: *canis* = dog and *caninus* = of dog;
used for plants considered as inferior. This
Agrostis is very small.

COMMON NAMES

ENGLISH: Velvet bent
FRENCH: Agrostis des chiens
GERMAN: Sumf straussgras

DISTRIBUTION

Native to Europe and temperate
Asia.

FIGURE 5.1
Inflorescence and part of the leaf
of *Agrostis canina*

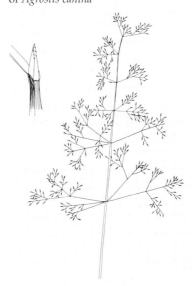

PLATE 5.1
Agrostis canina

A. PEETERS

PLATE 5.2
Agrostis canina

DESCRIPTION

MORPHOLOGICAL DESCRIPTION
Perennial plant, small-sized, hairless, without rhizomes. Stems often horizontal then raised (ascendent), often stoloniferous, 10–60 cm high. Leaf blade rolled when young. Leaves from the base have rolled, thin blades. Leaves of the stem have spreading but narrow blades. Blades always flexible, pale green to greyish green. Ligule long, generally acute. No auricles. Panicle-like, loose inflorescence, spreading at flowering and contracted afterwards. Spikelets 1-flowered, with a long bent awn on the lemma (*Agrostis capillaris*, *A. stolonifera* and *A. gigantea* have no awn in the spikelets except the *aristata* form of *A. capillaris*) (Figure 5.1) (Plates 5.1 and 5.2).

1000-SEED WEIGHT
0.08–0.10 g (small seeds).

CHROMOSOME NUMBER
2n = 14 (diploid).

ECOLOGICAL REQUIREMENTS

ALTITUDE – VEGETATION BELTS
From lowlands to high altitudes in mountain areas; from hill to alpine belts.

SOIL MOISTURE
Optimum on wet and very wet soils (Figure 5.2).

SOIL FERTILITY
Limited to nutrient-poor soils but often with high organic matter content, such as peaty soils. Adapted to very acid soils (Figures 5.2 and 5.3).

SOIL TYPE AND TOPOGRAPHY
Usually on siliceous soils: sand and schist, for example.

CLIMATE REQUIREMENTS
Wide climate range. Very resistant to cold but sensitive to drought.

FIGURE 5.2
Ecological optimum and range for soil pH and humidity of *Agrostis canina*

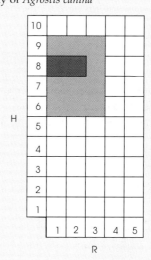

R: 1 = extremely acid and 5 = alkaline
H: 1 = extremely dry and 10 = permanently flooded

Complete key in Chapter 2, section 2.8.

FIGURE 5.3
Ecological optimum and range for nutrient availability and management of *Agrostis canina*

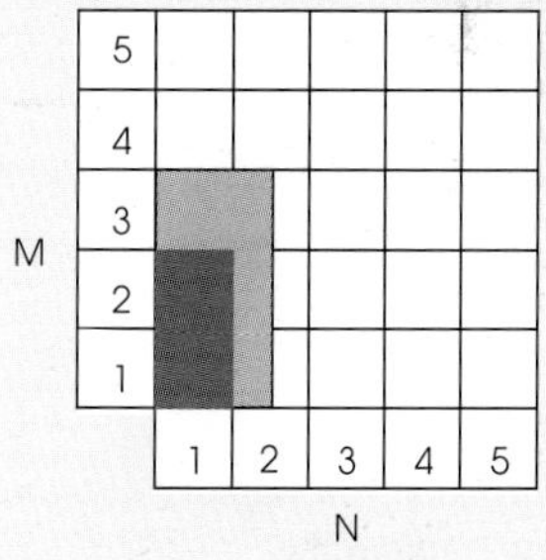

N: 1 = very low and 5 = very high
M: 1 = one defoliation per year and
5 = 5 defoliations per year or more

Complete key in Chapter 2, section 2.8.

TOLERANCE OF CUTTING AND GRAZING
Clear preference for cutting (hay) meadows (Figure 5.3).

SOCIABILITY – COMPETITIVENESS – PLANT COMMUNITIES

Poorly competitive species. Its small size and slow growth allow it to survive only on very poor soils or to colonize gaps in the cover of wet moors, or oligotrophic bogs. Associated with *Molinia caerulea, Ranunculus flammula* and *Carex* spp.

Mostly frequent in *Molinion caeruleae, Juncion acutiflori* and *Caricion fuscae*. Less frequent in *Bromion racemosi*.

COMPATIBILITY WITH OTHER FORAGE GRASSES AND LEGUMES

Incompatible with productive forage species.

INDICATOR VALUE

Wet soil, very poor in nutrients, especially in P.

AGRONOMIC CHARACTERISTICS

SOWING RATES

If sown, the ideal rate would probably be similar to *Agrostis capillaris*.

VARIETIES

Ornamental variety: Silver Needles.

FERTILIZATION

The communities in which *A. canina* grows are not fertilized.

DRY MATTER YIELDS AND SEASONALITY OF PRODUCTION

Very low production in the sites where it thrives spontaneously.

NUTRITIVE VALUE

Probably similar to *A. capillaris*.

ACCEPTABILITY

As hay, it is certainly well accepted.

DISEASES

In general, *Agrostis* spp. are very susceptible to diseases and in particular to snow mould caused by different pathogens, including *Microdochium* (= *Fusarium) nivale*, one of the most destructive.

However, there are considerable variations between ecotypes of *A. canina*, which range from very susceptible to moderately resistant to this disease. Other diseases infecting this *Agrostis* include rusts (*Puccinia coronata, P. graminis*) (Ellis and Ellis, 1997).

USE FOR PURPOSES OTHER THAN FORAGE PRODUCTION

The ornamental variety is sometimes used in gardens and parks.

MAIN QUALITIES

Reasonable quality compared with other species that are found in the same communities. Quality-wise, it is certainly better than *Carex* spp., *Juncus* spp. and *Molinia caerulea*.

MAIN SHORTCOMINGS

Very low productivity.

AGROSTIS CAPILLARIS L.

Agrostis capillaris is very close to *A. castellana* and some varieties sold as *A. capillaris* are actually *A. castellana*, e.g. the variety 'Highland bent'.

SYNONYM

A. pumila L., *A. tenuis* Sibth., *A. vulgaris* With.

ETYMOLOGY

LATIN: *agrostis* or GREEK: αγρωστις
Name of an undetermined grass but also GREEK: αγρωσ = *agros* = field; species that ornaments fields.

FIGURE 5.4
Inflorescence and part of the leaf of *Agrostis capillaris*

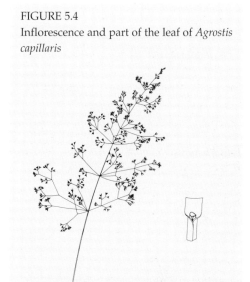

PLATE 5.3
Agrostis capillaris

PLATE 5.4
Agrostis capillaris

LATIN: *capillus* = hair; the inflorescence seems to be hairy.

COMMON NAMES

ENGLISH: Common bent, brown top
FRENCH: Agrostis commun
GERMAN: Gemeines straussgras, gewöhnliches straussgras, rotes straussgras

DISTRIBUTION

Native to Europe and western Asia. Has become subcosmopolitan in temperate regions.

DESCRIPTION

MORPHOLOGICAL DESCRIPTION
Perennial plant, small-sized, hairless, with short rhizomes. Stems often horizontal then raised (ascendent), often stoloniferous, 10–80 (–100) cm high. Blade rolled when young, thin, narrow (1–5 mm). Ligule short (0.5–2 mm), more broad than long, truncated, asymmetric (top of the ligule off-centre). No auricles. Panicle-like, loose inflorescence, usually spreading, ovate, rarely contracted even after flowering. Spikelets 1-flowered (Figure 5.4) (Plates 5.3 and 5.4).

PHYSIOLOGICAL PECULIARITIES
A. capillaris hybridizes with *Agrostis stolonifera*.

   One of the few grasses producing stolons and rhizomes simultaneously.

1000-SEED WEIGHT
0.06–0.10 g (small seeds).

CHROMOSOME NUMBER
2n = 28 (tetraploid).

ECOLOGICAL REQUIREMENTS

ALTITUDE – VEGETATION BELTS
Wide altitude range: from sand dunes or sea cliffs up to 3 000 m in the Alps (Rameau *et al.*, 1993); from hill to alpine belts.

FIGURE 5.5
Ecological optimum and range for soil pH and humidity of *Agrostis capillaris*

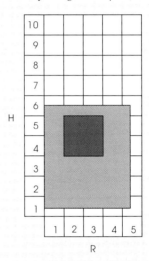

R: 1 = extremely acid and 5 = alkaline
H: 1 = extremely dry and 10 = permanently flooded

Complete key in Chapter 2, section 2.8.

FIGURE 5.6
Ecological optimum and range for nutrient availability and management of *Agrostis capillaris*

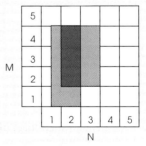

N: 1 = very low and 5 = very high
M: 1 = one defoliation per year and
5 = 5 defoliations per year or more

Complete key in Chapter 2, section 2.8.

SOIL MOISTURE
Optimum on normally drained or dry soils (Figure 5.5).

SOIL FERTILITY

Species typical of mesotrophic to oligotrophic conditions. Often indicative of a low P content in soil. Is often associated with *Festuca rubra*, which nevertheless prefers slightly richer (mesotrophic) soils (Figures 5.5 and 5.6).

SOIL TYPE AND TOPOGRAPHY

The optimum conditions of soil fertility are met mainly on sandy soils, schists, granites and gneiss, where the soil reaction is acid, but also on chalk and on dolomites where the soil reaction is basic. *A. capillaris* and *A. stolonifera* often have a complementary distribution in relation to topography. *A. capillaris* is more abundant on plateaux and on the upper part of slopes, while *Agrostis stolonifera* is more common on the lower part of slopes, namely on colluvium and in valleys. However, the distribution areas of the 2 species largely overlap.

CLIMATE REQUIREMENTS

Extremely resistant to summer heat and to winter cold. Its climate requirements are thus low. Present in Atlantic or continental climatic areas, at low or high altitudes. Tolerates shade, for instance, at the edges of forests.

TOLERANCE OF CUTTING AND GRAZING

Indifferent to the management regime (cutting or grazing). In lawns, tolerates frequent mowing (*circa* 20 cuts/year). Often dominates in extensively grazed grasslands on poor and dry soils (shallow or excessively drained soils). Co-dominant with *Festuca rubra* in mesotrophic and extensively grazed grasslands. Associated with *Festuca rubra*, *Trisetum flavescens* and *Anthoxanthum odoratum* in mainly cut grassland (Figure 5.6).

SOCIABILITY – COMPETITIVENESS – PLANT COMMUNITIES

Not a highly competitive species. Compatible with a high plant diversity (50–60 species/100 m^2).

Mostly frequent in the alliances of the *Arrhenatheretalia: Cynosurion cristati*, *Agrostis-Festuca* grasslands, *Arrhenatherion elatioris*, *Polygono-Trisetion flavescentis* namely; of the *Nardo-Callunetea: Violo-Nardion strictae*, *Nardion strictae*, *Deschampsia flexuosa* grasslands; and of the *Sedo-Scleranthetea*. Less frequent in *Mesobromion erecti*, *Koelerio-Phleion phleoidis* and dry intensive grasslands.

COMPATIBILITY WITH OTHER FORAGE GRASSES AND LEGUMES

Compatible with many other non-competitive forage species, but

TABLE 5.1

Examples of varieties of *Agrostis capillaris*

Variety	Seed breeder (country)
Allure	Joordens (NL)
Aros	DLF (DK)
Bardot	Barenbrug (NL)
Heriot	PBI (UK) – Barenbrug (NL)
Leivkin	Hellerud (N)
Ligretta	DSV (D)
Litenta	DSV (D)
Saboval	WPBS (UK)

TABLE 5.2

Annual yields (tonnes DM/ha) of *Agrostis capillaris* compared with *Lolium perenne* in Scotland, UK

| Species | Annual N fertilization (kg/ha) | | | | |
variety	0	120	240	360	480
L. perenne					
Perma	2.4	6.1	9.7	11.9	13.2
A. capillaris					
Highland	2.2	5.4	8.8	10.1	10.6
Saboval	2.0	5.6	8.0	9.8	10.0

Note: 6 cuts/year regime
Source: Frame, 1991

TABLE 5.3

Comparison of annual DM yields of 4 varieties of *Agrostis capillaris*

Variety	Yield (tonnes DM/ha)	Relative yield (%)
Leivkin	9.4	100
Allure	8.9	95
Litenta	8.7	93
Ligretta	7.9	84

Note: Average of 2 years and 5 sites in regimes of 3 or 4 cuts/year
Source: Lehmann *et al.*, 1992

suppressed by the most productive species. Can be mixed with *Festuca rubra, Trisetum flavescens* and *Trifolium repens.*

INDICATOR VALUE
Rather poor soil, especially in P.

AGRONOMIC CHARACTERISTICS

SOWING RATES
Pure: 10–20 kg/ha.
Mixed: 5 kg/ha in complex mixtures (8–10 species);
10 kg/ha in simple mixtures with *Festuca rubra* and *Trifolium repens.*

VARIETIES
A list of varieties is shown in Table 5.1.

FERTILIZATION
Very low requirements but modest production; N fertilization should be less than 100 kg/ha. An association with

legumes such as *Trifolium repens,* common bird's-foot trefoil (*Lotus corniculatus*) or black medick (*Medicago lupulina*) is preferable to pure-sown, fertilized swards. Its abundance is reduced by high fertilization, which benefits more competitive species such as *Festuca rubra, Holcus lanatus* and *Lolium perenne.*

DRY MATTER YIELDS AND SEASONALITY OF PRODUCTION
Average production ranking but proportionately more important than *Lolium perenne* in summer (Henderson, Edwards and Hammerton, 1962; Cowling and Lockyer, 1965). Yields were always lower than *Lolium perenne* in a 6 cuts/year regime, even without N fertilization (Table 5.2).

In Switzerland, the yields were also lower than other productive species. The differences in performance among varieties were significant (Table 5.3).

FIGURES 5.7–5.10

Annual yields of *Agrostis capillaris* compared with *Lolium perenne* and *Phleum pratense* in 3 cutting regimes and several nitrogen fertilization levels (average of 3 production years)

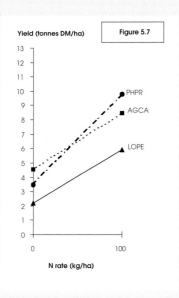

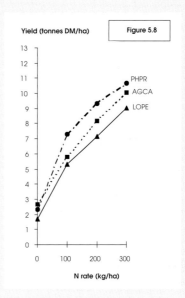

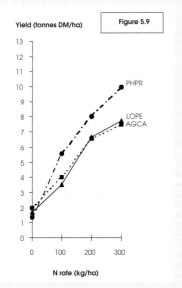

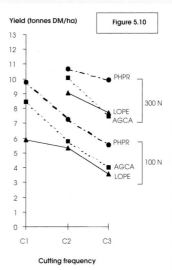

AGCA: *A. capillaris;* LOPE: *L. perenne;* PHPR: *P. pratense.* Fig. 5.7: 2 cuts/year (C1); Fig. 5.8: 4 cuts/year (C2); Fig. 5.9: 5-6 cuts/year (C3); Fig. 5.10: 100 and 300 kg N/ha

Source: Peeters and Decamps, 1999

In an experiment in Belgium (Peeters and Decamps, 1999), *A. capillaris* was more productive than *Lolium perenne* with infrequent cutting regimes (2 and 4 cuts/year) and at low fertilizer N rates. The yields of *Lolium perenne* were only higher in the 6 cuts/year regime and for high N rates (200 and 300 kg N/ha). Nevertheless, *A. capillaris* was less productive than *Phleum pratense*, except with no N (Figures 5.7 to 5.10). In all cases, the yields of *A. capillaris* were higher than *Agrostis stolonifera* on the well drained soil where the trial was carried out.

NUTRITIVE VALUE
Quality data are sparse.

In mixtures with *Lotus corniculatus* or greater lotus (*L. pedunculatus*) the digestible organic matter in the dry matter (DOMD) of *A. capillaris* was considerably lower than that of *Phleum pratense* but N contents were similar (Table 5.4).

In a 6 cuts/year regime, the average organic matter digestibility (OMD) of *A. capillaris* (70.2 percent) was clearly lower than that of *Lolium perenne* (79.6 percent) while in a 3 cuts/year regime, the difference between the 2 species widened (58.1 and 72.5 percent respectively) (Frame, 1991).

ACCEPTABILITY
Well accepted at a leafy stage by domestic animals. Very well accepted by deer in forest clearings.

ANIMAL PERFORMANCE
The animal performance results obtained on swards dominated by *A. capillaris* are very variable. For example, in the results of Common *et al.* (1991b) obtained in Scotland (UK), the *Agrostis/Festuca* community gave an increase of 17 percent of sheep grazing days/year in comparison with the *Molinia* community and an increase of 33 percent compared with the *Nardus* community, i.e. 2 500 grazing days/year for sheep of almost 50 kg live weight (LW).

The live weight gain (LWG) of sheep was about 54 g/day.

DISEASES
Generally very susceptible to diseases and in particular to snow mould caused by different pathogens, including *Microdochium (= Fusarium) nivale*, one of the most damaging.

However, there are great differences between varieties of *A. capillaris* (Smith, Jackson and Woolhouse, 1989). Other diseases like *Puccinia coronata*, *P. graminis* and *Helminthosporium* diseases (*Drechslera* spp.) can be observed (Ellis and Ellis, 1997).

TABLE 5.4
Annual DOMD and N content in the DM of *Agrostis capillaris* compared with *Phleum pratense*

	Year	Agrostis capillaris (%)	Phleum pratense (%)
DOMD	1	49.4	57.3
	2	48.6	56.6
	3	49.8	53.1
N	1	2.13	2.02
	2	1.92	1.92
	3	2.06	1.57

Note: The grasses were harvested in mixture with *Lotus* spp. in a 3 cuts/year regime
Source: Hopkins *et al.*, 1996

USE FOR PURPOSES OTHER THAN FORAGE PRODUCTION

Largely used for the creation of fine lawns in gardens and parks. It is seldom used in sports fields and in heavily trampled lawns because it does not tolerate trampling as well as *Lolium perenne*. It is associated in lawn mixtures with *Festuca ovina, F. rubra* and *Poa pratensis*. It can help to stabilize soils, for example on new ski slopes in mountain areas, on road and motorway side verges. It is also used for soil conservation and restoration. It is one of the best grasses for species-rich sward restoration.

MAIN QUALITIES

'Low-requirement' grass, well adapted to extensive husbandry systems. In mountains, together with *Festuca rubra*, it forms the basis of the best grazed grasslands. It is very compatible with *Trifolium repens* and produces a large part of its yield in summer. Hence, it contributes to sustaining production when many other species have low productivity.

MAIN SHORTCOMINGS

Not productive enough to be utilized in fertile lowlands. Moreover, in mixed species swards, it is suppressed by other more responsive species when a high N fertilization is applied.

AGROSTIS GIGANTEA ROTH

FIGURE 5.11
Inflorescence and part of the leaf of *Agrostis gigantea*

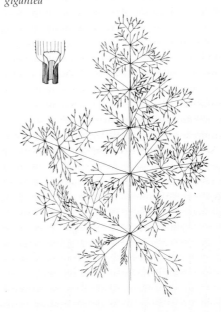

SYNONYM

A. alba auct. non L., *A. nigra* With., *A. stolonifera* L. ssp. *gigantea* Gaudin ex Schübler et Martens.
A. gigantea and *A. stolonifera* were wrongly labelled as *A. alba* for quite a long time.

ETYMOLOGY

LATIN: *agrostis* or GREEK: αγρωστις
Name of an undetermined grass but also
GREEK: αγρωσ = *agros* = field; species that ornaments fields.
LATIN: *giganteus* = giant; from Gigas: one of the giants who wanted to climb Mount Olympus to dethrone Jupiter. This *Agrostis* is very tall compared with the other species of the genus.

PLATE 5.5
Agrostis gigantea

PLATE 5.6
Agrostis gigantea

COMMON NAMES
ENGLISH: Black bent, red top
FRENCH: Agrostis géant
GERMAN: Riesen straussgras

DISTRIBUTION
Native to temperate and cold areas of the
northern hemisphere.

DESCRIPTION
MORPHOLOGICAL DESCRIPTION
Perennial plant, robust, hairless, with more
or less elongated rhizomes. Stems 30–120 cm
high, sometimes stoloniferous. Blade rolled
when young, 3–10 mm wide. Ligule long,
with a truncated and often dentate top. No
auricles. Panicle-like, loose inflorescence,
8–30 cm long, not or slightly contracted,
mainly on the top, after flowering. Spikelets
1-flowered (Figure 5.11) (Plates 5.5 and 5.6).

1000-SEED WEIGHT
0.08–0.11 g (small seeds).

CHROMOSOME NUMBER
2n = 42 (hexaploid).

ECOLOGICAL REQUIREMENTS

ALTITUDE – VEGETATION BELTS
From lowlands to high altitudes in mountain
areas; from hill to subalpine belts.

SOIL MOISTURE
Optimum on well drained, relatively dry
soils, but wide range; can grow on very wet
soils (Figure 5.12).

SOIL FERTILITY
Thrives on moderately to highly nutrient-
rich soils (Figures 5.12 and 5.13).

SOIL TYPE AND TOPOGRAPHY
Mainly prefers light, sandy soils.

CLIMATE REQUIREMENTS
Large climatic range. Good resistance to cold
and drought.

FIGURE 5.12
Ecological optimum and range for soil pH and
humidity of *Agrostis gigantea*

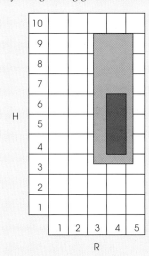

R: 1 = extremely acid and 5 = alkaline
H: 1 = extremely dry and
10 = permanently flooded

Complete key in Chapter 2, section 2.8.

FIGURE 5.13
Ecological optimum and range for nutrient
availability and management of *Agrostis
gigantea*

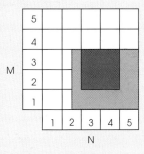

N: 1 = very low and 5 = very high
M: 1 = one defoliation per year and
5 = 5 defoliations per year or more

Complete key in Chapter 2, section 2.8.

TOLERANCE OF CUTTING AND GRAZING
Does not tolerate intensive grazing. Adapted
to infrequent cutting for hay production or
to extensive grazing in mountain areas.

Much more abundant in arable land and in disturbed ground than in grassland (Figure 5.13).

SOCIABILITY – COMPETITIVENESS – PLANT COMMUNITIES
Poorly competitive species in grassland, but in arable land it can be an invasive weed, especially on light soil.

Mostly frequent in *Arrhenatherion elatioris, Bromion racemosi* and other alliances of the *Molinio-Arrhenatheretea*, and in ruderal vegetation of the *Chenopodietea*. Less frequent in *Phragmitetalia*.

COMPATIBILITY WITH OTHER FORAGE GRASSES AND LEGUMES
Rarely mixed with other forage species, except in difficult areas. In Switzerland, it can be added to complex mixtures for grazing in mountain areas (Mosimann *et al.*, 1996a).

AGRONOMIC CHARACTERISTICS

SOWING RATES
Pure: 10–20 kg/ha.

Mixed: 5 kg/ha in complex mixtures with 8–10 species;
10 kg/ha in simple mixtures.

VARIETIES
A list of forage varieties is shown in Table 5.5.

FERTILIZATION
Important for intensive use.

DRY MATTER YIELDS AND SEASONALITY OF PRODUCTION
Average to high annual production (Table 5.6).

NUTRITIVE VALUE
Probably similar to *Agrostis capillaris*, or lower.

ACCEPTABILITY
Quite well accepted at a leafy stage by grazing animals. Well ingested as hay.

DISEASES
The diseases of this grass are not well known. The following diseases can

TABLE 5.5
Examples of varieties of *Agrostis gigantea*

Variety	Seed breeder (country)
BR 1408	Grassland Div (NZ)
Kita	ZNRO (PL)
Listra	DSV (D)
Streaker	CEBECO (NL)

TABLE 5.6
Annual yields (tonnes DM/ha) of 4 varieties of *Agrostis gigantea*

Variety	Average	Minimum	Maximum
Kita	13.4	8.5	18.0
Streaker	13.7	8.2	19.9
Listra	13.0	-	17.5
BR 1408	10.6	6.0	15.3

Note: Average of 2 years and 5 sites in regimes of 3 or 4 cuts/year
Source: Lehmann *et al.*, 1992

probably be observed since they also appear on *Agrostis stolonifera*: the snow mould caused by different pathogens including *Microdochium (= Fusarium) nivale*, one of the most damaging agents, *Puccinia coronata, P. graminis, Drechslera* spp., *Mastigosporium* leaf fleck (*Mastigosporium rubricosum*) and leaf blotch (*Rhynchosporium orthosporum*) (Ellis and Ellis, 1997).

MAIN QUALITIES
Fairly productive grass, very winter hardy and adapted to hay production. It is mainly of interest in mountain areas and possibly in other difficult situations.

MAIN SHORTCOMINGS
Not productive enough to be utilized in fertile lowlands. It does not tolerate intensive grazing and its nutritive value is rather poor.

Profiles of individual grass species

AGROSTIS STOLONIFERA L.

There are several ecotypes, botanical
varieties or subspecies in meadows,
on sand, on salty mud, as well as
in boggy places with freshwater.

SYNONYM
A. alba auct. non. L.
A. stolonifera and *Agrostis gigantea*
were wrongly labelled as *A. alba* for
a long period.

ETYMOLOGY
LATIN: *agrostis* or GREEK: αγρωστις
Name of an undetermined grass but also
GREEK: αγρωσ = *agros* = field; species that
ornaments fields.
LATIN: *stolo* = stolon and *ferre* = to carry. This
grass carries stolons.

FIGURE 5.14
Inflorescence and part of the leaf of *Agrostis
stolonifera*

PLATE 5.7
Agrostis stolonifera

PLATE 5.8
Agrostis stolonifera

COMMON NAMES

ENGLISH: Creeping bent
FRENCH: Agrostis stolonifère
GERMAN: Fioringras

DISTRIBUTION

Native to temperate and cold areas of the northern hemisphere.

DESCRIPTION

MORPHOLOGICAL DESCRIPTION
Perennial plant, small-sized, hairless, stoloniferous. Stems ascendent, 10–60 (–100) cm high. Blade rolled when young, narrow (1–6 mm). Ligule long (2–8 mm), longer than broad, lanceolate, with a rounded top. No auricles. Panicle-like, loose inflorescence, wide, contracted after flowering, 2–15 cm long. Spikelets 1-flowered (Figure 5.14) (Plates 5.7 and 5.8).

PHYSIOLOGICAL PECULIARITIES
A. stolonifera hybridizes with *Agrostis capillaris*.

1000-SEED WEIGHT
0.06–0.09 g (small seeds).

CHROMOSOME NUMBER
2n = 28 (tetraploid).
Other chromosome numbers have also been observed: 2n = 30, 32, 35, 42, 44 or 46.

ECOLOGICAL REQUIREMENTS

ALTITUDE – VEGETATION BELTS
Mostly in lowlands but can grow at quite high altitudes in mountains, up to 2 000 m in the Alps on river banks and fertile alluvium (Rameau *et al.*, 1993); from hill to subalpine belts.

SOIL MOISTURE
Has a clear preference for cool to wet soils. Can even thrive on the surface of ponds where its floating leaves sometimes cover areas of hundreds of square metres (Figure 5.15).

FIGURE 5.15
Ecological optimum and range for soil pH and humidity of *Agrostis stolonifera*

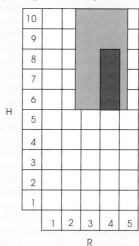

R: 1 = extremely acid and 5 = alkaline
H: 1 = extremely dry and
10 = permanently flooded

Complete key in Chapter 2, section 2.8.

FIGURE 5.16
Ecological optimum and range for nutrient availability and management of *Agrostis stolonifera*

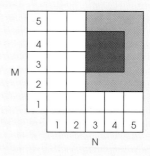

N: 1 = very low and 5 = very high
M: 1 = one defoliation per year and
5 = 5 defoliations per year or more

Complete key in Chapter 2, section 2.8.

SOIL FERTILITY
More demanding than *Agrostis capillaris*, it thrives on mesotrophic soils and often on

nutrient-rich or even very rich soils. The cool to wet soils where it is abundant often have a high organic matter content and some of them are very peaty. Tolerates a wide range of acidity, except extremely acid soils (Figures 5.15 and 5.16). The subspecies *maritima* is salt-tolerant.

SOIL TYPE AND TOPOGRAPHY
The ideal moisture conditions for this *Agrostis* are found on alluvium, colluvium, peat (pure or mineral-peaty), clay soil, impermeable schist, or even humus-rich sand if the water table is close to the surface. From a topographic point of view, the sites of *A. stolonifera* are often found in valleys, on shady hillsides, on slopes where springs appear and in wet plains.

CLIMATE REQUIREMENTS
Mostly adapted to wet climates but has wide distribution; frequently encountered in continental climates. Tolerates cold and also shade, especially in old orchards, but not drought.

TOLERANCE OF CUTTING AND GRAZING
Quite indifferent to the management regime, but its presence is less frequent in hay meadows. A mixed (cutting-grazing) regime or a silage cutting regime (3–4 cuts/year) is particularly suitable. Close grazing keeps it under control whereas it is favoured by long growing periods and a high cutting height (Figure 5.16).

SOCIABILITY – COMPETITIVENESS – PLANT COMMUNITIES
Poorly competitive, except in optimum growth conditions where it develops in large patches. It then grows outwards in waves, with its stolons spreading and other species are overwhelmed.

Mostly frequent in humid improved grasslands, *Cynosurion cristati, Arrhenatherion elatioris, Polygono-Trisetion flavescentis, Bromion racemosi,*

Deschampsion caespitosae, Agropyro-Rumicion and *Armerion maritimae* (*A. stolonifera* ssp. *maritima*).

COMPATIBILITY WITH OTHER FORAGE GRASSES AND LEGUMES
Compatible with *Lolium perenne* on wet and nutrient-rich soils.

INDICATOR VALUE
Relatively wet to very wet, nutrient-rich soil.

AGRONOMIC CHARACTERISTICS

SOWING RATES
Rarely sown in practice.
Pure: 10–20 kg/ha.
Mixed: 10 kg/ha with *Lolium perenne, Trifolium repens* and perhaps with *Alopecurus pratensis.*

VARIETIES
A list of varieties is shown in Table 5.7.

FERTILIZATION
Moderate requirements in pure stand. In mixture with *Lolium perenne*, the N fertilization can be increased to 300 kg/ha in permanent grassland.

DRY MATTER YIELDS AND SEASONALITY OF PRODUCTION
Tables 5.8 to 5.10 show that the dry matter (DM) yields of *A. stolonifera* are lower than those of *Lolium perenne* under optimum conditions (good drainage, high fertilization). It often gives 70–75 percent of the yield of *Lolium perenne*. However, on wet soils, the yield of *A. stolonifera* can be higher than *Lolium perenne* even at high fertilization. On well drained soil, *A. stolonifera* production can be slightly higher or equal to *Lolium perenne* when the N fertilization is low.

It is generally acknowledged that the summer production of *Agrostis* spp. is proportionately higher than *Lolium perenne*.

TABLE 5.7

Examples of varieties of *Agrostis stolonifera*

Variety	Seed breeder (country)
Kromi	DLF (DK)
Penncross	Tee 2 Green (USA)
Providence	Seed Research, Oregon (USA)
Regent	WVPB and Barenbrug (USA)

TABLE 5.8

Annual yields of *Agrostis stolonifera* compared with *Lolium perenne*

Species	Annual N fertilization (kg/ha)				Average yield (tonnes DM/ha)	Relative yield (%)
	0	200	400	800		
L. perenne	0.8	4.7	9.8	12.9	7.0	100
A. stolonifera	0.7	3.6	6.4	8.6	4.8	69

Note: 6 cuts/year regime, average of 3 years
Source: Sheldrick, Lavender and Martyn, 1990

TABLE 5.9

Dry matter yields of *Agrostis stolonifera* relative to *Lolium perenne* (= 100)

Author(s)	Year	Annual N Fertilization (kg/ha)	Relative yield (%)	Comments
Grubb	1968	196	58	62% cover of *A. stolonifera*
Morris and Thomas	1972	250-300	68	Well drained soil, not peaty. Average of 2 years.
			120	Podzol, peaty. 200–300 m of altitude. Average of 2 years.
Ivins	1974	variable	68	Synthesis: *Agrostis* in general
Haggar	1976	400	76	
Frame	1982	0	110	Average of 3 years
		240	88	"
		480	78	"
Smith and Allcock	1985	50-300	70	Average of N rates
Frame and Tiley	1988	360	76	Average of 3 years

Source: Synthesis of literature data by Sheldrick, Lavender and Martyn, 1990

TABLE 5.10

Annual yields (tonnes DM/ha) of *Agrostis stolonifera* compared with *Lolium perenne* in Scotland, UK

Species	Annual N fertilization (kg/ha)				
	0	120	240	360	480
L. perenne	2.4	6.1	9.7	11.9	13.2
A. stolonifera	2.6	6.3	8.5	10.0	10.5

Note: 6 cuts/year regime, average of 3 years
Source: Frame, 1991

The data in Table 5.11 show a higher growth of *A. stolonifera* compared with *Lolium perenne* in June and July. However, other data do not converge on this point. It seems that the soil moisture also has a major influence. On dry soil or during dry years, *A. stolonifera* would produce proportionately less biomass in summer than *Lolium perenne*; in normal conditions, the differences would be small; on wet soil, *A. stolonifera* would produce proportionately more in summer (Morris and Thomas, 1972; Charles *et al.*, 1978; Sheldrick *et al.*, 1990) but this last point remains controversial.

On well drained soils, poor in organic matter, in Belgium *A. stolonifera* was one of the less productive species among the 11 species tested (Peeters and Decamps, 1999). Its yields were higher than *Poa trivialis* but clearly lower than *A. capillaris*, even at high N fertilization. A low water supply, particularly in summer, was certainly responsible for this poor performance. At a high N rate (300 kg/ha), the yield of *A. stolonifera* was 73 percent of *Lolium perenne* in a 4 cuts/year regime and 66 percent for 6 cuts/year. It surpassed the yield of *Lolium perenne* only at low fertilization (0 and 100 kg N/ha) of the 2 cuts/year regime. It never exceeded the yield of *Phleum pratense* (Figures 5.17 to 5.20).

The growth of *A. stolonifera* was clearly slower than *Lolium perenne* in spring before the senescence of *L. perenne* in early June (Table 5.12). At the end of May, the difference of accumulated biomass between the 2 species ranged from 2–4 tonnes DM/ha.

NUTRITIVE VALUE
In the work of Thomas and Morris (1973), the digestibility (DOMD) of *A. stolonifera* was low, as low as *Festuca rubra*, and thus considerably lower than *Lolium perenne*; they suggested the N contents should be similar to *L. perenne* despite a significantly

lower production. These results were confirmed by Frame (1991) for the OMD values but the N content of *A. stolonifera* was higher than *Lolium perenne* in his trials. Haggar (1976) noted better DOMD values for *A. stolonifera* since they were similar to those of *Lolium perenne* (Table 5.13).

A very comprehensive trial in the UK compared *A. stolonifera* and *Lolium perenne* (6 cuts/year, 4 fertilization rates) on a site favourable to *A. stolonifera* (Table 5.14).

It can be concluded from these data that the digestibility of *A. stolonifera* is lower than *Lolium perenne*, but that the difference of about 3 percent is rather small. In the same trial, N contents were clearly higher for *A. stolonifera*.

A. stolonifera digestibility seems to decline very fast at the beginning of spring but then the rate of decline becomes much slower than *Lolium perenne* (Haggar, 1976). The digestibility will always be lower by almost 5 percent than *Lolium perenne* during the first growth cycle up to the beginning of June.

In Belgium, a difference of 5 percent digestibility was noted at the beginning of April, but this difference was reduced as the season advanced. At the beginning of June, the digestibility of the 2 species was similar (Figure 5.21).

The mineral nutrient contents (P, K, Ca, Mg) do not seem to differ from those in *Lolium perenne*.

ACCEPTABILITY
Well ingested by animals.

DISEASES
Generally speaking, the *Agrostis* spp. are very susceptible to diseases, and in particular to snow mould caused by different pathogens, including *Microdochium* (= *Fusarium*) *nivale*, one of the most destructive. However, there are considerable variations between varieties

TABLE 5.11

Yields (tonnes DM/ha) of *Agrostis stolonifera* compared with *Lolium perenne* in England, UK

Species		Cutting dates		
	5 June	1 August	5 September	Total
L. perenne	5.15	1.95	0.70	7.80
A. stolonifera	5.43	4.79	0.74	10.96

Note: 3 cuts/year, annual N fertilization: 120 kg/ha
Source: Haggar, 1976

TABLE 5.12

Evolution of yields (tonnes DM/ha) of *Agrostis stolonifera* compared with *Lolium perenne* during the first growth cycle in the springs of 1991–93 in Belgium

			Dates			
1991	8 April	15 April	24 April	13 May	27 May	11 June
L. perenne	1.0	1.8	2.8	6.0	8.7	12.9
A. stolonifera	1.0	1.5	2.2	3.4	5.0	7.2
1992	15 April	29 April	7 May	19 May	27 May	9 June
L. perenne	0.9	2.8	4.7	7.8	10.1	9.9
A. stolonifera	1.4	2.4	3.3	5.7	7.3	9.5
1993	16 April	26 April	4 May	17 May	25 May	1 June
L. perenne	1.7	3.8	5.4	7.7	9.2	9.7
A. stolonifera	1.2	2.7	3.8	5.6	7.4	8.4

Note: N fertilization during the first growth cycle in spring: 100 kg/ha
Source: Peeters and Decamps, 1994

TABLE 5.13

Annual digestibility of *Agrostis stolonifera* compared with *Lolium perenne* according to several authors

Authors	A. stolonifera (%)	L. perenne (%)	Comments
Thomas and Morris (1973)	52.3	58.2	DOMD, 1 cut/month, average of 5 sites and 2 years
Frame (1991)	71.2	79.6	OMD, 6 cuts/year, average of 2 years and 1 site
Frame (1991)	62.6	72.5	OMD, 3 cuts/year, average of 1 year and 1 site
Haggar (1976)	65.4	65.4	DOMD (D-value) 1 cut/month, average of 1 year and 1 site

TABLE 5.14

Annual digestibility (DOMD) of *Agrostis stolonifera* compared with *Lolium perenne*

Species			Annual N rates (kg N/ha)		
	0	200	400	800	Average
L. perenne	66.4	66.3	67.3	66.2	66.6
A. stolonifera	61.4	64.9	64.1	64.2	63.6

Note: 6 cuts/year regime, average of 3 years
Source: Sheldrick, Lavender and Martyn, 1990

FIGURES 5.17–5.20

Annual yields of *Agrostis stolonifera* compared with *Lolium perenne* and *Phleum pratense* in 3 cutting regimes and several nitrogen fertilization levels (average of 3 production years)

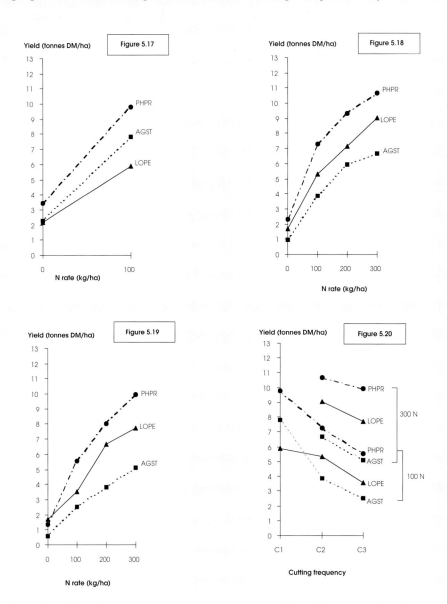

AGST: *A. stolonifera*; LOPE: *L. perenne*; PHPR: *P. pratense*. Fig. 5.17: 2 cuts/year (C1); Fig. 5.18: 4 cuts/year (C2); Fig. 5.19: 5–6 cuts/year (C3); Fig. 5.20: 100 and 300 kg N/ha

Source: Peeters and Decamps, 1999

of *A. stolonifera*. Other diseases such as *Puccinia coronata, P. graminis, Drechslera* spp., *Mastigosporium rubricosum* and *Rhynchosporium orthosporum* can be observed (Ellis and Ellis, 1997).

USE FOR PURPOSES OTHER THAN FORAGE PRODUCTION
Can be used for lawns, especially in the 'greens' of golf courses. It can also be used for stabilizing wet soils on new ski slopes in mountain areas, for establishing field margins along river sides and for species-rich sward restoration in wet habitats.

MAIN QUALITIES
Can contribute to the yield of permanent grassland on fertile and cool soils because of its good summer productivity. Its quality and its acceptability to stock are rather good. Has a slower decline of digestibility than *Lolium perenne* in summer. It is very cold-resistant and thus it can dominate *Lolium perenne* if the latter has suffered from a severe winter.

MAIN SHORTCOMINGS
Not sufficiently productive to be sown regularly.

FIGURE 5.21
Evolution of the digestibility of *Agrostis stolonifera* compared with *Lolium perenne* during an uninterrupted growth period in spring (100 kg N/ha) (linear regression equations and determination coefficients shown)

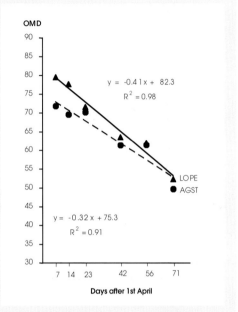

AGST: *A. stolonifera*; LOPE: *Lolium perenne*

Source: Peeters, 1992

ALOPECURUS GENICULATUS L.

ETYMOLOGY
GREEK: αλωπεχ = alopex = fox and ουρα = oura = tail; the spike is like a fox's tail.
LATIN: *geniculatus* = that has knees.

COMMON NAMES
ENGLISH: Marsh foxtail, floating foxtail
FRENCH: Vulpin genouillé
GERMAN: Knick-fuchsschwanz

DISTRIBUTION
Native to Europe and western Asia.

DESCRIPTION

MORPHOLOGICAL DESCRIPTION
Annual or perennial plant, small-sized, hairless (except spikelets), stoloniferous. Stems first creeping, rooted at nodes, then raised (ascendent stolons), 15–60 cm high.

FIGURE 5.22
Inflorescence and part of the stem and leaf of *Alopecurus geniculatus*

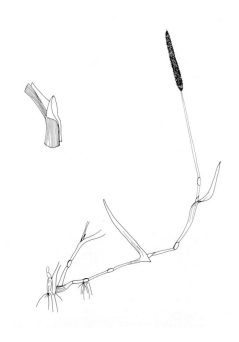

PLATE 5.9
Alopecurus geniculatus

PLATE 5.10
Alopecurus geniculatus

Blade rolled when young, rather short, wide (3–5 mm), green to greyish green, with clear veins. Sheath of the upper leaf slightly bulging. Ligule long. No auricles. Spike-like inflorescence, much shorter and thinner than *Alopecurus pratensis*, hairy-silky, dark green to black. Spikelets 1-flowered. Glumes just united at the base, lengthy ciliate on the back. Lemma awned (Figure 5.22) (Plates 5.9 and 5.10).

CHROMOSOME NUMBER
2n = 28 (tetraploid).

ECOLOGICAL REQUIREMENTS

ALTITUDE – VEGETATION BELTS
From lowlands to high altitudes in mountain areas; from hill to subalpine belts.

SOIL MOISTURE
On wet to very wet soils often found in small dips or in hollows of microreliefs, e.g. tractor tracks (Figure 5.23).

SOIL FERTILITY
Moderately demanding (mesotrophic) to very demanding (eutrophic) from the point of view of soil nutrient availability. Thus, distinguishable from the ecology of *Agrostis canina*, which has the same preferences for soil moisture, but thrives at a much lower nutrient availability level (Figures 5.23 and 5.24). Salt-tolerant.

SOIL TYPE AND TOPOGRAPHY
Often on clay or peat soils, rich in nutrients.

CLIMATE REQUIREMENTS
Wide distribution in temperate climates. Cold-resistant. Drought-sensitive.

TOLERANCE OF CUTTING AND GRAZING
Clear preference for grazed grasslands or for mixed (cutting-grazing) management. Tolerates stock trampling very well (Figure 5.24).

FIGURE 5.23
Ecological optimum and range for soil pH and humidity of *Alopecurus geniculatus*

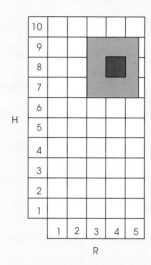

R: 1 = extremely acid and 5 = alkaline
H: 1 = extremely dry and
10 = permanently flooded

Complete key in Chapter 2, section 2.8.

FIGURE 5.24
Ecological optimum and range for nutrient availability and management of *Alopecurus geniculatus*

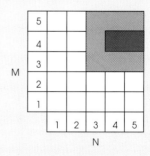

N: 1 = very low and 5 = very high
M: 1 = one defoliation per year and
5 = 5 defoliations per year or more

Complete key in Chapter 2, section 2.8.

SOCIABILITY – COMPETITIVENESS – PLANT COMMUNITIES
Poorly competitive, though often found as sole occupant of microsites.

Mostly frequent in *Agropyro-Rumicion, Bromion racemosi, Deschampsion caespitosae* and humid improved grasslands.

COMPATIBILITY WITH OTHER FORAGE GRASSES AND LEGUMES
Incompatible with productive forage species.

INDICATOR VALUE
Wet, nutrient-rich soil, often on heavy or trampled soils.

AGRONOMIC CHARACTERISTICS

FERTILIZATION
The communities in which *A. geniculatus* is found can have been heavily fertilized.

DRY MATTER YIELDS AND SEASONALITY OF PRODUCTION
Very low production in the sites where it thrives spontaneously.

NUTRITIVE VALUE
Probably quite low quality.

ACCEPTABILITY
Probably very low or even nil after heading.

TOXICITY
Comparable with *Alopecurus pratensis* and, once the anti-quality factor in that species is identified, it will be useful to evaluate its presence in *A. geniculatus*.

ANIMAL PERFORMANCE
Very low.

MAIN QUALITIES
Almost no forage value but it tends not to invade the sites favourable to more valuable forage species.

MAIN SHORTCOMINGS
Very low yield and quality.

ALOPECURUS PRATENSIS L.

ETYMOLOGY
GREEK: αλωπεχ = alopex = fox and oυρα = oura = tail; the spike is like a fox's tail.
LATIN: *pratensis* = from meadows.

COMMON NAMES
ENGLISH: Meadow foxtail, common foxtail
FRENCH: Vulpin des prés
GERMAN: Wiesen-fuchsschwanz

DISTRIBUTION
Native to temperate regions of Europe and Asia. Has become subcosmopolitan in temperate regions.

DESCRIPTION

MORPHOLOGICAL DESCRIPTION
Perennial plant, robust, hairless (except spikelets), rhizomatous. Stems erect,

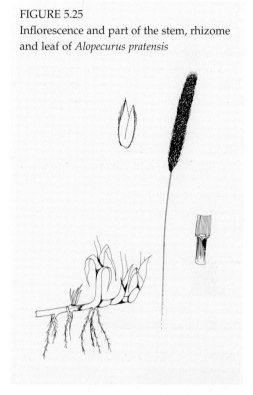

FIGURE 5.25
Inflorescence and part of the stem, rhizome and leaf of *Alopecurus pratensis*

PLATE 5.11
Alopecurus pratensis

PLATE 5.12
Alopecurus pratensis

PLATE 5.13
Alopecurus pratensis

40–120 (–170) cm high. Blade rolled when young, large (3–7 mm), quite coarse, dark green, with irregular ribs. Ligule of average size, variable, truncated, oblique at the base. No auricles. Spike-like, cylindrical inflorescence, narrowed to a fairly sharp point at the top, hairy-silky, dark green to black. Spikelets 1-flowered. Glumes united in the lower third, lengthily ciliate on the back. Lemma awned (Figure 5.25) (Plates 5.11 to 5.13).

PHYSIOLOGICAL PECULIARITIES
Very early flowering.

1000-SEED WEIGHT
0.60–0.99 g (average-sized seeds).

CHROMOSOME NUMBER
2n = 28 (tetraploid).

ECOLOGICAL REQUIREMENTS

ALTITUDE – VEGETATION BELTS
Widespread mainly in the lowlands. In the Alps, grows up to 1 200 m (Caputa, 1967). From hill to mountain belts.

SOIL MOISTURE
Thrives on cool and wet soils (Figure 5.26).

SOIL FERTILITY
Optimum on nutrient-rich to very rich soils. Grows on slightly acid to alkaline soils (Figures 5.26 and 5.27).

SOIL TYPE AND TOPOGRAPHY
Widespread mainly on clay soils. Other types of soils can be suitable, provided they are wet and nutrient-rich: alluvium, colluvium and peat soils.

CLIMATE REQUIREMENTS
Wide climate range except dry climates such as Mediterranean. Extremely resistant to cold and to long snowy periods. Drought-sensitive. Tolerates shade, for example in old orchards.

FIGURE 5.26
Ecological optimum and range for soil pH and humidity of *Alopecurus pratensis*

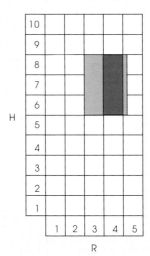

R: 1 = extremely acid and 5 = alkaline
H: 1 = extremely dry and
10 = permanently flooded

Complete key in Chapter 2, section 2.8.

FIGURE 5.27
Ecological optimum and range for nutrient availability and management of *Alopecurus pratensis*

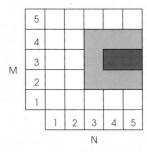

N: 1 = very low and 5 = very high
M: 1 = one defoliation per year and
5 = 5 defoliations per year or more

Complete key in Chapter 2, section 2.8.

TOLERANCE OF CUTTING AND GRAZING
Favoured by late spring utilization.
Optimum intermediate between hay cutting
(2 cuts/year) and exclusive grazing (6–7
grazings/year). When cut 3–4 times/year –
a management frequency that occurs in
grasslands cut only for silage or in
grasslands alternately cut and grazed – it
can develop almost pure populations on rich
and wet soils. If the first use is intended as a
grazing, this may be difficult because the
soil of the optimum site for *A. pratensis* may
not be sufficiently firm at the beginning of
spring, thus grazing is often delayed. When
the soil is sufficiently drained to allow
grazing, the stems of *A. pratensis* are already
tall and so are little grazed. The species is

thus able to continue its growth cycle,
encouraging development of rhizomes and
making seed production possible. This
allows *A. pratensis* to develop in the sward
to the detriment of other species, which are
later growing and better grazed by the
animals. Little by little, the sward can then
be overwhelmed by *A. pratensis*. Longer
growing periods, as in extensive hay
production, are not that beneficial because
the growth is so fast in the spring that leaf
senescence often occurs well before harvest.
Other species can then dominate it. In silage
production, *A. pratensis* has just enough time
to mature and to dominate the other species
before the first harvest (Figure 5.27).
In Belgium (Figures 5.28 and 5.29), the

FIGURES 5.28 AND 5.29
Proportion (% FM) of *Alopecurus pratensis* in mixture with *Lolium perenne* or *Phleum pratense* in 3
cutting regimes and 4 nitrogen fertilizer levels, after 5 years of production

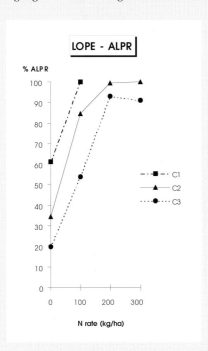 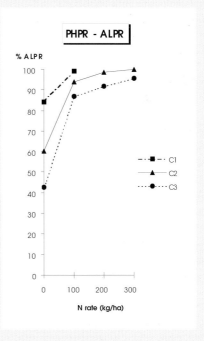

ALPR: *A. pratensis*; LOPE: *L. perenne*; PHPR: *P. pratense*. C1 = 2 cuts/year, C2 = 4 cuts/year,
C3 = 5–6 cuts/year

Source: Peeters, Decamps and Janssens, 1999

percentage by fresh matter (% FM) of *A. pratensis* was measured in an experiment where it was either in mixture with *Phleum pratense* or with *Lolium perenne*. *A. pratensis* dominated these 2 species particularly in 2 main situations: when cutting was infrequent and the N fertilization was high.

SOCIABILITY – COMPETITIVENESS – PLANT COMMUNITIES
Highly competitive species when conditions are optimum. There are generally very few other species on *A. pratensis* grasslands. Beneath fruit trees, it can often be found in conjunction with cow parsley (*Anthriscus sylvestris*). In grasslands alternately used for cutting and grazing and managed at medium intensity, *A. pratensis* often appears in conjunction with *Rumex acetosa* and *Ranunculus acris*, which also thrive on cool, wet and fertile soils. *Poa trivialis* and *Holcus lanatus* are also frequently associated with this species.

Mostly frequent in *Arrhenatherion elatioris*, *Bromion racemosi* and other alliances of the *Molinio-Arrhenatheretea*, in *Filipendulion ulmariae* and humid improved grasslands.

COMPATIBILITY WITH OTHER FORAGE GRASSES AND LEGUMES
In Switzerland, it is used in mixtures suitable for cutting and for wet, cold areas, as well as in high-altitude grasslands. It is associated with *Lolium perenne* (early varieties), *Poa pratensis*, *Festuca rubra*, *F. pratensis* and *Trifolium repens*.

INDICATOR VALUE
Wet, nutrient-rich soil, often with high organic matter content. Mainly managed for silage or mixed use for hay and grazing. In mainly grazed swards, reveals a late first management or a low stocking rate in spring.

AGRONOMIC CHARACTERISTICS
SOWING RATES
Pure: 20–25 kg/ha;
         70–100 kg/ha for coated seeds, i.e.

coated with polymers in order to make the sowing easier.
Mixed: 10–15 kg/ha in mixture with *Lolium perenne, Festuca pratensis, Trifolium repens, T. pratense*;
5 kg/ha in complex mixtures with 8–10 species.

VARIETIES
A list of forage varieties is shown in Table 5.15. Ornamental varieties: Aureovariegatus, Aureus.

FERTILIZATION
High level necessary for intensive use.

DRY MATTER YIELDS AND SEASONALITY OF PRODUCTION
In France, annual DM yields reached nearly 15 tonnes/ha (Table 5.16).

In Switzerland with a 5 cuts/year regime and 200 kg N/ha, high yields were obtained at about 450 m altitude (Table 5.17). The other 2 varieties tested in this trial, Lipex and Alko, had average yields very similar to those of Vulpera, reaching 98 and 95 percent, respectively, of its yield.

Few experiments have compared the yield of *A. pratensis* with productive grasses. In Belgium, in a 2 cuts/year regime and with low fertilization, it produced as much as *Lolium perenne*, but less than species better adapted to this cutting regime i.e. *Arrhenatherum elatius, Phleum pratense* and *Dactylis glomerata*. In a 4 cuts/year regime, its yield was overtaken by *Dactylis glomerata* but it produced approximately the same yield as *Holcus lanatus* and *Phleum pratense*. In a 6 cuts/year regime, it ranked among the most productive species, its yield being slightly lower than those of *Dactylis glomerata* and *Phleum pratense* (Figures 5.30 to 5.33).

A. pratensis growth starts earlier than *Lolium perenne* in spring but then in the first fortnight of May, *L. perenne* accumulates much more biomass. That is partially due to an early senescence of the leaves of *A.*

TABLE 5.15
Examples of varieties of *Alopecurus pratensis*

Variety	Seed breeder (country)
Alko	Steinach (D)
Lipex	DSV (D)
Vulpera	FAL (CH)

TABLE 5.16
Annual yields (tonnes DM/ha) of *Alopecurus pratensis* at different nitrogen rates in Normandy, France

| | Annual N fertilization (kg/ha) | | | | |
	0	80	160	240	320
Annual yields	10.1	10.9	11.9	13.9	14.7

Note: 5 cuts/year, average of 2 years
Source: Laissus, 1979

TABLE 5.17
Annual yields (tonnes DM/ha) of *Alopecurus pratensis*, variety Vulpera, obtained at 2 sites in Switzerland over two years

| Variety | Year 1 | | Year 2 | | |
	Site 1	Site 2	Site 1	Site 2	Average
Vulpera	10.6	12.1	11.3	11.6	11.4

Note: 5 cuts/year, annual N fertilization: 200 kg/ha
Source: Mosimann *et al.*, 1996b

pratensis that are often strongly attacked by a *Mastigosporium* leaf fleck (Table 5.18).

NUTRITIVE VALUE
A. pratensis is generally acknowledged to be not very digestible. Indeed, the leaves are physically quite coarse. Its great earliness affects its digestibility, which is already low when the other grasses in a mixed sward reach the optimum utilization stage. This negative judgement is somewhat tempered by some authors; Tingle and van Adrichem (1974) discovered in British Columbia (Canada) that *A. pratensis* was more digestible and had a higher N content at a yield slightly lower than *Phleum pratense*. The digestibility measured *in vivo* in nylon bags for *A. pratensis* cut at early flowering stages (cuts 1 and 2), then at the vegetative stage (cut 3) reached the following values: 76–78 percent at cut 1, 68–70 percent at cut 2 and 58–62 percent at cut 3 (van Adrichem and Tingle, 1975). These digestibility values were higher than those of 15 other grasses evaluated at the same place and cut at the same physiological stage (Tingle and Elliott, 1975).

In a large-scale experiment (7 sites) organized in Canada, *A. pratensis* had a higher digestibility (DDM) (70.6 percent) than *Phleum pratense* (52.5 percent) and 5 other forage grasses, when cut at a similar physiological stage. *A. pratensis* is richer in N, P, Ca and Mg than *Phleum pratense* (Rode

TABLE 5.18

Evolution of yields (tonnes DM/ha) of *Alopecurus pratensis* compared with *Lolium perenne* during the first growth cycle in the springs of 1991–93 in Belgium

				Dates		
1991	8 April	15 April	24 April	13 May	27 May	11 June
L. perenne	1.0	1.8	2.8	6.0	8.7	12.9
A. pratensis	1.3	2.4	3.1	4.8	6.2	6.6
1992	15 April	29 April	7 May	19 May	27 May	9 June
L. perenne	0.9	2.8	4.7	7.8	10.1	9.9
A. pratensis	1.6	3.0	5.0	6.8	7.8	7.3
1993	16 April	26 April	4 May	17 May	25 May	1 June
L. perenne	1.7	3.8	5.4	7.7	9.2	9.7
A. pratensis	2.0	4.0	5.9	6.9	7.4	7.0

Note: N fertilization during the first growth cycle in spring: 100 kg/ha
Source: Peeters and Decamps, 1994

and Pringle, 1986; Broersma, 1991).

However, it accumulates more nitrate than *Phleum pratense* and *Phalaris arundinacea*, especially with high N fertilization (Kline and Broersma, 1983; Broersma and Mir, 1990). Many authors insist on the importance of cutting *A. pratensis* at the beginning of heading (Pringle and Kline, 1982; Schröder and Adolf, 1993; Kozlowski and Golinska, 1994).

Daccord (1988) observed a declining rate of digestibility (OMD), similar to *Dactylis glomerata*, during the first growth cycle (-1.8 percent per week) but the digestibility of *A. pratensis* was constantly about 2 percent lower than that of *D. glomerata*.

In Belgium, *A. pratensis* digestibility (OMD) declined considerably quicker than that of *Lolium perenne* (-3.5 *versus* -2.9 percent per week) (Figure 5.34) or of *Dactylis glomerata*. It was also constantly lower than those of 8 other grasses tested.

ACCEPTABILITY

Poorly ingested by grazing animals, even as hay, if they have the choice of other species. However, Rode and Pringle (1986) did not note a significant difference in DM or digestible DM ingestion between *A. pratensis* and *Phleum pratense* by young steers grazing paddocks where these species were sown in

pure stands. The voluntary DM intake was even slightly higher for *A. pratensis* (9.1 versus 8.6 kg/day).

TOXICITY

There is apparently an anti-quality factor in the herbage which has not yet been identified (see below).

ANIMAL PERFORMANCE

Even though its chemical composition and digestibility are superior, animal performance is inferior when fed *A. pratensis* instead of *Phleum pratense*. Rode and Pringle (1986) used young steers to compare both species and noted a higher stock-carrying capacity for *A. pratensis* (501 versus 443 grazing days/ha), though the daily LWGs were lower on *A. pratensis* paddocks (0.79 versus 1.13 kg LWG/day). Despite the higher stock-carrying capacity of *A. pratensis*, the LW production was lower than on *Phleum pratense* paddocks (391 versus 502 kg LWG/ha/year). After the grazing period, the 2 lots of animals received an identical feed of *Phleum pratense* and clover for 56 days. The steers that had previously grazed *A. pratensis* continued to have lower growth rates (0.54 kg LWG/day) than steers that had grazed *Phleum pratense* (0.64 kg LWG/day) (Rode, 1986). This is not

FIGURES 5.30–5.33

Annual yields of *Alopecurus pratensis* compared with *Lolium perenne* and *Phleum pratense* in 3
cutting regimes and several nitrogen fertilization levels (average of 3 production years)

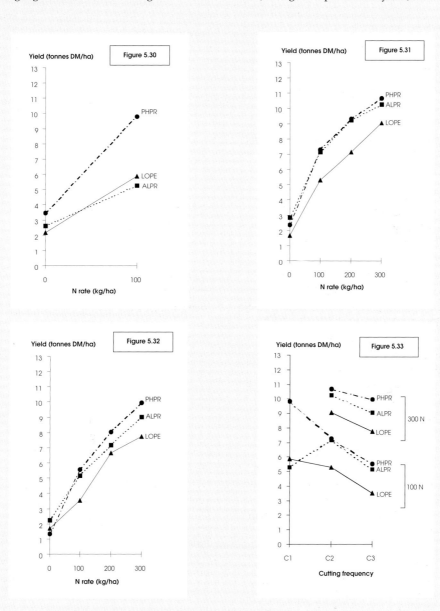

ALPR: *A. pratensis;* LOPE: *L. perenne;* PHPR: *P. pratense.* Fig. 5.30: 2 cuts/year (C1); Fig. 5.31: 4 cuts/year (C2);
Fig. 5.32: 5–6 cuts/year (C3); Fig. 5.33: 100 and 300 kg N/ha

Source: Peeters and Decamps, 1999

consistent with the compensatory growth
principle, since the animals that had a lower
growth rate at grazing should theoretically
have fattened up quicker during the second
phase of the experiment, suggesting that *A.
pratensis* contains some anti-quality factor or
factors that affect animal performance
adversely. However, no alkaloid, cyanogenic
glycoside or nitro-toxin have been detected
in this species (Broersma, 1991).

DISEASES
Massively attacked by *Mastigosporium album*.
These attacks appear quickly at the
beginning of a growing period. Other
diseases have also been observed: ergot
(*Claviceps purpurea*), *Rhynchosporium secalis*
and brown stripe (*Cercosporidium graminis*)
(Ellis and Ellis, 1997).

*USE FOR PURPOSES OTHER THAN FORAGE
PRODUCTION*
Sometimes used for establishing species-rich
meadows and field margins on wet soils,
rich in organic matter.

MAIN QUALITIES
Productive species which starts growing
very early in spring. It could thus be used to
extend the grazing period at that time of the
year. It is rich in N and minerals. When
harvested at leafy or early heading growth
stage, its digestibility is satisfactory. It can be
a useful species in cold areas and on wet
and nutrient-rich soils.

MAIN SHORTCOMINGS
The digestibility is constantly inferior to
other forage grasses in spring due to its
great earliness. However, at equivalent
growth stages, its digestibility is better than
many other species, including *Phleum*

FIGURE 5.34
Evolution of the digestibility of *Alopecurus
pratensis* compared with *Lolium perenne*
during an uninterrupted growth period in
spring (100 kg N/ha) (linear regression
equations and determination coefficients
shown)

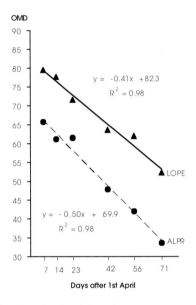

ALPR: *A. pratensis*; LOPE: *L. perenne*

Source: Peeters, 1992

pratense. Animal performance when fed *A.
pratensis* does not reflect its DM yield
or its chemical composition.
An anti-quality factor, as yet unknown,
seems to affect its performance.
Moreover, this grass has a low
acceptability by animals when mixed
with other species in the sward, partially
due to its lower leaf:stem ratio. Also these
other grasses are constantly at a less
advanced physiological stage during the
first growth period.

ANTHOXANTHUM ODORATUM L.

ETYMOLOGY
GREEK: ανθος = anthos = flower and ζανθος = xanthos = yellow, the colour of the panicle after flowering.
LATIN: *odoratus* = fragrant, scented.

COMMON NAMES
ENGLISH: Sweet vernal-grass
FRENCH: Flouve odorante
GERMAN: Gemeines ruchgras

DISTRIBUTION
Native to Europe, temperate Asia and Northern Africa. Has become sub-cosmopolitan in temperate regions.

DESCRIPTION

MORPHOLOGICAL DESCRIPTION
Perennial plant, small-sized, hairy, fragrant

FIGURE 5.35
Inflorescence and part of the leaf of *Anthoxanthum odoratum*

PLATE 5.14
Anthoxanthum odoratum

PLATE 5.15
Anthoxanthum odoratum

(because of coumarin content), caespitose. Stems erect, 10–60 (–80) cm high. Leaf blade rolled when young, wide (3–5 mm). Ligule rather long, sharp, often purple-spotted. Auricles transformed to hairs located on the top of the sheath. Panicle-like inflorescence, 3–7 (–10) cm high, more or less compact and then spike-like. Spikelets with 1 single fertile flower and 2 sterile rudiments. Glumes hairless or hairy. Lemma awned (Figure 5.35) (Plates 5.14 and 5.15).

PHYSIOLOGICAL PECULIARITIES
Maximum number of leaves per tiller: 4 (Sydes, 1984).

1000-SEED WEIGHT
0.40–0.63 g (small seeds).

CHROMOSOME NUMBER
2n = 20 (tetraploid).
Another chromosome number has also been observed: 2n = 10.

ECOLOGICAL REQUIREMENTS

ALTITUDE – VEGETATION BELTS
From coastal areas up to 3 000 m in the Alps (Rameau *et al.*, 1993); from hill to alpine belts.

SOIL MOISTURE
Optimum on dry soils but also frequent on well drained or even slightly wet soils (Figure 5.36).

SOIL FERTILITY
Restricted to nutrient-poor soils especially those low in P. More abundant on acid soils than on alkaline soils (Figures 5.36 and 5.37).

SOIL TYPE AND TOPOGRAPHY
Thrives on a very large range of soil textures.

CLIMATE REQUIREMENTS
Wide climate range. Good resistance to drought. Tolerant of wetness. Very resistant to cold and heat.

FIGURE 5.36
Ecological optimum and range for soil pH and humidity of *Anthoxanthum odoratum*

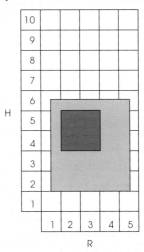

R: 1 = extremely acid and 5 = alkaline
H: 1 = extremely dry and
10 = permanently flooded

Complete key in Chapter 2, section 2.8.

FIGURE 5.37
Ecological optimum and range for nutrient availability and management of *Anthoxanthum odoratum*

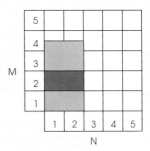

N: 1 = very low and 5 = very high
M: 1 = one defoliation per year and
5 = 5 defoliations per year or more

Complete key in Chapter 2, section 2.8.

TOLERANCE OF CUTTING AND GRAZING
Clear preference for hay meadows but not absent from extensively grazed grasslands (Figure 5.37).

SOCIABILITY – COMPETITIVENESS – PLANT COMMUNITIES

Poorly competitive though can be a component of species-rich grasslands.

Mostly frequent in *Agrostis-Festuca* grasslands, *Cynosurion cristati, Arrhenatherion elatioris, Polygono-Trisetion flavescentis, Violo-Nardion strictae, Nardion strictae, Juncion acutiflori, Molinion caeruleae, Poion alpinae, Mesobromion erecti* and *Koelerio-Phleion phleoidis*.

COMPATIBILITY WITH OTHER FORAGE GRASSES AND LEGUMES

Incompatible with productive forage species.

INDICATOR VALUE

Very poor soil, especially in P. Extensive management.

AGRONOMIC CHARACTERISTICS

SOWING RATES

In commercial wild flora mixtures of 20–30 constituents, which are sown at 30–50 kg/ha, *A. odoratum* is often included at up to 6 percent by seed number.

FERTILIZATION

The communities in which *A. odoratum* is found are usually not fertilized.

DRY MATTER YIELDS AND SEASONALITY OF PRODUCTION

One of the less productive grasses in a 6 cuts/year regime, but when cut less frequently (3 cuts/year), it achieved high yields, similar to *Lolium perenne* (Table 5.19). Thomas and Morris (1973) also recorded high performances of *A. odoratum* compared with *Lolium perenne* at 5 locations of increasing altitude (between 49 and 303 m) in the North Pennines (UK). *Lolium perenne* was slightly more productive (5.5 versus 4.9 tonnes DOM/ha) at low altitude (49 m), but the advantage decreased above 200 m (2.6 versus 2.4 tonnes DOM/ha at 303 m altitude). The sites were of low soil fertility, especially the higher sites.

It can be concluded from these experiments that *A. odoratum* can equal or outyield *Lolium perenne* at low N fertilization and in general at nutrient-poor sites. Also, infrequent cutting regimes are more favourable than frequent cutting regimes.

NUTRITIVE VALUE

The digestibility of *A. odoratum* is moderately high. Its average OMD (74.6 percent) was lower than that of *Lolium perenne* (79.6 percent) or *Holcus lanatus* (75.8 percent), but similar to that of *Cynosurus cristatus* (74.9 percent) and clearly higher than *Agrostis* spp., *Festuca rubra* and *Poa pratensis* in a 6 cuts/year regime (Frame, 1991); these rankings were also confirmed in a 3 cuts/year regime. In the experiment of Thomas and Morris (1973), the average DOMD of *A. odoratum* (57.5 percent) was slightly lower

TABLE 5.19

Annual yields (tonnes DM/ha) of *Anthoxanthum odoratum* compared with *Lolium perenne* in Scotland, UK

	Annual N fertilization (kg/ha)				
	0	120	240	360	480
6 cuts/year **(average of 3 years)**					
L. perenne	2.4	6.1	9.7	11.9	13.2
A. odoratum	2.6	5.3	7.8	9.5	9.7
3 cuts/year **(1 year of experiment)**					
L. perenne	3.8	6.0	10.8	14.7	15.8
A. odoratum	5.2	6.7	10.5	14.4	15.1

Source: Frame, 1991

than *Lolium perenne* (58.2 percent) but higher than *Cynosurus cristatus* (55.5 percent), *Agrostis stolonifera* (52.2 percent) and *Festuca rubra* (51.9 percent). With the same fertilization, *A. odoratum* had a higher N content than *Lolium perenne* but its mineral contents were lower (Thomas and Morris, 1973; Frame, 1991).

ACCEPTABILITY
Well accepted by cattle.

TOXICITY
Contains an alkaloid, coumarin, which is toxic at a high rate. However, in practice, this substance does not seem to create any problems.

DISEASES
Puccinia graminis, P. poae-nemoralis, Claviceps purpurea and *Drechslera dematioidea* are among the diseases observed on *A. odoratum* (Ellis and Ellis, 1997).

USE FOR PURPOSES OTHER THAN FORAGE PRODUCTION
Rather attractive grass that is regularly incorporated in wild flora mixtures. It is even sometimes used in lawns for the pleasant smell that is produced during mowing. It can be used to flavour white alcohol, as is *Hierochloe odorata*, another grass that contains coumarin, and which is placed in bottles of vodka in Poland (Zubrowka: bison vodka).

MAIN QUALITIES
Well suited to poor soils and extensive systems for hay production. Its production and composition are much better than is generally believed.

MAIN SHORTCOMINGS
Not very flexible to management and disappears quickly if the fertilization and defoliation frequency are increased. Production and quality are not sufficiently satisfactory to make this species a valuable forage plant.

ARRHENATHERUM ELATIUS (L.)
BEAUV. EX J. ET C. PRESL

FIGURE 5.38
Inflorescence and part of the leaf of
Arrhenatherum elatius

There are 2 subspecies. The subspecies
elatius thrives in hay meadows.
The subspecies *bulbosum* (Willd.) Hyl.
(Syn.: var. *bulbosum* (Willd.) St-Amans.)
is rare in grassland and is a weed in
arable land. It has a more southern
and Western European distribution than
subspecies *elatius*.

SYNONYM
Avena elatior L.

ETYMOLOGY
GREEK: αρρην = arrhen = male and αθηρ =
ather = spike beard; male flower lengthily
aristate.
LATIN: *elatio* = height and *elatus* = high, tall.

PLATE 5.16
Arrhenatherum elatius

PLATE 5.17
Arrhenatherum elatius

PLATE 5.18
Arrhenatherum elatius

COMMON NAMES
ENGLISH: Tall oat-grass, false oat-grass
FRENCH: Fromental
GERMAN: Glatthafer

DISTRIBUTION
Native to Europe, western Asia and Northern Africa. Has become subcosmopolitan in temperate regions.

DESCRIPTION

MORPHOLOGICAL DESCRIPTION
Perennial plant, robust, hairless to more or less hairy, caespitose. Stems erect, 60–150 (–180) cm high. Blade rolled when young, large, with translucent lines, often slightly narrowed on the upper third, hairless or hairy. Ligule almost 1.5 mm long, denticulate, truncated. No auricles. Panicle-like inflorescence, 10–30 cm long, more or less spreading. Spikelets 2–3-flowered. The lower flower normally male with a long bent awn on the lemma, the second hermaphrodite with usually a shorter awn on the lemma, the upper sterile (Figure 5.38) (Plates 5.16 to 5.18).

1000-SEED WEIGHT
2.50–3.53 g (large seeds).

CHROMOSOME NUMBER
2n = 28 (tetraploid).
Another chromosome number has also been observed: 2n = 14.

BIOLOGICAL FLORA
Pfitzenmeyer (1962).

ECOLOGICAL REQUIREMENTS

ALTITUDE – VEGETATION BELTS
Disappears at altitude, from 1 500 m in the Alps (Rameau *et al.*, 1993), where it is replaced as a dominant grass in hay meadows by *Trisetum flavescens*; from hill to mountain belts.

SOIL MOISTURE
Optimum on well drained and dry soils. Very

FIGURE 5.39
Ecological optimum and range for soil pH and humidity of *Arrhenatherum elatius*

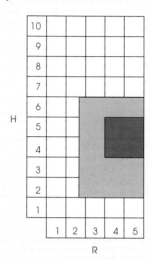

R: 1 = extremely acid and 5 = alkaline
H: 1 = extremely dry and
10 = permanently flooded

Complete key in Chapter 2, section 2.8.

FIGURE 5.40
Ecological optimum and range for nutrient availability and management of *Arrhenatherum elatius*

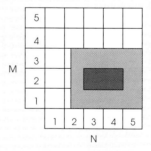

N: 1 = very low and 5 = very high
M: 1 = one defoliation per year and
5 = 5 defoliations per year or more

Complete key in Chapter 2, section 2.8.

sensitive to soil moisture; disappears with winter floodings. Never found on peat soils (Figure 5.39).

SOIL FERTILITY

Prefers average to high soil fertility. Does not tolerate extreme soil acidity; widespread on neutral to alkaline soils (Figures 5.39 and 5.40).

SOIL TYPE AND TOPOGRAPHY

Not very demanding in soil texture requirements but does best on medium-heavy (loam, sandy-clay) and light (sandy-loam, sand) soils.

CLIMATE REQUIREMENTS

Very resistant to summer drought. Although well adapted to continental climates, dislikes late spring frosts. Does not tolerate shade. In oceanic climates, prefers south-facing slopes where the microclimate is warmer.

TOLERANCE OF CUTTING AND GRAZING

Typical species of hay meadows (optimum management 2 cuts/year or 1 late cut and 1 grazing/year). Eliminated where exclusively grazed (Figure 5.40).

In Belgium (Figures 5.41 and 5.42), the % FM of *A. elatius* was measured in an experiment where it was in mixture with *Phleum pratense*, and also with *Lolium perenne*. *A. elatius* dominated *Lolium perenne* and *Phleum pratense* in the 2 and 4 cuts/year regimes, particularly when the N fertilization was high. In a frequent cutting regime (5 or 6 cuts/year), *A. elatius* was totally dominated by *Lolium perenne*. The mixture with *Phleum pratense* was better balanced.

FIGURES 5.41 AND 5.42

Proportion (% FM) of *Arrhenatherum elatius* in mixture with *Lolium perenne* or *Phleum pratense* in 3 cutting regimes and 4 levels of nitrogen fertilization, after 5 years of production

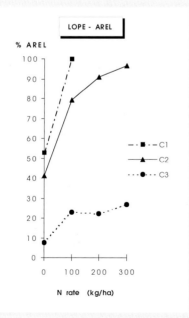
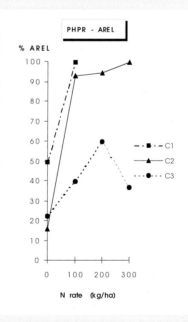

AREL: *A. elatius*; LOPE: *L. perenne*; PHPR: *P. pratense*. C1 = 2 cuts/year; C2 = 4 cuts/year; C3 = 5–6 cuts/year

Source: Peeters, Decamps and Janssens, 1999

SOCIABILITY – COMPETITIVENESS – PLANT COMMUNITIES

Competitive species when conditions are suitable. The subspecies *bulbosum* is a troublesome weed but the subspecies *elatius* that thrives in grassland is not likely to invade arable crops.

Mostly frequent in *Arrhenatherion elatioris*. Less frequent in *Mesobromion erecti* and *Koelerio-Phleion phleoidis*.

COMPATIBILITY WITH OTHER FORAGE GRASSES AND LEGUMES

Used in forage mixtures with *Dactylis glomerata, Festuca pratensis* and *Trifolium pratense* for cutting in Central Europe for moderately intensive to extensive systems (Mosimann *et al.*, 1996a).

INDICATOR VALUE

Sward mainly cut for hay.

AGRONOMIC CHARACTERISTICS

SOWING RATES

Pure: 30–40 kg/ha;
           60–70 kg/ha for coated seeds, i.e. coated with polymers in order to make the sowing easier.

Mixed: 10–20 kg/ha with *Dactylis glomerata, Festuca pratensis, Trifolium pratense;*
           15 kg/ha with lucerne (*Medicago sativa*).

Difficult to sow because of the presence of a long awn on the seed. Awned seeds form balls in the seeder. The use of coated seeds helps to solve the problem.

VARIETIES

A list of forage varieties is shown in Table 5.20. Ornamental variety: *A. elatius* ssp. *bulbosum*: Variegatum.

FERTILIZATION

Moderate requirements.

DRY MATTER YIELDS AND SEASONALITY OF PRODUCTION

In Switzerland, yields were high and the differences between varieties small (Table 5.21).

In dry and warm climates, the superiority of *A. elatius* over other productive forage species is clearly apparent (Tables 5.22 and 5.23). In cool and wet temperate climates, *A. elatius* is also one of the most productive species, particularly with infrequent or moderately frequent cutting regimes (2–4 cuts/year) (Peeters and Decamps, 1999) (Figures 5.43 to 5.46). In a 2 cuts/year regime, it was by far the most productive and ahead of *Phleum pratense* and *Dactylis glomerata*. In a 4 cuts/year regime, *Dactylis glomerata* equalled it, but it always produced more than *Phleum pratense* and *Lolium perenne* at high N levels. In the 6 cuts/year regime, its yield was comparable to *Dactylis glomerata*, *Phleum pratense*, *Holcus lanatus*, *Alopecurus pratensis* and *Festuca rubra* and superior to *Lolium perenne*. It is generally believed that it regrows very little after the first cut but this opinion is not justified. The growth of *A. elatius* is faster than *Lolium perenne* in spring. It is particularly able to accumulate much more biomass at the end of the first growth period (beginning of June) than *Lolium perenne* (Table 5.24).

NUTRITIVE VALUE

Good quality, probably similar though perhaps a little lower than that of *Phleum pratense*. Contains more carotene than *Lolium perenne* (Moon, 1939 cited in Spedding and Diekmahns, 1972). The OMD of *A. elatius* declined very rapidly (-4.3 percent per week) during the first growth period in spring, even though its digestibility was similar to *Lolium perenne* at the beginning of growth (Figure 5.47). The qualities of the 2 species became differentiated after about 15 May. When harvesting for silage production, the digestibilities of the 2 species are thus almost identical.

ACCEPTABILITY

Poorly ingested by cattle in grazing because of its bitter taste. However, this taste disappears during drying for hay (Mosimann *et al.*, 1998).

TABLE 5.20
Examples of varieties of *Arrhenatherum elatius*

Variety	Seed breeder (country)
Arel	Steinach (D)
Arone	Steinach (D)
Grano*	ZG (D)
Modus *	Oseva (CZ)
Tualitin	CEBECO (NL)

* awnless variety

TABLE 5.21
Comparison of annual yields (tonnes DM/ha) of 3 varieties of *Arrhenatherum elatius*

Variety	Average	Minimum	Maximum
Arel	14.4	9.8	19.2
Tualitin	14.0	8.6	18.4
Grano	13.8	8.6	17.8

Note: Average of 2 years and 5 sites with regimes of 3 or 4 cuts/year
Source: Lehmann *et al.*, 1992

TABLE 5.22
Comparison of annual yields (tonnes DM/ha) of *Arrhenatherum elatius* with 3 other forage grasses in Umbria, Italy at 170 m altitude

Grasses	Annual yield
Arrhenatherum elatius	6.9
Lolium perenne	4.6
Dactylis glomerata	4.0
Phleum pratense	3.7

Note: Averages of 4 years, including the year of sowing
Source: Bianchi and Ciriciofolo, 1978

TABLE 5.23
Comparison of annual yields (tonnes DM/ha) of *Arrhenatherum elatius* with 5 other forage grasses at Troyan, Bulgaria, at 350 m altitude

Grasses	Annual yield
Arrhenatherum elatius	13.3
Phalaris bulbosa	13.0
Festuca arundinacea	12.9
Dactylis glomerata	12.7
Bromus inermis	12.2
Lolium perenne	10.6

Note: Averages of 3 years, including year of sowing; 2 cuts/year in year of sowing, 3 cuts/year in full harvest years; annual N fertilization: 120 kg/ha
Source: Lingorsky, 1994

FIGURES 5.43–5.46

Annual yields of *Arrhenatherum elatius* compared with *Lolium perenne* and *Phleum pratense* in 3 cutting regimes and several nitrogen fertilization levels (average of 3 production years)

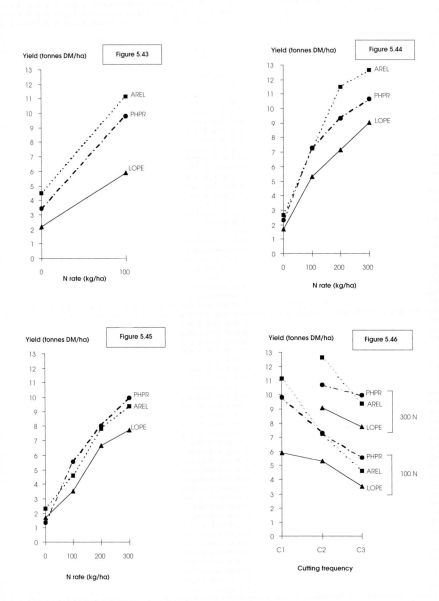

AREL: *A. elatius*; LOPE: *L. perenne*; PHPR: *P. pratense*. Fig. 5.43: 2 cuts/year (C1); Fig. 5.44: 4 cuts/year (C2); Fig. 5.45: 5–6 cuts/year (C3); Fig. 5.46: 100 and 300 kg N/ha

Source: Peeters and Decamps, 1999

TABLE 5.24

Evolution of yields (tonnes DM/ha) of *Arrhenatherum elatius* compared with *Lolium perenne* during the first growth period in the springs of 1991–93 in Belgium

	Dates					
1991	8 April	15 April	24 April	13 May	27 May	11 June
L. perenne	1.0	1.8	2.8	6.0	8.7	12.9
A. elatius	1.6	2.6	3.6	6.1	8.7	13.4
1992	15 April	29 April	7 May	19 May	27 May	9 June
L. perenne	0.9	2.8	4.7	7.8	10.1	9.9
A. elatius	2.1	4.0	6.2	10.7	12.1	12.6
1993	16 April	26 April	4 May	17 May	25 May	1 June
L. perenne	1.7	3.8	5.4	7.7	9.2	9.7
A. elatius	2.0	4.4	7.4	10.6	11.6	12.1

Note: N fertilization during the first growth cycle in spring: 100 kg/ha
Source: Peeters and Decamps, 1994

DISEASES

A rust, *Puccinia brachypodii* var. *arrhenatheri*, is frequently observed. A smut, *Ustilago avenae*, that develops on cultivated oats is almost always present in the panicles of *A. elatius* (Ellis and Ellis, 1997). *A. elatius* can be struck down by the bacterial disease, *Xanthomonas translucens* pv. *arrhenatheri*, which causes the plants to wither.

USE FOR PURPOSES OTHER THAN FORAGE PRODUCTION

Can be used to stabilize dry slopes and road verges for example. In set-asides, provides a tall and permeable cover that is favourable to wildlife reproduction and shelter. Suited for runoff control in field margins. The ornamental variety is sometimes used in mixed borders.

MAIN QUALITIES

Very productive grass, of good forage quality, capable of being used for hay or silage. It is particularly adapted to dry conditions. It is well suited to infrequent (2 cuts/year for hay production) or moderate (3–4 cuts/year for silage production for instance) cutting regimes.

MAIN SHORTCOMINGS

Badly ingested at a fresh stage by animals. It is eliminated by regular grazing.

FIGURE 5.47

Evolution of the digestibility of *Arrhenatherum elatius* compared with *Lolium perenne* during an uninterrupted growth period in spring (100 kg N/ha) (linear regression equations and determination coefficients shown)

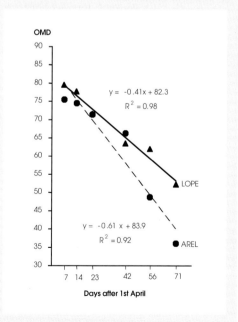

$$y = -0.41x + 82.3$$
$$R^2 = 0.98$$

$$y = -0.61 x + 83.9$$
$$R^2 = 0.92$$

Days after 1st April

AREL: *A. elatius;* LOPE: *L. perenne*

Source: Peeters, 1992

AVENULA PRATENSIS (L.) DUM.

SYNONYM
Avena pratensis L., *Helictotrichon pratense* (L.)
Besser, *Avenochloa pratensis* (L.) Holub

ETYMOLOGY
LATIN: *aveo* = I desire; grass attractive for cattle.
LATIN: *pratensis*: from meadows.

COMMON NAMES
ENGLISH: Meadow oat-grass
FRENCH: Avoine des prés
GERMAN: Rauher wiesenhafer

AVENULA PUBESCENS (HUDS.) DUM.

SYNONYM
Avena pubescens Huds., *Helictotrichon
pubescens* (Huds.) Pilger, *Avenochloa pubescens*
(Huds.) Holub

ETYMOLOGY
LATIN: *aveo* = I desire; grass attractive for cattle.
LATIN: *pubescere* = to come into
adolescence and *pubescens* = with short and
soft hairs.

COMMON NAMES
ENGLISH: Hairy oat-grass
FRENCH: Avoine pubescente
GERMAN: Flaumiger wiesenhafer

DISTRIBUTION
A. pubescens: Native to Europe except its
southern part, and to median Asia.

FIGURE 5.48
Inflorescences and parts of the stems and of
the leaves of *Avenula pubescens* and *Avenula
pratensis*

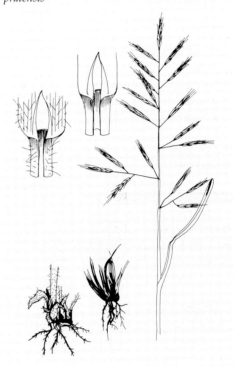

A. pratensis: Native to Europe and western
Asia.

DESCRIPTION

MORPHOLOGICAL DESCRIPTION
Perennial plants, average-sized, hairless
(*A. pratensis*) or hairy (*A. pubescens*).
Stems erect, 30–100 cm high. Blade
folded when young, hairless and rough
(*A. pratensis*) or hairy (*A. pubescens*). Ligule
2–5 (–8) mm long on the upper leaves,
oblong. No auricles. Panicle-like
inflorescence, long (10–20 cm), loose.
Spikelets 2–4-flowered, including
2–3 aristate flowers, lemma with a
long bent awn (*A. pubescens*) or spikelets
with 3–6 aristate flowers, lemma with a long
bent awn (*A. pratensis*) (Figure 5.48).

PHYSIOLOGICAL PECULIARITIES

Maximum number of leaves per tiller: 4.0 (*A. pratensis*) and 3.7 (*A. pubescens*) (Sydes, 1984).

1000-SEED WEIGHT

A. pubescens: 2.6 g (large seeds).

CHROMOSOME NUMBER

A. pratensis: 2n = 126 (octodecaploid). Other chromosome numbers have also been observed: 2n = 84 or 147.
A. pubescens: 2n = 14 (diploid).

BIOLOGICAL FLORA

Dixon (1991).

ECOLOGICAL REQUIREMENTS

ALTITUDE – VEGETATION BELTS

From the lowlands to high altitudes in mountain areas; from hill to alpine belts.

SOIL MOISTURE

Restricted to dry soils (Figures 5.49 and 5.50).

SOIL FERTILITY

Species of nutrient-poor soils, often calcareous (Figures 5.49 to 5.52).

SOIL TYPE AND TOPOGRAPHY

Encountered on a large range of soil textures. Often on sun-oriented slopes or on excessively drained soils.

CLIMATE REQUIREMENTS

Wide distribution in temperate climates. Prefers sunny and warm microclimates. Resistant to drought and winter cold.

TOLERANCE OF CUTTING AND GRAZING

Species of cut (*A. pubescens*) or extensively grazed (*A. pratensis*) grasslands (Figures 5.51 and 5.52).

FIGURES 5.49 AND 5.50

Ecological optimum and range for soil pH and humidity of *Avenula pratensis* and *Avenula pubescens*

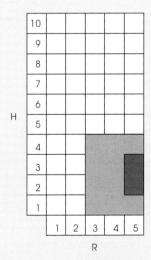

A. pratensis

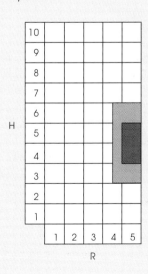

A. pubescens

R: 1 = extremely acid and 5 = alkaline
H: 1 = extremely dry and 10 = permanently flooded

Complete key in Chapter 2, section 2.8.

FIGURES 5.51 AND 5.52

Ecological optimum and range for nutrient availability and management of *Avenula pratensis*
and *Avenula pubescens*

A. pratensis

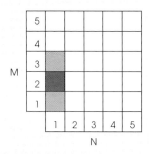

A. pubescens

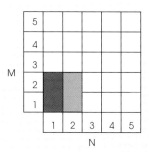

N: 1 = very low and 5 = very high
M: 1 = one defoliation per year and 5 = 5 defoliations per year or more

Complete key in Chapter 2, section 2.8.

*SOCIABILITY – COMPETITIVENESS – PLANT
COMMUNITIES*
Poorly competitive. Not found in large
populations.

A. pratensis
Mostly frequent in *Mesobromion erecti*
and other alliances of the *Festuco-Brometea*,
and in *Violo-Nardion strictae*.

A. pubescens
Mostly frequent in *Mesobromion erecti*
and other alliances of the *Festuco-Brometea*,
and in *Arrhenatherion elatioris*. Less frequent in
Cynosurion cristati and *Violo-Nardion strictae*.

*COMPATIBILITY WITH OTHER FORAGE
GRASSES AND LEGUMES*
Incompatible with productive forage species.

INDICATOR VALUE
A. pratensis
Dry or very dry, nutrient-poor soil. Extensive
grazing.

A. pubescens
Dry or very dry, nutrient-poor soil. Sward cut
for hay.

AGRONOMIC CHARACTERISTICS

FERTILIZATION
The communities in which *A. pratensis*
and *A. pubescens* are found are not
fertilized.

*DRY MATTER YIELDS AND SEASONALITY OF
PRODUCTION*
Low production on the sites where they thrive
spontaneously.

NUTRITIVE VALUE
Quality not known but probably poor to
average.

ACCEPTABILITY
Ingested as hay.

MAIN QUALITIES
Both species adapted to dry situations and to
extensive utilization systems especially in
mountain areas.

MAIN SHORTCOMINGS
These 2 *Avenula* species have a low
productivity and their quality is probably
low.

BRIZA MEDIA L.

FIGURE 5.53
Inflorescence and part of stem and leaf of
Briza media

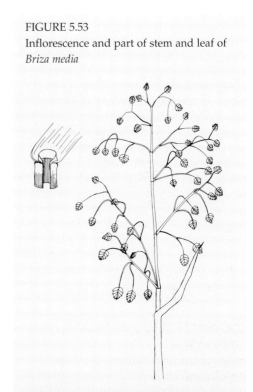

ETYMOLOGY
GREEK: βριθο = britho = I rock,
I balance; spikelets very mobile
and very easily swayed by wind.
LATIN: *medius* = which is in the middle,
by comparison with *B. maxima*
(*maximus* = the largest) and *B. minor*
(*minor* = smaller).

COMMON NAMES
ENGLISH: Common quaking grass,
totter grass, quaker grass
FRENCH: Amourette commune
GERMAN: Gewöhnliches zittergras

DISTRIBUTION
Native to Europe and western Asia.

PLATE 5.19
Briza media

PLATE 5.20
Briza media

DESCRIPTION

MORPHOLOGICAL DESCRIPTION
Perennial plant, small-sized, hairless, tufted with short stolons. Stems 10–60 cm high. Blade rolled when young, large, rough. Ligule short (0.5–1.5 mm), rounded or truncate on the top. No auricles. Panicle-like inflorescence, very characteristic, spreading, pyramidal. Spikelets 5–9-flowered, hanging, very easily swayed by wind, ovate to sub-orbicular (heart-shaped), 4–7 mm long (Figure 5.53) (Plates 5.19 and 5.20).

1000-SEED WEIGHT
0.40–0.64 g (small seeds).

CHROMOSOME NUMBER
2n = 14 (diploid) or 28 (tetraploid).

BIOLOGICAL FLORA
Dixon (2002).

ECOLOGICAL REQUIREMENTS

ALTITUDE – VEGETATION BELTS
From the lowlands up to more than 2 000 m in mountains (Caputa, 1967); from hill to subalpine belts.

SOIL MOISTURE
Optimum on dry soils (Figure 5.54).

SOIL FERTILITY
Only on nutrient-poor soils, especially those low in P. Adapted well to both very acid and alkaline soils, probably because of the existence of different ecotypes (Cooper, 1976). Less frequent on neutral soils (Figures 5.54 and 5.55).

SOIL TYPE AND TOPOGRAPHY
Encountered on a large range of soil textures.

CLIMATE REQUIREMENTS
Large climate range. Good resistance to drought and cold.

FIGURE 5.54
Ecological optimum and range for soil pH and humidity of *Briza media* (two distinct ecotypes)

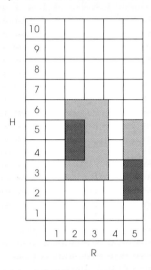

R: 1 = extremely acid and 5 = alkaline
H: 1 = extremely dry and
10 = permanently flooded

Complete key in Chapter 2, section 2.8.

FIGURE 5.55
Ecological optimum and range for nutrient availability and management of *Briza media*

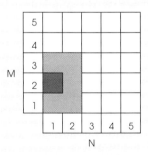

N: 1 = very low and 5 = very high
M: 1 = one defoliation per year and
5 = 5 defoliations per year or more

Complete key in Chapter 2, section 2.8.

TOLERANCE OF CUTTING AND GRAZING
Species of cut or extensively grazed grasslands (Figure 5.55).

SOCIABILITY – COMPETITIVENESS – PLANT COMMUNITIES
Poorly competitive. Component of species-rich grasslands. Does not constitute large populations.

Mostly frequent in *Mesobromion erecti, Koelerio-Phleion phleoidis, Xerobromion erecti, Violo-Nardion strictae, Cynosurion cristati, Agrostis-Festuca* grasslands, *Arrhenatherion elatioris, Polygono-Trisetion flavescentis, Molinion caeruleae* and *Juncion acutiflori*. Less frequent in *Nardion strictae*.

COMPATIBILITY WITH OTHER FORAGE GRASSES AND LEGUMES
Incompatible with productive forage species.

INDICATOR VALUE
Very poor soil, especially in P. Extensive management.

AGRONOMIC CHARACTERISTICS

VARIETIES
Ornamental varieties: Limouzi, Lutescens.

FERTILIZATION
The communities in which *B. media* is found are not fertilized.

DRY MATTER YIELDS AND SEASONALITY OF PRODUCTION
Low production in the sites where it thrives spontaneously.

NUTRITIVE VALUE
Quality not known but probably poor.

ACCEPTABILITY
Ingested as hay.

DISEASES
Attacked by several non-specialized fungi including *Puccinia graminis* (Ellis and Ellis, 1997).

USE FOR PURPOSES OTHER THAN FORAGE PRODUCTION
Very attractive inflorescence because of the original shape of its spikelets. Graceful species which can be incorporated in multi-species wild flora mixtures. The wild type or the ornamental varieties are sometimes used in mixed borders.

MAIN QUALITIES
Adapted to dry situations and to extensive systems, especially in mountain areas.

MAIN SHORTCOMINGS
Not very productive in spontaneous swards and its quality is probably rather low.

BROMUS CATHARTICUS Vahl

SYNONYM
B. willdenowii Kunth, *B. unioloides* Bonpl. and Kunth, *B. schraderi* Kunth, *Ceratochloa cathartica* (Vahl) Herter, *C. unioloides* (Hum., Bonpl. and Kunth) Beauv.

ETYMOLOGY
GREEK: βρομος = bromos = food, grass eaten by cattle; also Greek name of oats.
Could also derive from GREEK: βρομω = bromo = I rumble.
It was believed that this plant protected against thunder.
GREEK: καθαρτικος = katartikos = purgative. In Chile, the rhizomes are used as a purgative.

COMMON NAMES
ENGLISH: Rescue grass, prairie grass, schraders brome grass
FRENCH: Brome purgatif, brome cathartique
GERMAN: Purgier trespe, plattähren trespe, ahrengrasähnliche trespe

DISTRIBUTION
Native to South America. Cultivated and naturalized in diverse warm temperate regions of Europe, North America, Australia and New Zealand.

DESCRIPTION

MORPHOLOGICAL DESCRIPTION
Annual or perennial, robust plant. Stems erect, 30–90 cm high. Leaf blade rolled when young, large. Leaf sheath covered by fine hairs. Ligule long, truncate, slightly dentate. No auricles. Panicle-like inflorescence, stiff, with erect branches. Spikelets flat, with 6–12 aristate flowers (Figure 5.56).

FIGURE 5.56
Inflorescence and part of the stem and leaf of *Bromus catharticus*

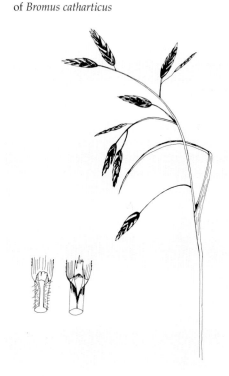

PHYSIOLOGICAL PECULIARITIES
Compared with Westerwold ryegrass (*Lolium multiflorum* ssp. *westerwoldicum*), it has lower tiller and leaf populations, larger tillers and more live leaf per tiller. The roots are more evenly distributed in the soil profile and reach a greater depth (Hume, 1991). Highly self-fertile, it may be self- or cross-pollinated. It is an alternative species, the production of fertile stems being very important in the year of sowing, comparable with that of *Lolium multiflorum* ssp. *westerwoldicum*; heading is important at each regrowth including in autumn (Betin and Mansat, 1979).

1000-SEED WEIGHT
11.5–13 g (large seeds).

CHROMOSOME NUMBER
2n = 42 (hexaploid).

ECOLOGICAL REQUIREMENTS

ALTITUDE – VEGETATION BELTS
Sown in the lowlands.

SOIL MOISTURE
Very sensitive to high soil moisture.

SOIL FERTILITY
Reacts well to nutrient availability. Requires high N fertilization for optimum growth.

SOIL TYPE AND TOPOGRAPHY
Adapted to fertile, free-draining soils.

CLIMATE REQUIREMENTS
Adapted to warm climates. Rather cold-sensitive. Very resistant to heat and to a certain extent to drought.

TOLERANCE OF CUTTING AND GRAZING
Sensitive to trampling, especially in wet soil. Well adapted to cutting. Suited to strict rotational grazing but not continuous stocking.

SOCIABILITY – COMPETITIVENESS – PLANT COMMUNITIES
Very competitive.

COMPATIBILITY WITH OTHER FORAGE GRASSES AND LEGUMES
Compatible with *Trifolium pratense* and *Medicago sativa.*

AGRONOMIC CHARACTERISTICS

SOWING RATES
Pure: 60–80 kg/ha.
Mixed with *Trifolium pratense* or *Medicago sativa*: 40 kg/ha

Difficult to sow because of the presence of a long awn on the seed. De-awned seed can be used.

VARIETIES
A list of forage varieties is shown in Table 5.25.

FERTILIZATION
High N rate required to express its full potential in pure stand.

DRY MATTER YIELDS AND SEASONALITY OF PRODUCTION
The yield in the year of sowing can be relatively high if sown in spring. In following years, the production is very high, similar to *Festuca arundinacea*, reaching 15–20 tonnes DM/ha in optimum conditions.

In Scotland (UK), *B. catharticus* was more productive than *Lolium perenne* in the first 2 production years in regimes of 6 and 4 cuts/year. In the 4 cuts/year regime, it had a similar production to *Lolium multiflorum* in the first 2 years and was more productive than *Phleum pratense* and *Dactylis glomerata* in the first year but not the second one (Table 5.26).

In Britanny (F), *B. catharticus* was highly productive (Table 5.27). Its yield was comparable with those of *Festuca arundinacea* and *Dactylis glomerata*, but it was higher in the year of sowing. Its winter and early spring production was important in this oceanic climate, the yield reaching 4–5 tonnes DM/ha in April. The summer yield was also high, comparable with that of *Festuca arundinacea*. Its persistency was better than *Lolium multiflorum* but lower than that of the other species.

Under good water supply, *B. catharticus* can continue to produce at high temperatures when *Lolium multiflorum* stops growing. In autumn, it continues to grow much later than *Dactylis glomerata*; in an oceanic climate there is almost no break of growth if the winter is mild (Betin and Gillet, 1983).

In a large multisite experiment (17 sites) in France (Betin and Gillet, 1983), the performances of *B. catharticus* were compared with those of *Lolium multiflorum*, *Festuca arundinacea* and *Dactylis glomerata*. In the year of sowing, *B. catharticus* was more productive than the 3 other species. In the

TABLE 5.25
Examples of varieties of *Bromus catharticus*

Variety	Seed breeder (country)
Anabel	Loiseau Semences (F)
Baladin	GIE Semunion (F)
Banco	Limagrain Genetics (F)
Bellegarde	INRA (F)
Cabro	Sisforaggera (I)
Decibel	RAGT (F)
Luprime	INRA (F)
Meribel	INRA (F)

TABLE 5.26
Annual yields (tonnes DM/ha) of *Bromus catharticus* **compared with** *Lolium perenne,*
Lolium multiflorum, Phleum pratense **and** *Dactylis glomerata*

Species (variety)	Year 1	Year 2	Year 3
6 cuts/year			
B. catharticus (Matua)	13.4	12.6	10.0
L. perenne (average of Talbot and Fantoom)	13.1	12.0	10.2
4 cuts/year (experiment 1)			
B. catharticus (Matua)	18.3	15.7	12.9
L. perenne (average of Talbot and Fantoom)	12.0	13.7	12.3
4 cuts/year (experiment 2)			
B. catharticus (Matua)	17.0	13.1	12.8
L. multiflorum (Rvp)	16.9	13.6	16.0
P. pratense (Scots)	12.1	13.3	14.5
D. glomerata (S26)	10.5	15.2	15.4

Note: N fertilization during the first growth cycle in spring: 350–360 kg/ha
Source: Frame and Morrison, 1991

TABLE 5.27
Annual yields (tonnes DM/ha) of *Bromus catharticus* **compared with** *Lolium multiflorum,*
Lolium perenne, Dactylis glomerata **and** *Festuca arundinacea*

Species (variety)	A0 (3 cuts)	A1 (6 cuts)	A2 (5 cuts)	A3 (5 cuts)
L. multiflorum (Tetrone)	6.9	16.0	9.9	-
L. perenne (Vigor)	7.5	15.8	9.9	13.0
D. glomerata (Lucifer)	6.1	19.1	13.4	15.6
F. arundinacea (Clarine)	6.3	17.1	14.5	16.0
B. catharticus (average of Bellegarde and Delta)	8.1	17.6	13.1	16.5

Note: 60 kg N/ha before each cut, 4 years of experiment (A0 to A3)
Source: Simon, Le Corre and Coppenet, 1983

following years, the yield of the first early cuts were similar for *B. catharticus* and *Lolium multiflorum*. The summer growth and the annual yield were comparable for *B. catharticus*, *Festuca arundinacea* and *Dactylis glomerata*. *B. catharticus* established much faster than *Festuca arundinacea* and *Dactylis glomerata* and was more persistent than *Lolium multiflorum*.

NUTRITIVE VALUE

It has a high content of soluble carbohydrates that makes it very suitable for silage-making.

In Scotland (UK), *B. catharticus* had good digestibility values (OMD), but they were lower than those of *Lolium perenne* (Frame and Morrison, 1991). In a 6 cuts/year regime, the OMD ranged from 0.75–0.76 for *B. catharticus* and from 0.78–0.80 for *Lolium perenne*, while in a 4 cuts/year regime, from 0.71–0.74 and from 0.76–0.78 respectively. In another 4 cuts/year experiment, the OMD values of *B. catharticus* were slightly lower than those of *Lolium multiflorum* or *Phleum pratense*, but higher than those of *Dactylis glomerata*. The N contents were very similar for *B. catharticus* and *Lolium perenne*, but the P content of *B. catharticus* was lower.

In Britanny (F), the N content was high even at heading, the spikelets containing more N than the leaves. The energy content was intermediate between *Festuca arundinacea* and *Dactylis glomerata*. The P contents were good and the Na contents were very high, higher than those of *Lolium perenne* (Simon, Le Corre and Coppenet, 1983).

In the multisite French experiment (Betin and Gillet, 1983), the digestibility (DMD) of *B. catharticus* was similar to *Lolium multiflorum*. The content of soluble carbohydrates was comparable with *Festuca arundinacea* and the N content similar to *Lolium multiflorum*.

ACCEPTABILITY

Good intake as hay or silage. Cattle can graze it up to an advanced stage since the loose panicle head is acceptable (Gillet, 1980). However, awns of *B. catharticus* and other *Bromus* spp. (though not *B. inermis*, which is awnless) can provoke irritation in the mouth and throat of cattle.

DISEASES

The following diseases have been noted: black rust (*Puccinia graminis* ssp. *graminis*), yellow rust (*P. striiformis*), powdery mildew (*Blumeria* [= *Erysiphe*] *graminis*), *Drechslera tritici* ssp. *repentis*, *Ramularia* leaf spot (*Ramulispora bromi*), *Rhynchosporium secalis*, *Ustilago bromivora* on the panicle, *Claviceps purpurea*, choke (*Epichloë typhina*) (Raynal *et al.*, 1989; Ellis and Ellis, 1997). *B. catharticus* can be adversely affected by *Xanthomonas translucens* pv. *graminis*, the so-called bacterial wilt which is sometimes responsible for the plant's disappearance.

MAIN QUALITIES

Adapted to dry areas. It can provide early spring growth and has good summer growth. It establishes easily and vigorously and is very competitive against weeds. It is very productive in the year of sowing. Highly productive and mainly adapted to cutting. Very compatible with *Trifolium pratense* and *Medicago sativa*. Has a good digestibility and a high soluble carbohydrate content. Well accepted by grazing animals.

MAIN SHORTCOMINGS

Very sensitive to soil wetness. Early spring growth can be delayed by late frost. Does not tolerate trampling especially in wet conditions. Produces new stems after each cut and these stems are not very leafy. Not well adapted to continuous stocking. Probably has low P content. Average persistency.

BROMUS ERECTUS HUDS.

SYNONYM
Bromopsis erecta (Huds.) Fourr.,
Zerna erecta (Huds.) S.F. Gray

ETYMOLOGY
GREEK: βρομος = bromos = food,
grass eaten by cattle; also Greek
name of oats.
Could also derive from
GREEK: βρομω = bromo = I rumble.
It was believed that this plant
protected against thunder.
LATIN: *erectus* = tall, erected, straight.

COMMON NAMES
ENGLISH: Upright brome
FRENCH: Brome dressé
GERMAN: Aufrechte trespe

FIGURE 5.57
Inflorescence and part of the stem and leaf
of *Bromus erectus*

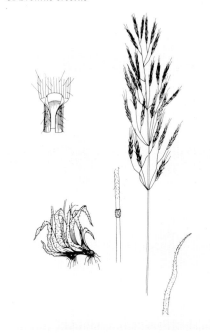

PLATE 5.21
Bromus erectus

P & W GUINCHART, BIOS

PLATE 5.22
Bromus erectus

A. PEETERS

DISTRIBUTION
Native to Europe except the northern part, and to Northern Africa. Introduced to North America.

DESCRIPTION

MORPHOLOGICAL DESCRIPTION
Perennial plant, robust, hairy, caespitose. Stems erect, 30–100 cm high. Leaf blade folded when young, large and spreading on the upper leaves (up to 6 mm wide), usually rolled (2–3 mm) on the lower leaves. Ligule short, truncate, slightly dentate. No auricles. Panicle-like inflorescence, stiff, with erect branches. Spikelets with 5–9 aristate flowers, 2–3 cm long. Lemma with awn almost half the lemma length (Figure 5.57) (Plates 5.21 and 5.22).

1000-SEED WEIGHT
4.0–5.3 g (large seeds).

CHROMOSOME NUMBER
2n = 42 (hexaploid) or 56 (octoploid).

ECOLOGICAL REQUIREMENTS

ALTITUDE – VEGETATION BELTS
Distributed mainly in the lowlands but present up to almost 2 200 m in the Alps (Rameau *et al.*, 1993); from hill to subalpine belts.

SOIL MOISTURE
Restricted to dry and very dry soils (Figure 5.58).

SOIL FERTILITY
Only present on nutrient-poor soils (Figures 5.58 and 5.59). Optimum on alkaline soils but present also on acid soils.

SOIL TYPE AND TOPOGRAPHY
Thrives on stony, medium-deep to shallow calcareous soils.

CLIMATE REQUIREMENTS
Well distributed in wet temperate, warm

FIGURE 5.58
Ecological optimum and range for soil pH and humidity of *Bromus erectus*

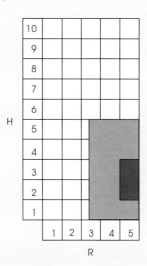

R: 1 = extremely acid and 5 = alkaline
H: 1 = extremely dry and
10 = permanently flooded

Complete key in Chapter 2, section 2.8.

FIGURE 5.59
Ecological optimum and range for nutrient availability and management of *Bromus erectus*

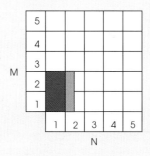

N: 1 = very low and 5 = very high
M: 1 = one defoliation per year and
5 = 5 defoliations per year or more

Complete key in Chapter 2, section 2.8.

temperate and Mediterranean climates. Very resistant to drought.

TOLERANCE OF CUTTING AND GRAZING
Tolerant of cutting but sensitive to grazing
(Figure 5.59).

*SOCIABILITY – COMPETITIVENESS – PLANT
COMMUNITIES*
Competitive. Can constitute dense
species-poor swards, if the sward cover
is not managed. Under cutting there is
often a high number of species and
the brome dominance is smaller.
Such swards often evolve into an
Arrhenatherum elatius grassland,
especially if fertilized.
    Mostly frequent in *Mesobromion erecti,
Xerobromion erecti* and *Koelerio-Phleion
phleoidis.* Less frequent in *Arrhenatherion
elatioris* and *Molinion caeruleae.*

*COMPATIBILITY WITH OTHER FORAGE
GRASSES AND LEGUMES*
Spontaneous populations include very few
valuable forage species. However, *Dactylis
glomerata* and *Poa pratensis* are regularly
noted.

INDICATOR VALUE
Dry to very dry, nutrient-poor soil. Extensive
management.

AGRONOMIC CHARACTERISTICS

FERTILIZATION
The communities in which *B. erectus* grows
are rarely fertilized.

*DRY MATTER YIELDS AND SEASONALITY OF
PRODUCTION*
Yields of *B. erectus* communities are low and
very variable (Table 5.28).

NUTRITIVE VALUE
Poor.

ACCEPTABILITY
Moderately acceptable to sheep, e.g. less than
Arrhenatherum elatius (highly acceptable) but
better than *Festuca arundinacea* (poorly
acceptable) or *Brachypodium pinnatum*
(refused) (Troxler, Jans and Floch, 1990).

*USE FOR PURPOSES OTHER THAN FORAGE
PRODUCTION*
Can be used for species-rich sward
restoration in very dry conditions, on
calcareous soils namely.

MAIN QUALITIES
Well adapted to dry and very dry situations
where it grows spontaneously. The
communities where it is dominant provide
valuable production in unfavourable
conditions where it is difficult to obtain
good quality forage.

MAIN SHORTCOMINGS
Poor nutritive value. A low level of
fertilization helps to deplete it to the benefit
of *Arrhenatherum elatius* in cut swards, and of
Dactylis glomerata and *Poa pratensis* in cut or
grazed swards.

TABLE 5.28
Examples of annual yields (tonnes DM/ha) of *Bromus erectus* communities

Authors	Yield	Comments
Coppel and Etienne (1992)	3.6	'Average' community in the Pre-Alps (F)
	4.0	Community on sandstone
	1.1	Community in the area of white oak forests
	5.0	Community fertilized with 60 kg N/ha
Hubert (1978)	almost 1.0	Causses (F)
Garde (1990)	almost 1.5	Haute Provence (F)
Santilocchi (1989)	almost 1.5	Umbria (I)

BROMUS HORDEACEUS L.

FIGURE 5.60
Inflorescence and part of the leaf of *Bromus hordeaceus*

SYNONYM
B. mollis L.

ETYMOLOGY
GREEK: βρομος = bromos = food, grass eaten by cattle; also Greek name of oats.
Could also derive from GREEK: βρομω = bromo = I rumble.
It was believed that this plant protected against thunder.
LATIN: *hordeaceus* = like a hordeum; *hordeum* = barley.
Hordeum could itself come from *hordus* = heavy; the bread made with barley is heavy.

COMMON NAMES
ENGLISH: Soft brome, lop grass
FRENCH: Brome mou
GERMAN: Weiche trespe

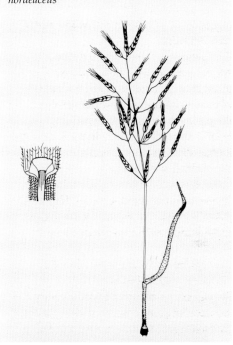

PLATE 5.23
Bromus hordeaceus

PLATE 5.24
Bromus hordeaceus

DISTRIBUTION

Native to Europe, western Asia and Northern Africa. Has become subcosmopolitan in temperate regions.

DESCRIPTION

MORPHOLOGICAL DESCRIPTION

Annual or biennial plant, medium-sized, extremely hairy, caespitose. Stems erect, 15–100 cm high. Leaf blade rolled when young, large, floppy, very hairy. Ligule short, truncate, slightly dentate. No auricles. Panicle-like inflorescence, loose, spreading during flowering and contracted afterwards, almost 10 cm long. Spikelets with 7–11 flowers, softly hairy, quite large, lemma with awn of same length (Figure 5.60) (Plates 5.23 and 5.24).

PHYSIOLOGICAL PECULIARITIES

Short-living plant which produces large amounts of seeds compared with the total biomass production.

1000-SEED WEIGHT

Almost 3 g (large seeds).

CHROMOSOME NUMBER

2n = 28 (tetraploid).

ECOLOGICAL REQUIREMENTS

ALTITUDE – VEGETATION BELTS

From the lowlands to high altitudes in mountain areas; from hill to subalpine belts.

SOIL MOISTURE

In cutting meadows, optimum on normally drained soils. In grazed grassland, only on dry soils (Figure 5.61).

SOIL FERTILITY

Species of soils moderately rich to rich in nutrients (mesotrophic to slightly eutrophic species). Mainly on slightly acid to neutral soils (Figures 5.61 and 5.62).

FIGURE 5.61

Ecological optimum and range for soil pH and humidity of *Bromus hordeaceus*

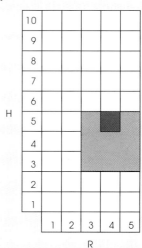

R: 1 = extremely acid and 5 = alkaline
H: 1 = extremely dry and
10 = permanently flooded

Complete key in Chapter 2, section 2.8.

FIGURE 5.62

Ecological optimum and range for nutrient availability and management of *Bromus hordeaceus*

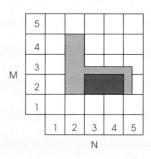

N: 1 = very low and 5 = very high
M: 1 = one defoliation per year and
5 = 5 defoliations per year or more

Complete key in Chapter 2, section 2.8.

SOIL TYPE AND TOPOGRAPHY

Adapted to large range of soil textures, from sand to clay.

CLIMATE REQUIREMENTS
All temperate climates suitable. Species adapted to variations in temperature and to drought because of regeneration from seeds buried in the soil.

TOLERANCE OF CUTTING AND GRAZING
Species typical of hay meadows where it mixes with *Arrhenatherum elatius*, *Poa trivialis*, *Dactylis glomerata* and *Alopecurus pratensis*. Late cutting enables *B. hordeaceus* to produce seeds. Its low tiller density in the sward encourages seed germination and formation of *B. hordeaceus* seedlings after the cut. Does not tolerate trampling but in extensively grazed dry grasslands where the stocking rate is low, it survives because the open sward allows seed germination and in addition, a certain number of plants escape from grazing and trampling (Figure 5.62).

SOCIABILITY – COMPETITIVENESS – PLANT COMMUNITIES
Poorly competitive. Can develop large populations but these are not very dense because tillering is limited. Associated with many other species in mesotrophic cutting grassland. Favoured by organic manure application.

Mostly frequent in *Arrhenatherion elatioris* and *Cynosurion cristati*.

COMPATIBILITY WITH OTHER FORAGE GRASSES AND LEGUMES
Appears in temporary late-cut grasslands sown with *Phleum pratense*, *Festuca pratensis*, *Trifolium pratense*, *T. repens* and sometimes *Lolium perenne*. In these grasslands, it is considered as an undesirable plant.

INDICATOR VALUE
Sward cut for hay or extensive grazing on dry soil.

AGRONOMIC CHARACTERISTICS

FERTILIZATION
B. hordeaceus grows well in communities that are often fertilized with organic and/or mineral fertilizers.

DRY MATTER YIELDS AND SEASONALITY OF PRODUCTION
Moderate production.

NUTRITIVE VALUE
Acknowledged to be of bad quality because the leaf:stem ratio is low. However, in hay production its large seeds could provide significant amounts of energy.

ACCEPTABILITY
Well ingested as hay or silage. Often refused during grazing because of stemminess caused by earliness of heading.

DISEASES
Attacked by *Puccinia recondita* ssp. *bromina*.

MAIN QUALITIES
Helps to sustain yield in old leys managed by cutting, where the sward has deteriorated. Appears spontaneously in these grasslands, often mixed with *Poa trivialis*. Can also be an important component of permanent grasslands used for cutting.

MAIN SHORTCOMINGS
Nutritive value low since plants mature rapidly.

BROMUS INERMIS Leyss.

SYNONYM
Zerna inermis (Leyss.) Lindm., *Bromopsis inermis* (Leyss.) Holub

ETYMOLOGY
GREEK: βρομος = bromos = food, grass eaten by cattle; also Greek name of oats.
Could also derive from *GREEK*: βρομω = bromo = I rumble.
It was believed that this plant protected against thunder.
LATIN: *inermis* = not armed, harmless. This brome has no awns or short awns. Awns of other brome species can provoke irritations in the mouth and the throat of cattle.

COMMON NAMES
ENGLISH: Smooth brome, Hungarian brome, awnless brome
FRENCH: Brome inerme
GERMAN: Unbegrannte trespe

FIGURE 5.63
Inflorescence and part of the stem, rhizome and leaf of *Bromus inermis*

PLATE 5.25
Bromus inermis

PLATE 5.26
Bromus inermis

PLATE 5.27
Bromus inermis

DISTRIBUTION
Native to temperate Europe and Asia.

DESCRIPTION

MORPHOLOGICAL DESCRIPTION
Perennial plant, robust, hairless, with long
rhizomes. Stems erect, 30–100 (–160) cm
high. Leaf blade rolled when young, large
(5–12 mm), rough. Ligule very short
(1 mm), truncate. No auricles. Panicle-like
inflorescence, oblong, with erect branches,
half-verticillate. Spikelets with 5–8 flowers
exceptionally awned (awn: 4 mm
maximum) (Figure 5.63) (Plates 5.25 to
5.27).

1000-SEED WEIGHT
3.0–4.0 g (large seeds).

CHROMOSOME NUMBER
2n = 42 (hexaploid) or 56 (octoploid).
Other chromosome numbers have also been
observed: 2n = 28 or 76.

ECOLOGICAL REQUIREMENTS

ALTITUDE – VEGETATION BELTS
Mostly in the lowlands; from hill to
mountain belts.

SOIL MOISTURE
Very resistant to drought but moisture-
sensitive (Figure 5.64).

SOIL FERTILITY
Prefers average to high soil fertility. Does
not tolerate soil acidity (Figures 5.64 and
5.65).

SOIL TYPE AND TOPOGRAPHY
Adapted to light-textured soils.

CLIMATE REQUIREMENTS
Adapted to dry, temperate regions, e.g. very
abundant in Hungary. Does not tolerate
shade. Very resistant to cold.

FIGURE 5.64
Ecological optimum and range for soil pH and
humidity of *Bromus inermis*

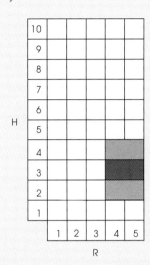

R: 1 = extremely acid and 5 = alkaline
H: 1 = extremely dry and
10 = permanently flooded

Complete key in Chapter 2, section 2.8.

FIGURE 5.65
Ecological optimum and range for nutrient
availability and management of *Bromus
inermis*

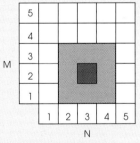

N: 1 = very low and 5 = very high
M: 1 = one defoliation per year and
5 = 5 defoliations per year or more

Complete key in Chapter 2, section 2.8.

TOLERANCE OF CUTTING AND GRAZING
Well adapted to cutting; tolerates extensive,
but not intensive, grazing. Does not with-
stand frequent defoliation (Figure 5.65).

SOCIABILITY – COMPETITIVENESS – PLANT COMMUNITIES
Mostly frequent in *Agropyretalia, Cirsio-Brachypodion pinnati* and *Arrhenatherion elatioris*.

COMPATIBILITY WITH OTHER FORAGE GRASSES AND LEGUMES
Compatible with *Trifolium pratense* and *Medicago sativa*.

AGRONOMIC CHARACTERISTICS

SOWING RATES
Pure: 20–25 kg/ha.
Mixed with *Trifolium pratense* or *Medicago sativa*: 10 kg/ha.

VARIETIES
There are very few forage varieties in Europe. Some examples of varieties used in North America are: Barton, Baylor, Beacon, Blair, Bromex, Fox, Manchar, Saratoga. Grasslands Tiki is a New Zealand variety. Ornamental variety: Skinner's Gold.

FERTILIZATION
High level required to realize its potential.

DRY MATTER YIELDS AND SEASONALITY OF PRODUCTION
High yields, often of the same order as *Dactylis glomerata*.

NUTRITIVE VALUE
At early growth stages, this brome is of good quality, but since it must be defoliated at a late maturity stage in spring to aid its persistence, its quality is then often moderate to low. Mixed with *Medicago sativa*, the quality of the association is improved.

ACCEPTABILITY
Intake a little better than *Dactylis glomerata* (Marten and Jordan, 1974).

ANIMAL PERFORMANCE
Provides lamb LWGs of the same order as *Dactylis glomerata* (Marten and Jordan, 1974).

USE FOR PURPOSES OTHER THAN FORAGE PRODUCTION
Can be used for stabilizing soils in dry conditions for instance in new ski pistes (at moderate altitude) and on road or motorway verges.

MAIN QUALITIES
Can be used in difficult climate areas, very cold in winter or very warm in summer, for instance, continental steppe areas in the USA, Canada, Ukraine and the Russian Federation. Productive and compatible with *Medicago sativa*.

MAIN SHORTCOMINGS
Not suitable for intensive cutting management. Lacks persistence unless defoliated late in spring and so its nutritive value is then moderate.

CYNOSURUS CRISTATUS L.

FIGURE 5.66
Inflorescence and part of the leaf of
Cynosurus cristatus

ETYMOLOGY
GREEK: κυνος = cynos = of dog and ουρα =
oura = tail; the panicle is like a dog's tail.
LATIN: *cristatus* = crested; the panicle is like a
cockscomb.

COMMON NAMES
ENGLISH: Crested dog's tail
FRENCH: Crételle
GERMAN: Wiesen kammgras

DISTRIBUTION
Native to Europe and the Caucasus region.

DESCRIPTION
MORPHOLOGICAL DESCRIPTION
Perennial plant, small-sized, hairless,
caespitose. Stems erect, 20–70 (–90) cm high.
Leaf blade folded when young but can also

PLATE 5.28
Cynosurus cristatus

PLATE 5.29
Cynosurus cristatus

be rolled during elongation. Leaf blade narrow (1–3 mm), with well marked veins, shiny on the lower surface. Ligule short (2 mm maximum), truncate. No auricles. Sheath of the lower leaves pale yellow or brownish yellow. Spike-like panicle, narrow, dense, erect, 2–10 cm long, with very short branches. Two types of spikelets, fertile with 3–7 flowers and sterile, very reduced in size (Figure 5.66) (Plates 5.28 and 5.29).

1000-SEED WEIGHT
0.55–0.80 (small seeds).

CHROMOSOME NUMBER
2n = 14 (diploid).

BIOLOGICAL FLORA
Lodge (1959).

ECOLOGICAL REQUIREMENTS

ALTITUDE – VEGETATION BELTS
From the lowlands up to high altitudes in mountain areas; from hill to subalpine belts.

SOIL MOISTURE
Optimum on normally drained soils (Figure 5.67).

SOIL FERTILITY
Large range of soil nutrient availability except for the richest and the poorest soils (mesotrophic species with large range). Grows mainly on acid to neutral soils (Figures 5.67 and 5.68).

SOIL TYPE AND TOPOGRAPHY
Preference for clay, loam and sandy-clay soils.

CLIMATE REQUIREMENTS
Thrives in all temperate climates. Very resistant to cold.

TOLERANCE OF CUTTING AND GRAZING
Species typical of extensively or moderately intensively grazed grasslands. Disappears with high stocking rates associated with high

FIGURE 5.67
Ecological optimum and range for soil pH and humidity of *Cynosurus cristatus*

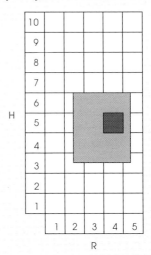

R: 1 = extremely acid and 5 = alkaline
H: 1 = extremely dry and
10 = permanently flooded

Complete key in Chapter 2, section 2.8.

FIGURE 5.68
Ecological optimum and range for nutrient availability and management of *Cynosurus cristatus*

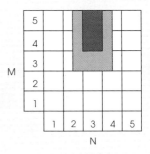

N: 1 = very low and 5 = very high
M: 1 = one defoliation per year and
5 = 5 defoliations per year or more

Complete key in Chapter 2, section 2.8.

fertilization. Absent in cutting grassland but tolerates occasional cuts in grazed grassland (Figure 5.68).

Poorly competitive. Tillering limited and so tufts not very dense. Considerable populations are sometimes noted but are usually encountered in small patches. In extensive grasslands, associated with *Festuca rubra* and *Agrostis capillaris*. In moderately intensive grasslands, associated with *Lolium perenne*. Slow to establish.

Mostly frequent in *Cynosurion cristati*. Less frequent in *Arrhenatherion elatioris* and *Polygono-Trisetion flavescentis*.

COMPATIBILITY WITH OTHER FORAGE GRASSES AND LEGUMES
In Switzerland, used in complex mixtures for the creation of permanent grasslands in high-altitude zones where *Lolium perenne* is at the limit of its adaptation (Mosimann *et al.*, 1996a). *Poa pratensis, Festuca rubra, F. pratensis, Phleum pratense, Lolium perenne*, black bent (*Agrostis gigantea*), *Trifolium repens* and *Lotus corniculatus* are the main components of these mixtures.

INDICATOR VALUE
Mainly grazed sward.

AGRONOMIC CHARACTERISTICS

SOWING RATES
Pure: 20–30 kg/ha.
Mixed: 5 kg/ha for the creation of grazed permanent grasslands with *Lolium*

perenne, Poa pratensis, Festuca rubra, Phleum pratense and *Trifolium repens* (Mosimann *et al.*, 1996a).

VARIETIES
Southlands (NZ).

FERTILIZATION
The communities in which *C. cristatus* grows are moderately fertilized, but some are not fertilized at all.

DRY MATTER YIELDS AND SEASONALITY OF PRODUCTION
Moderate production, clearly lower than *Lolium perenne* at high N fertilization, but equivalent at low or moderate fertilization (Table 5.29). Generally speaking, it produced less than *Holcus lanatus, Festuca rubra, Agrostis stolonifera* and *Poa pratensis*. Its production was of the same order as *Agrostis capillaris*. A similar ranking of species was obtained in a 4 cuts/year regime and 360 kg N/ha by Frame (1989).

Mixed with *Trifolium repens*, *C. cristatus* produced less than *Festuca rubra* and *Lolium perenne* but more than *Holcus lanatus, Poa pratensis* or *Agrostis stolonifera*. Among grasses, it is one of the most compatible with *Trifolium repens* (Frame, 1990).

In a trial network of 5 sites at increasing altitude gradients under low fertility conditions in the North Pennines (UK), *C. cristatus* produced less DOM (2.9 tonnes/ha) than *Lolium perenne* (3.6 tonnes/ha), as much as *Festuca rubra* (2.7 tonnes/ha) and more

TABLE 5.29

Annual yields (tonnes DM/ha) of *Cynosurus cristatus* compared with *Lolium perenne* in Scotland, UK

	Annual N fertilization (kg/ha)				
	0	120	240	360	480
6 cuts /year (average of 3 years)					
L. perenne	2.4	6.1	9.7	11.9	13.2
C. cristatus	2.6	5.5	8.2	9.5	11.0
3 cuts/year (1-year experiment)					
L. perenne	3.8	6.0	10.8	14.7	15.8
C. cristatus	4.1	6.3	9.7	12.4	12.0

Source: Frame, 1991

than *Agrostis stolonifera* (2.6 tonnes/ha)
(Thomas and Morris, 1973).

NUTRITIVE VALUE
Good digestibility, lower than *Lolium perenne*
but similar to *Holcus lanatus* and *Anthoxanthum
odoratum* (Thomas and Morris, 1973; Frame,
1989; 1991). *C. cristatus* is richer in N than
Lolium perenne at the same N fertilization level.
Mineral (P, K, Ca, Mg) contents are very similar.

ACCEPTABILITY
Very well accepted at a leafy stage.
However, a high proportion of tillers of *C.
cristatus* head even under grazing and so it
is necessary to top these rejected stems to
ensure quality regrowths.

DISEASES
Few known diseases. *Puccinia graminis*
sometimes infects plants (Ellis and Ellis, 1997).

*USE FOR PURPOSES OTHER THAN FORAGE
PRODUCTION*
Constituent of wild flora mixtures for
neutral (pH 6–7) and calcareous soils
(Crofts and Jefferson, 1999).

MAIN QUALITIES
Good production when grazed on rather
poor soils and with low N fertilization.
Nutritive value very good at leafy stage.
Resistant to drought and very resilient
to cold. Can play a secondary role in
extensive or moderately intensive
grasslands.

MAIN SHORTCOMINGS
Not productive enough at high rates
of N fertilization and may disappear
in these conditions. In late spring,
often produces many flowering stems
that are refused by grazing animals.

DACTYLIS GLOMERATA L.

After the studies of Stebbins and Zohary
(1959), Jones, Carroll and Borrill (1961)
and Borrill and Carroll (1969), the species
D. glomerata contains 3 ploidy levels:
diploid (2n = 14), tetraploid (2n = 28)
and hexaploid (2n = 42). Seventeen
subspecies can be defined: 14 diploids
and 3 tetraploids. Six other minor
tetraploid taxa and one diploid taxon have
also been described, with restricted
geographical ranges (Mousset, 2000).
The subspecies *hispanica* can be tetra- or
hexaploid. It is an important forage
species in the Mediterranean basin.
The subspecies described here
is *glomerata* (tetraploid). It is the most
important forage taxon for temperate areas.
The subspecies *aschersoniana* which
is widespread in Eastern Europe could
also have a good production potential
(Mousset, 2000).

FIGURE 5.69
Inflorescence and part of the leaf
of *Dactylis glomerata*

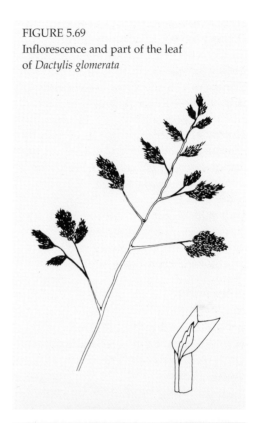

PLATE 5.30
Dactylis glomerata

PLATE 5.31
Dactylis glomerata

ETYMOLOGY
GREEK: δακτυλος = dactylos = finger.
The branches of the panicle are like fingers.
LATIN: *glomeratus* = clustered in a ball.
The spikelets are clustered in balls at the top
of the branches of the panicle.

COMMON NAMES
ENGLISH: Cocksfoot, orchard grass
FRENCH: Dactyle vulgaire, dactyle aggloméré
GERMAN: Gemeines knäuelgras

DISTRIBUTION
Native to Europe, western Asia and
Northern Africa. Has become
subcosmopolitan in temperate regions.

DESCRIPTION

MORPHOLOGICAL DESCRIPTION
Perennial plant, robust, hairless, caespitose.
Stems erect, 20–120 cm high, compressed at
the base. Leaf blade folded when young,
often accordion-pleated in the top third,
large (4–12 mm), quite stiff, long, with no
visible nerve, hulled at the top. Ligule very
large, irregular, white. No auricles.
Inflorescence in stiff panicle, spreading or
dense, erect, with basal branches without
spikelets over a long part. Spikelets 3–6-
flowered, 5–6 mm long, in compact clusters
(Figure 5.69) (Plates 5.30 and 5.31).

1000-SEED WEIGHT
0.8–1.4 g (average-sized seeds).

CHROMOSOME NUMBER
2n = 28 (tetraploid).

BIOLOGICAL FLORA
Beddows (1959).

ECOLOGICAL REQUIREMENTS

ALTITUDE – VEGETATION BELTS
From the lowlands up to 2 400 m in the Alps
(Rameau *et al.*, 1993); from hill to alpine belts.

FIGURE 5.70
Ecological optimum and range for soil pH and
humidity of *Dactylis glomerata*

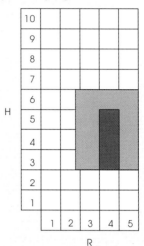

R: 1 = extremely acid and 5 = alkaline
H: 1 = extremely dry and
10 = permanently flooded

Complete key in Chapter 2, section 2.8.

FIGURE 5.71
Ecological optimum and range for nutrient
availability and management of *Dactylis glomerata*

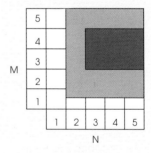

N: 1 = very low and 5 = very high
M: 1 = one defoliation per year and
5 = 5 defoliations per year or more

Complete key in Chapter 2, section 2.8.

SOIL MOISTURE
Optimum on normally drained to dry soils.
Can grow on slopes and on very dry
shallow soils. Dislikes excessive moisture
(Figure 5.70).

SOIL FERTILITY
Wide range of adaptability for nutrient availability. Can grow on rather poor soils but is much more productive on fertile soils. Its strong root system can develop over a large volume of soil and thus absorb more nutrients than the other species of the same community. Large range for soil pH but prefers slightly acid or alkaline soils (Figures 5.70 and 5.71).

SOIL TYPE AND TOPOGRAPHY
Indifferent to soil texture. Very rare or absent from peat soils.

CLIMATE REQUIREMENTS
Large climate range. Frost-, heat- and drought-resistant. Markedly thermophilous: appreciates sun-oriented slopes (warm microclimates). However, also tolerates shade very well, for instance, in old orchards.

TOLERANCE OF CUTTING AND GRAZING
Widely distributed in cut as well as in grazed grasslands, from very extensive (1 hay cut and aftermath grazed) to very intensive systems (7 grazing rotations or intensive continuous grazing). Being sensitive to trampling, pure-sown *D. glomerata* swards deteriorate if grazing is too intensive or if the soil is too wet when it is grazed (Figure 5.71).

In a Belgian trial in which *D. glomerata* was mixed with *Phleum pratense* or *Lolium perenne*, *D. glomerata* largely dominated these 2 species, particularly when N fertilization was high. *Lolium perenne* was more competitive to *D. glomerata* than *Phleum pratense* (Figures 5.72 and 5.73).

SOCIABILITY – COMPETITIVENESS – PLANT COMMUNITIES
Very competitive except in the first year after sowing. After 2 to 3 years, it can become the dominant species in the mixtures under cutting. Less competitive in well grazed mixed swards.

Mostly frequent in *Cynosurion cristati*, intensive grasslands, *Arrhenatherion elatioris*, *Polygono-Trisetion flavescentis* and *Mesobromion erecti*. Less frequent in *Agropyro-Rumicion*.

COMPATIBILITY WITH OTHER FORAGE GRASSES AND LEGUMES
Mainly mixed with *Medicago sativa* or with *Trifolium pratense* in cutting grassland. The *Medicago sativa-D. glomerata* association is rarely well balanced and in most cases, *Medicago sativa* will dominate the mixture.

Sometimes mixed with *Lolium perenne* in grazing mixtures adapted to dry areas. However, this is not recommended since the highly acceptable *Lolium perenne* is overgrazed and *D. glomerata*, being undergrazed, becomes tufted and lignified and so is rejected by stock unless they are forced to graze the sward short – a management that causes *D. glomerata* to disappear. *D. glomerata* is much less compatible with *Trifolium repens* than *Lolium perenne*, *Festuca pratensis* or *Phleum pratense* and it tends to stifle the *Trifolium*. The Ladino types are better adapted to *D. glomerata* mixtures than the smaller types of *Trifolium repens*. In Switzerland, *D. glomerata* is included in several complex mixtures including *Poa pratensis* that has similar ecological requirements, particularly for drought resistance (Mosimann *et al.*, 1996a).

AGRONOMIC CHARACTERISTICS

SOWING RATES
Pure: 15–20 kg/ha (though up to 30–40 kg/ha).
Mixed:10–5 kg/ha with *Medicago sativa* or *Trifolium pratense*;
4–6 kg/ha in complex mixtures with *Festuca pratensis*, *Lolium perenne*, *Phleum pratense* and *Trifolium repens*.

In Switzerland, *D. glomerata* is used in 2 main types of mixtures: a cutting mixture with *Medicago sativa* or 'long persistent'

FIGURES 5.72 AND 5.73

Proportion (% FM) of *Dactylis glomerata* in mixture with *Lolium perenne* or *Phleum pratense* in 3
cutting regimes and 4 nitrogen fertilizer levels, after 5 years of production

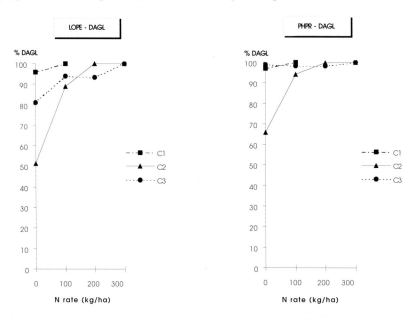

DAGL: *D. glomerata;* LOPE: *L. perenne;* PHPR: *P. pratense.* C1 = 2 cuts/year; C2 = 4 cuts/year;
C3 = 5–6 cuts/year

Source: Peeters, Decamps and Janssens, 1999

Trifolium pratense and in mixtures with *T.
repens* and other grasses for alternately cutting
and grazing. Elsewhere, *D. glomerata* is mainly
sown pure or mixed with *Medicago sativa*.

VARIETIES
A list of forage varieties is shown in Table 5.30.
Ornamental variety: Variegata.

FERTILIZATION
Needs high fertilization for intensive use,
but less than *Lolium perenne* for equivalent
DM production.

*DRY MATTER YIELDS AND SEASONALITY OF
PRODUCTION*
Very high annual production and one of the
most productive grasses in temperate

regions. Yields of 15–20 tonnes DM/ha are
frequently noted in both Europe and the
eastern USA. Stays productive even on
rather poor soils and at low fertilization.
Production is well spread over the growing
season and it is one of the grasses that
produces the most in the summer season. In
Belgium, *D. glomerata* was the most
productive grass among 10 other species in
cutting regimes of 4–6 cuts/year and at high
annual N rates (200–300 kg N/ha). It was
much more productive than *Lolium perenne*
and was only markedly overtaken by *Phleum
pratense* in the infrequent cutting regime (2
cuts/year) (Figures 5.74 to 5.77). *D. glomerata*
accumulates biomass at least as fast as
Lolium perenne during the first growth cycle
and at most of the cutting dates, its yield

TABLE 5.30
Examples of varieties of *Dactylis glomerata*

Varieties	Seed breeder (country)
Early to mid-early varieties	
Amba	DLF (DK)
Loke	Svalöf-Weibull (S)
Oberweihst	ZG (D)
Padania	ISCF-Lodi (I)
Reda	FAL (CH)
Late to mid-late varieties	
Abertop	WPBS (UK)
Athos	Green Genetics (F)
Baraula	Barenbrug (NL)
Cabrett	Carneau (F)
Greenly	RAGT (F)
Lidalgo	DSV (D)
Ludovic	INRA (F)
Lupré	INRA (F)
Prato	RAC (CH)
Vaillant	Verneuil Semunion (F)

TABLE 5.31
Evolution of yields (tonnes DM/ha) of *Dactylis glomerata* compared with *Lolium perenne* during the first growth cycle in the springs of 1991–93

	Dates					
1991	8 April	15 April	24 April	13 May	27 May	11 June
L. perenne	1.0	1.8	2.8	6.0	8.7	12.9
D. glomerata	1.5	2.6	3.6	6.3	8.7	11.6
1992	15 April	29 April	7 May	19 May	27 May	9 June
L. perenne	0.9	2.8	4.7	7.8	10.1	9.9
D. glomerata	1.2	3.0	4.9	8.0	10.3	11.3
1993	16 April	26 April	4 May	17 May	25 May	1 June
L. perenne	1.7	3.8	5.4	7.7	9.2	9.7
D. glomerata	2.3	4.6	7.1	9.9	12.5	11.2

Note: N fertilization during the first growth cycle in spring: 100 kg/ha
Source: Peeters and Decamps, 1994

was higher than *L. perenne*. In uninterrupted growth, it reached maximum yields of about 12.5 tonnes DM/ha (Table 5.31).

NUTRITIVE VALUE
D. glomerata has a lower digestibility than *Lolium perenne*. Peeters (1992) noted an average difference of 7 percent between digestibilities (OMD) of the 2 species from the beginning of April to mid-June. The rate of digestibility decline of *D. glomerata* (-0.41 percent per day) was similar to that of *Lolium perenne* (-0.41) (Figure 5.78). Spedding

and Diekmahns (1972) showed an average digestibility difference of 4 percent between the 2 species. *D. glomerata* is rather poor in soluble carbohydrates and quite high in cellulose and lignin compared with *Lolium perenne*. Its N and mineral content are similar to other productive grasses.

ACCEPTABILITY
D. glomerata is well accepted at a leafy stage. Mixed with *Lolium perenne* and other later-growing species, it is often too tall and lignified when the animals come to graze the

FIGURES 5.74–5.77
Annual yields of *Dactylis glomerata* compared with *Lolium perenne* and *Phleum pratense* in 3
cutting regimes and different nitrogen fertilization levels (average of 3 production years)

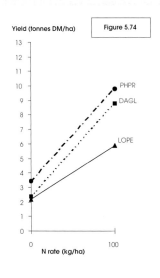

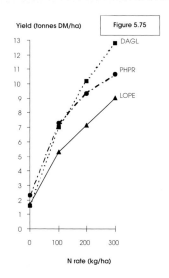

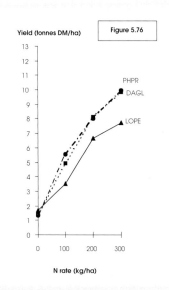

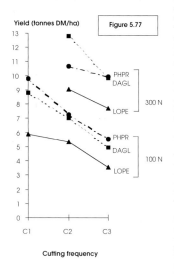

DAGL: *D. glomerata;* LOPE: *L. perenne;* PHPR: *P. pratense.* Fig. 5.74: 2 cuts/year (C1); Fig. 5.75: 4 cuts/year
(C2); Fig. 5.76: 5–6 cuts/year (C3); Fig. 5.77: 100 and 300 kg N/ha

Source: Peeters and Decamps, 1999

paddock. The latter then graze the shorter and more digestible species first and neglect *D. glomerata* which thus continues to deteriorate quality-wise. Under these circumstances, *D. glomerata* develops large tufts which are considerably taller than the rest of the grazing area and are rejected by cattle. It is therefore necessary to top these rejected plants in order to encourage quality regrowth.

ANIMAL PERFORMANCE

Fribourg *et al.* (1979) cited daily LWGs of steers of 1.07 kg/day in spring and 0.60 kg/day in summer in southeast USA, and a total maximum production of 560 kg LWG/ha.year for a stocking rate of 3.7 steers/ha with a final LW of 392 kg/steer. These results were obtained in an area where summer drought is intense and reduces the grazing period to 140 days/year. In more humid areas, where the use of *D. glomerata* in grazing is less justified, animal LWGs can be higher and sometimes similar to *Lolium perenne*.

DISEASES

Highly susceptible to disease, 2 of which are very important: *Mastigosporium rubricosum* and *Puccinia striiformis* var. *dactylidis*; 2 others are less detrimental, *Rhynchosporium orthosporum* and *Cercosporidium graminis*. *Mastigosporium rubricosum* is favoured by mild temperatures and high air humidity and it badly affects forage yield and quality. If 10 percent of the leaf area is affected by this disease, the dry matter yield falls by 30 percent and the soluble sugar content by 50 percent (O'Rourke, 1976). Varieties react very differently to the disease and this is also true for resistance to *Puccinia striiformis* var. *dactylidis*. *D. glomerata* is fairly resistant to snow mould, more so than *Lolium perenne*, but less so than *Phleum pratense*. This disease, which appears with heavy snow, is caused by a

FIGURE 5.78

Evolution of the digestibility of *Dactylis glomerata* compared with *Lolium perenne* during an uninterrupted growth period in spring (100 kg N/ha) (linear regression equations and determination coefficients shown)

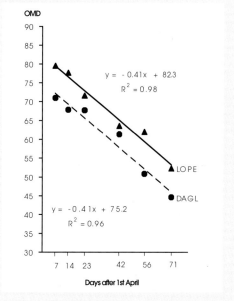

DAGL: *D. glomerata*; LOPE: *L. perenne*

Source: Peeters, 1992

combination of fungi including *Microdochium* (= *Fusarium*) *nivale* and *Typhula incarnata*. It results in plant disappearance in winter.

Other diseases appearing on this grass are: *Puccinia graminis* ssp. *graminicola*, *Uromyces dactylidis*, *Blumeria* (= *Erysiphe*) *graminis*, *Rhynchosporium secalis*, *Claviceps purpurea* and halo spot (*Pseudoseptoria stomaticola*). *Epichloë typhina* can develop spectacularly on *D. glomerata* stems during elongation and is very damaging for seed production (Raynal *et al.*, 1989; Ellis and Ellis, 1997). The variability in disease resistance of *D. glomerata* ecotypes and selections must be an important criterion in plant breeding.

USE FOR PURPOSES OTHER THAN FORAGE PRODUCTION

Sometimes used to stabilize slopes in new ski pistes in the mountains or in disused industrial sites (slag heaps for instance). In set-asides, provides a tall and permeable cover that is favourable to wildlife reproduction and shelter. Regularly used in field margins for controlling runoff and erosion. The ornamental variety is sometimes used in mixed borders.

MAIN QUALITIES

Productive grass. It is drought-resistant and produces markedly more in summer than *Lolium perenne*. Cold-resistant to a good degree. Well adapted to cutting and can be combined with *Medicago sativa*. At a leafy stage, it is well accepted by animals though less so than *Lolium perenne*. Very competitive, especially two to three years after sowing and very persistent in a cutting regime or in alternate cutting-grazing systems. Mainly suited to dry areas.

MAIN SHORTCOMINGS

Poorly adapted to exclusive grazing. In permanent grassland and mixed with *Lolium perenne*, it is less acceptable than *Lolium perenne* and so develops into large fibrous tufts that are refused by cattle. Despite important differences between varieties, it is often badly attacked by a disease complex, mainly *Mastigosporium rubricosum* and *Puccinia striiformis* var. *dactylidis* which are very detrimental to acceptability and feeding quality. Not very compatible with *Trifolium repens*.

DESCHAMPSIA CAESPITOSA (L.) BEAUV.

Many forms acknowledged as ecotypes or subspecies.

SYNONYM
Aira caespitosa L.

ETYMOLOGY
Dedicated to Dr Deschamps (eighteenth century), from St-Omer, a French botanist. *LATIN*: *caespes* or *cespes* (genitive: *cespitis*) = grass tuft, sod. This *Deschampsia* forms a large tuft.

COMMON NAMES
ENGLISH: Tufted hair-grass
FRENCH: Canche cespiteuse
GERMAN: Rasenschmiele

DISTRIBUTION
Native to temperate and cold areas of the northern hemisphere and to the mountain areas of tropical Africa.

FIGURE 5.79
Inflorescence and part of the leaf of
Deschampsia caespitosa

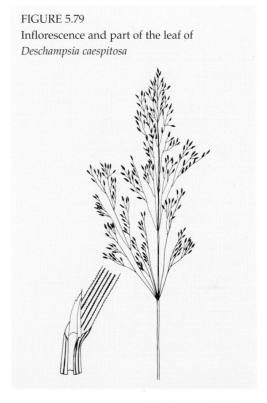

PLATE 5.32
Deschampsia caespitosa

PLATE 5.33
Deschampsia caespitosa

PLATE 5.34
Deschampsia caespitosa

DESCRIPTION

MORPHOLOGICAL DESCRIPTION
Perennial plant, robust, hairless, caespitose.
Stems erect, 30–150 cm high. Leaf blade
folded when young, 2–5 mm, large, rough,
often very long, strongly veined on the
upper part, sharp, dark green. Ligule
narrowly lanceolate, very acute, very long,
irregular, white. No auricles. Panicle-like
ample inflorescence, spreading even after
flowering, up to 40 cm long. Spikelets
2-flowered (rarely 3–4 fertile flowers),
3–6 mm long (Figure 5.79) (Plates 5.32 to 5.34).

1000-SEED WEIGHT
0.20–0.30 g (small seeds).

CHROMOSOME NUMBER
2n = 26.
Another chromosome number has also been
observed: 2n = 28.

BIOLOGICAL FLORA
Davy (1980).

ECOLOGICAL REQUIREMENTS

ALTITUDE – VEGETATION BELTS
From the lowlands up to 2 700 m in the Alps
(Rameau *et al.*, 1993); from hill to alpine belts.

SOIL MOISTURE
Optimum on wet and boggy soils (Figure
5.80).

SOIL FERTILITY
Optimum on moderately rich soils,
being a mesotrophic to slightly oligotrophic
species. Excluded from the richer soils.
Mainly on moderately acid to acid soils but
very large range for soil pH (Figures 5.80
and 5.81).

SOIL TYPE AND TOPOGRAPHY
Mainly distributed on peat soils but also
widespread on clay soils.

FIGURE 5.80
Ecological optimum and range for soil pH and
humidity of *Deschampsia caespitosa*

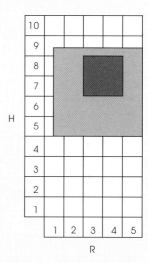

R: 1 = extremely acid and 5 = alkaline
H: 1 = extremely dry and
10 = permanently flooded

Complete key in Chapter 2, section 2.8.

FIGURE 5.81
Ecological optimum and range for nutrient
availability and management of *Deschampsia
caespitosa*

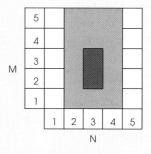

N: 1 = very low and 5 = very high
M: 1 = one defoliation per year and
5 = 5 defoliations per year or more

Complete key in Chapter 2, section 2.8.

CLIMATE REQUIREMENTS
Widespread in the climates of temperate
areas. Very resistant to cold and to

a long snowy period. Tolerates shade
very well.

TOLERANCE OF CUTTING AND GRAZING
Grows in grazed grassland as well as in
cutting grassland. Rejected by grazing
animals and tends to invade the sward
progressively if conditions are suitable.
Better controlled by cutting but its presence
makes hay-making more difficult (Figure
5.81).

*SOCIABILITY – COMPETITIVENESS – PLANT
COMMUNITIES*
Very competitive. If not controlled,
constitutes dense tufts and patches. When
conditions are suitable, quickly invades
little-used or neglected grasslands. Even
with regular grazing, cultural or chemical
means are often necessary to control it. A
common technique consists of grinding the
tufts close to the soil surface.
　　Mostly frequent in *Filipendulion ulmariae,
Deschampsion caespitosae, Molinion caeruleae,
Cynosurion cristati, Arrhenatherion elatioris,
Bromion racemosi* and humid improved
grasslands.

*COMPATIBILITY WITH OTHER FORAGE
GRASSES AND LEGUMES*
Incompatible with productive forage species.
Drainage and improvement of soil fertility
discourage it.

INDICATOR VALUE
Soil often wet, high probability of
fluctuating water table close to the soil
surface (pseudo-gley).

AGRONOMIC CHARACTERISTICS

VARIETIES
Ornamental varieties: Bronzeschleier, Fairy's
Joke = Vivipara, Goldgehänge, Goldschleier

= Golden Veil, Goldstaub, Goldtau = Golden
Dew, Ladywood Gold, Northern Lights,
Schottland, Tardiflora, Tauträger,
Waldschatt.
　　Some varieties are used for reclamation
of disturbed sites, e.g. Barcampsia
(Barenbrug, NL).

FERTILIZATION
The communities in which *D. caespitosa*
grows are hardly fertilized or not fertilized
at all.

*DRY MATTER YIELDS AND SEASONALITY OF
PRODUCTION*
Moderate to high yield.

NUTRITIVE VALUE
Very low digestibility. High lignin content.

ACCEPTABILITY
Unacceptable to grazing stock.

DISEASES
Three rusts: *Puccinia coronata*, a specific
species (*P. deschampsiae*) and *P. graminis*. A
Mastigosporium leaf fleck (*Mastigosporium
deschampsiae*). Occasionally, *Ustilago
striiformis* (Ellis and Ellis, 1997).

*USE FOR OTHER PURPOSES THAN FORAGE
PRODUCTION*
Used in gardens as decorative tufts in
borders and beneath the trees.
　　Constituent of wild flora mixtures for
damp acid or neutral soils (Crofts and
Jefferson, 1999).

MAIN QUALITIES
None for forage.

MAIN SHORTCOMINGS
Unacceptable to stock; ultra-poor nutritive
quality.

DESCHAMPSIA FLEXUOSA (L.) TRIN.

FIGURE 5.82
Inflorescence and part of the leaf
Deschampsia flexuosa

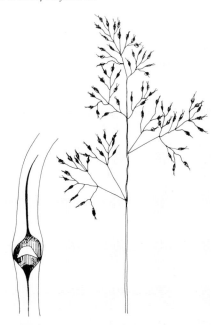

SYNONYM
Aira flexuosa L.

ETYMOLOGY
Dedicated to Dr Deschamps
(eighteenth century), from St-Omer,
a French botanist.
LATIN: *flexuosus* = tortuous, winding; the
branches of the panicle are tortuous.

COMMON NAMES
ENGLISH: Wavy hair-grass
FRENCH: Canche flexueuse
GERMAN: Geschlängelte schmiele

DISTRIBUTION
Native to temperate and cold areas of the
northern hemisphere.

PLATE 5.35
Deschampsia flexuosa

PLATE 5.36
Deschampsia flexuosa

DESCRIPTION

MORPHOLOGICAL DESCRIPTION
Perennial plant, moderate sized, hairless, turf forming, sometimes with short rhizomes. Stems erect, (20–) 30–60 (–80) cm high. Needle-like leaf blade, dark green. Blade of the upper leaves sometimes flat. Ligule quite short (1–3 mm), truncate. No auricles. Panicle-like, spreading inflorescence, with long flexuous branches, ovale shape, contracted after flowering. Spikelets with 2 fertile flowers, almost 5 mm long, variegated with purple and silvery white. Lemma with a bent, very striking awn (Figure 5.82) (Plates 5.35 and 5.36).

1000-SEED WEIGHT
0.30–0.40 g (small seeds).

CHROMOSOME NUMBER
2n = 28 (tetraploid).
Other chromosome numbers have also been observed: 2n = 14, 26 or 56.

BIOLOGICAL FLORA
Scurfield (1954).

ECOLOGICAL REQUIREMENTS

ALTITUDE – VEGETATION BELTS
From the lowlands up to 2 700 m in the Alps (Rameau *et al.*, 1993); from hill to alpine belts.

SOIL MOISTURE
Optimum on normally drained to dry soils (Figure 5.83).

SOIL FERTILITY
Restricted to soils that are very poor in nutrients (oligotrophic species) and acid to extremely acid. The accumulation of organic matter in these soils is often high because of low mineralization (Figures 5.83 and 5.84).

SOIL TYPE AND TOPOGRAPHY
Large range for soil texture.

FIGURE 5.83
Ecological optimum and range for soil pH and humidity of *Deschampsia flexuosa*

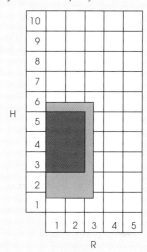

R: 1 = extremely acid and 5 = alkaline
H: 1 = extremely dry and
10 = permanently flooded

Complete key in Chapter 2, section 2.8.

FIGURE 5.84
Ecological optimum and range for nutrient availability and management of *Deschampsia flexuosa*

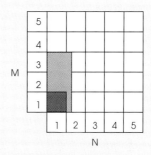

N: 1 = very low and 5 = very high
M: 1 = one defoliation per year and
5 = 5 defoliations per year or more

Complete key in Chapter 2, section 2.8.

CLIMATE REQUIREMENTS
Widespread in the climates of the temperate area. Very resistant to cold

and to a long snowy period. Tolerates shade very well.

TOLERANCE OF CUTTING AND GRAZING
In the lowlands, mainly restricted to forest on infertile acid soils. In the highlands, grows principally on grasslands usually poorly grazed by sheep (Figure 5.84).

SOCIABILITY – COMPETITIVENESS – PLANT COMMUNITIES
Poorly competitive in grasslands and in moors. In woodlands, can constitute large patches but, in grasslands, it mixes with other oligotrophic species like *Molinia caerulea*, *Sieglingia decumbens* and *Nardus stricta*.

Mostly frequent in *Deschampsia flexuosa* grasslands, *Violo-Nardion strictae* and *Nardion strictae*.

COMPATIBILITY WITH OTHER FORAGE GRASSES AND LEGUMES
Incompatible with productive forage species.

INDICATOR VALUE
Extremely poor soil, especially in P. Extensive management.

AGRONOMIC CHARACTERISTICS

VARIETIES
Ornamental variety: Tatra Gold = Aurea.

FERTILIZATION
The communities in which *D. flexuosa* grows are not fertilized.

DRY MATTER YIELDS AND SEASONALITY OF PRODUCTION
Low production in sites where it is spontaneous.

NUTRITIVE VALUE
Low.

ACCEPTABILITY
Ingested by animals but of low acceptability except by deer in forest clearings.

DISEASES
A specific rust species (*Uromyces airae-flexuosae*) is occasionally found (Ellis and Ellis, 1997).

USE FOR PURPOSES OTHER THAN FORAGE PRODUCTION
Constituent of wild flora mixtures for acid soils (Crofts and Jefferson, 1999). The ornamental variety is sometimes used in borders, beneath trees. In Scandinavia, it is used for covering wood roofs.

MAIN QUALITIES
Contributes to the production of infertile, high-altitude grasslands.

MAIN SHORTCOMINGS
Poorly productive and of low nutritional value.

ELYMUS REPENS (L.) GOULD

FIGURE 5.85
Inflorescence and part of the stem, rhizome and leaf of *Elymus repens*

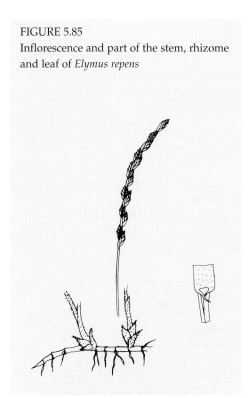

SYNONYM
Agropyron repens (L.) Beauv., *Elytrigia repens* (L.) Desv. ex Nevski

ETYMOLOGY
GREEK: ελυμος = elymos = undetermined cereal and ελμο = eluo = I envelop, I wrap up. In the spikelet, the glumes wrap up the flowers.
LATIN: *repens* = creeping. This grass produces many rhizomes creeping below the soil surface.

COMMON NAMES
ENGLISH: Couch, twitch, quack grass
FRENCH: Chiendent commun
GERMAN: Gemeine quecke

PLATE 5.37
Elymus repens

PLATE 5.38
Elymus repens

DISTRIBUTION

Native to Europe, western Asia and Northern Africa. Has become subcosmopolitan in temperate areas.

DESCRIPTION

MORPHOLOGICAL DESCRIPTION
Perennial plant, robust, more or less densely hairy, very rhizomatous. Stems erect, 30–120 cm high. Long or very long rhizomes. Leaf blade rolled when young, large, flat, green or bluish green, rough, more or less hairy on the upper part. Sheath of the lower leaves often very densely hairy. Ligule very short. Auricles narrow, quite short, sharp, variable. Spike-like inflorescence, long, with densely over-lapping spikelets. Spikelets with 5–10 flowers, often aristate, applied to the stem by their broader side. Glumes and lemma aristate or pointed. Very variable species (Figure 5.85) (Plates 5.37 and 5.38).

PHYSIOLOGICAL PECULIARITIES
E. repens is one of the latest grasses to mature, heading after *Phleum pratense*.

1000-SEED WEIGHT
2.0–2.5 g (average-sized seeds).

CHROMOSOME NUMBER
2n = 42 (hexaploid).
Other chromosome numbers have also been observed: 2n = 28, 35 or 56.

BIOLOGICAL FLORA
Palmer and Sagar (1963).

ECOLOGICAL REQUIREMENTS

ALTITUDE – VEGETATION BELTS
From the lowlands to high altitudes in mountain areas; from hill to alpine belts.

SOIL MOISTURE
Large range from dry to wet soils (Figure 5.86).

FIGURE 5.86
Ecological optimum and range for soil pH and humidity of *Elymus repens*

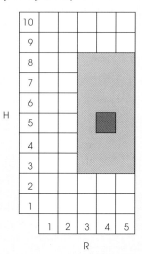

R: 1 = extremely acid and 5 = alkaline
H: 1 = extremely dry and
10 = permanently flooded

Complete key in Chapter 2, section 2.8.

FIGURE 5.87
Ecological optimum and range for nutrient availability and management of *Elymus repens*

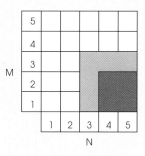

N: 1 = very low and 5 = very high
M: 1 = one defoliation per year and
5 = 5 defoliations per year or more

Complete key in Chapter 2, section 2.8.

SOIL FERTILITY
Optimum on nutrient-rich, neutral soils (eutrophic species) (Figures 5.86 and 5.87). Salt-tolerant.

SOIL TYPE AND TOPOGRAPHY
Large range of soil textures: can be
abundant on peat soil or on shallow mineral
soils if they are nutrient-rich, but optimum
on good soils.

CLIMATE REQUIREMENTS
Widespread in the climates of temperate
areas. Very resistant to cold, heat and
drought.

TOLERANCE OF CUTTING AND GRAZING
Not very common in exclusively grazed
grasslands. Very sensitive to defoliation
frequency. Optimum of 1 or 2 cuts/year.
Species very frequent and abundant in
heavily fertilized temporary grassland for

cutting, especially in cold areas (mountains,
Scandinavia) (Figure 5.87).
    In Belgium (Figures 5.88 and 5.89), the
fresh weight percentage of *E. repens* was
measured in an experiment where it was
either mixed with *Phleum pratense*, or with
Lolium perenne. *E. repens* did not dominate
these 2 species except in the 2 cuts/year
regime with 100 kg N/ha and in the 4
cuts/year regime with an N rate higher or
equal to 200 kg/ha; however, it was clearly
dominated in the frequent cutting regime,
even by *Phleum pratense*.

*SOCIABILITY – COMPETITIVENESS – PLANT
COMMUNITIES*
Competitive even though its initial

FIGURES 5.88 AND 5.89
Proportion (% FM) of Elymus repens in mixture with *Lolium perenne* or *Phleum pratense* in 3
cutting regimes and 4 levels of nitrogen fertilization, after 5 years of production

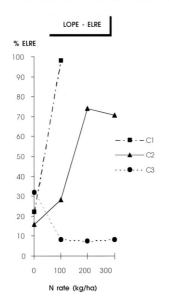
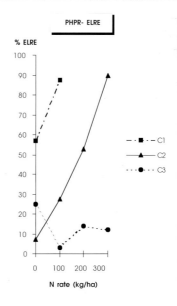

ELRE: *E. repens;* LOPE: *L. perenne;* PHPR: *P. pratense.* C1 = 2 cuts/year; C2 = 4 cuts/year; C3 = 5–6 cuts/year

Source: Peeters, Decamps and Janssens, 1999

establishment in a sward is low, but once established, it can become dominant. Often associated in temporary grassland with *Phleum pratense, Poa trivialis* and *Festuca pratensis*.

On a fertile soil and in permanent swards cut twice a year, becomes co-dominant with *Arrhenatherum elatius*. Favoured in rotations of cereals and cutting leys. During the cereal cropping phase, the soil cultivations cut and spread the rhizomes and the cutting frequency used (1 cut/year) is perfectly suited to it. Unless controlled, e.g. by herbicide, the density of rhizomes is thus high at the time of grassland sowing. *E. repens* can then produce more than three-quarters of the biomass in the grassland sward.

Mostly frequent in *Arrhenatherion elatioris*, intensive grasslands, *Agropyro-Rumicion crispi, Convolvulo-Agropyrion repentis,* other ruderal and weed alliances.

Less frequent in *Mesobromion erecti* and *Koelerio-Phleion phleoidis*.

COMPATIBILITY WITH OTHER FORAGE GRASSES AND LEGUMES

Compatible mainly with *Phleum pratense*, which has the same cutting frequency and soil fertility requirements, but also with *Festuca pratensis, Trifolium pratense* and, to a lesser extent, *Lolium perenne. Lolium multiflorum* can dominate it and control it in short-term swards.

INDICATOR VALUE

In grazed swards, only in recently sown mixtures. Elsewhere, sward cut for hay or silage and heavily fertilized.

AGRONOMIC CHARACTERISTICS

SOWING RATES

Pure: 30–50 kg/ha;
          100 kg/ha for coated seeds.

TABLE 5.32

Annual yields (tonnes DM/ha) of *Elymus repens* compared with a standard sward mixture in three cutting regimes in Finland

Cutting frequency	1974 (normal year)		1975 (dry year)	
	E. repens	Grass-clover	*E. repens*	Grass-clover
4 cuts/year	11.7	12.4	5.9	5.4
3 cuts/year	10.4	11.2	6.7	7.3
2 cuts/year	11.5	12.8	8.5	9.0

Note: annual N fertilization: 270 kg/ha
Source: Pulli, 1983

TABLE 5.33

Annual yields (tonnes DM/ha) of cutting grassland dominated by *Elymus repens* (79%) compared with a meadow dominated by *Lolium perenne* (89%) in 3 cutting regimes in Ardenne, Belgium

Cutting frequency and annual N fertilization	Yields	
	E. repens	*L. perenne*
4 cuts/year 325 kg N/ha	13.5	14.2
3 cuts/year 250 kg N/ha	14.2	14.0
2 cuts/year 175 kg N/ha	16.3	15.6

Source: Peeters, 1989

Mixed: 10–20 kg/ha with *Phleum pratense,*
Arrhenatherum elatius or *Festuca*
pratensis.

Never sown in practice. Conceivably, it
could be incorporated in cutting mixtures
for silage intended for suckler cows or
heifers, especially in cold climate areas.

VARIETIES

No bred variety but wild ecotype seeds
available. Neuteboom (1975, 1980) and
Taylor and Aarssen (1988) remarked upon
the huge morphological diversity in the
species concerning plant colour, leaf
hairiness, length of the spikelet awn, size of
leaves and stems, weight and thickness of
rhizomes and heading dates. Breeding
opportunities obviously exist in this
species.

FERTILIZATION

Needs a high fertilization to express its
potential.

DRY MATTER YIELDS AND SEASONALITY OF
PRODUCTION

Low or high production according to the
situation and the ecotype. The yield of *E.*
repens sward can be equivalent, slightly higher
or slightly lower to the best mixtures used in a
region (Tables 5.32 and 5.33). It can be even
higher in dry years (Kreil, Kaltofen and
Wacker, 1964), though not always necessarily
so. However, the establishment of *E. repens*
after sowing pure can be very low. In a trial in
Belgium with 10 other grasses, *E. repens* was
one of the less productive species, outyielding
only *Poa trivialis* and *Agrostis stolonifera*. This
was due to a very low sward density even in
harvest year 3 and even with high fertilization
(Figures 5.90 to 5.93). However, the *E. repens*
yields reached 93, 91 and 81 percent of *Lolium*
perenne yields in cutting regimes of 2, 4 and 6
cuts/year, respectively, with 100, 300 and 300
kg N/ha annually.

E. repens growth was clearly slower than
Lolium perenne growth, especially at the

beginning of the season. Then, the growth
accelerated but the *E. repens* production
rarely exceeded that of *Lolium perenne* at the
end of the first growth cycle (Table 5.34).

E. repens produces the major part of its
yield in June–July (Table 5.35).

NUTRITIVE VALUE

Digestibility is good to very poor. *E. repens*
has a digestibility similar to *Lolium perenne*
in spring. The rate of digestibility reduction
is even slower at the beginning of the
season, but on average it is quite similar
(-0.50 versus -0.41 percent per day for *Lolium*
perenne) (Peeters, 1992) (Figure 5.94).

Narashimhalu, Winter and Kunelius
(1982) found that the digestibility of *E. repens*
silage is lower than that of *Lolium perenne*
and *L. multiflorum* and similar to *Dactylis*
glomerata, while Salo, Nykanen and
Ormunen (1975) and Pulli (1976) noted it is
close to that of *Phleum pratense* and *Festuca*
pratensis. It is higher than that of *Bromus*
inermis (Marten, Sheaffer and Wyse, 1987).

The fibre contents are often higher than
those from better forage grasses. However,
Marten, Sheaffer and Wyse (1987) noted that
the cell wall content, as assessed by neutral
detergent fibre (NDF), was lower than
Bromus inermis (Table 5.36).

Peeters *et al.* (1991) observed a higher
crude fibre (CF) content for *E. repens* than for
Phleum pratense, Lolium perenne and *L.*
multiflorum during an uninterrupted growth
period from the beginning of May to the end
of July.

E. repens is poor in soluble carbohydrates
but its contents are also more stable with
advancing maturity than those of other good
forage grasses (Table 5.37).

The N content is almost always higher
than that in productive grasses cultivated
under the same conditions (Peeters *et al.*, 1991).
Neuteboom (1981) showed that the differences
in crude protein (CP) contents between *E.*
repens and *Lolium perenne* are greater when the
forage is young and N contents are high. This

TABLE 5.34

Dry matter yield (tonnes/ha) evolution of *Elymus repens* compared with *Lolium perenne* during the first growth cycle in the springs of 1991–93 in Belgium

				Dates		
1991	8 April	15 April	24 April	13 May	27 May	11 June
L. perenne	1.0	1.8	2.8	6.0	8.7	12.9
E. repens	0.2	0.8	1.2	2.6	4.7	7.1
1992	15 April	29 April	7 May	19 May	27 May	9 June
L. perenne	0.9	2.8	4.7	7.8	10.1	9.9
E. repens	0.5	1.5	2.5	5.3	7.4	10.1
1993	16 April	26 April	4 May	17 May	25 May	1 June
L. perenne	1.7	3.8	5.4	7.7	9.2	9.7
E. repens	0.5	2.4	4.3	6.3	8.3	9.1

Note: N fertilization during the first growth cycle in spring: 100 kg/ha
Source: Peeters and Decamps, 1994

TABLE 5.35

Seasonal yields (% basis) of *Elymus repens* compared with *Lolium perenne*

	April–May (%)	June–July (%)	August–September (%)
E. repens	32	47	21
L. perenne	41	43	16

Source: Peeters and Decamps, 1994

TABLE 5.36

Cell wall contents (NDF as % of DM) of bred and wild ecotypes of *Elymus repens* compared with *Bromus inermis* at 4 cutting dates

	18 May (%)	1 June (%)	15 June (%)	29 June (%)
E. repens (bred)	41.6	57.5	69.9	69.6
E. repens (wild)	46.4	59.2	66.4	65.8
B. inermis	49.2	65.1	66.7	65.8

Source: Marten, Sheaffer and Wyse, 1987

TABLE 5.37

Sugar + fructosan contents (% of DM) of *Elymus repens* compared with *Festuca pratensis* and *Phleum pratense* at 3 cutting dates

	13 June (%)	17 June (%)	20 June (%)
E. repens	9.8	9.7	9.6
F. pratensis	12.4	12.4	11.2
P. pratense	14.5	11.8	9.2

Source: Salo, Nykanen and Ormunen, 1975

FIGURES 5.90–5.93
Annual yields of *Elymus repens* compared with *Lolium perenne* and *Phleum pratense* in 3 cutting regimes and several nitrogen fertilization levels (average of 3 production years)

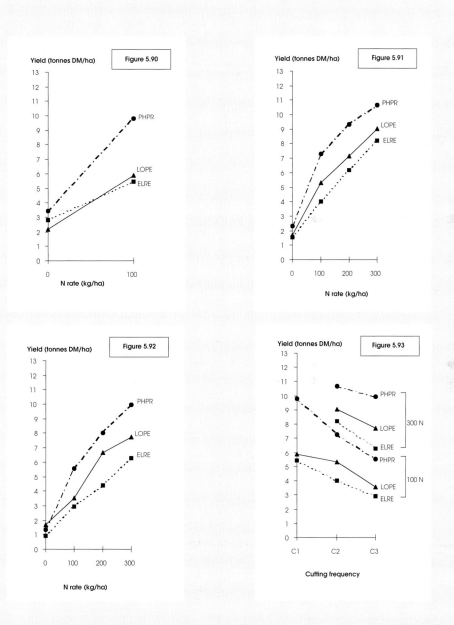

ELRE: *E. repens;* LOPE: *L. perenne;* PHPR: *P. pratense.* Fig. 5.90: 2 cuts/year (C1); Fig. 5.91: 4 cuts/year (C2); Fig. 5.92: 5–6 cuts/year (C3); Fig. 5.93: 100 and 300 kg N/ha

Source: Peeters and Decamps, 1999

superiority of N content of *E. repens*, compared with *Phleum pratense* and *Lolium perenne*, lasts from the beginning of May to the end of July (Peeters *et al.*, 1991).

ACCEPTABILITY
E. repens silage is well accepted by stock. According to Narashimhalu, Winter and Kunelius (1982), the intake of pre-wilted silage of *E. repens* is similar to that from *Dactylis glomerata* and *Lolium perenne*. The hay is also well ingested.

Fresh *E. repens* is better ingested than *Bromus inermis* but less well than *Medicago sativa* (Marten, Sheaffer and Wyse, 1987). Nevertheless when grazed, *E. repens* is avoided in swards with more acceptable grasses present, the roughness of the hairs being the probable reason.

DISEASES
E. repens is attacked by a great number of diseases but very few are specific. It is susceptible to several rusts, including *Puccinia coronata*, *P. graminis* and *P. recondita* ssp. *agropyrina*, which is the most common rust on this species. It is attacked by *Ustilago hypodites* and by *Blumeria* (= *Erysiphe*) *graminis*. It can suffer from snow mould, which appears under heavy snow cover and is caused by a complex of fungi including *Microdochium* (= *Fusarium*) *nivale* and *Typhula incarnata*. It can lead to the disappearance of plants in winter. *Rhynchosporium secalis*, *Drechslera tritici-repentis* and other *Drechslera* spp., as well as *Claviceps purpurea* can also be observed (Ellis and Ellis, 1997).

USE FOR PURPOSES OTHER THAN FORAGE PRODUCTION
Rhizomes were traditionally used for making hard brushes and, in the Alps, for filtering milk. Rhizomes contain starch and for this reason were harvested during periods of famine to make a flour, mixed or not with wheat flour and cooked as pancakes. Rhizomes have also many

medicinal properties: diuretic (Girre, 2001), emollient, vermifuge, febrifuge and depurative (Brüschweiler, 1999).

MAIN QUALITIES
Productive grass with a good digestibility and a high N content. Extremely resistant to cold and tolerates drought. Well accepted as hay and silage. Can contribute to yield in cutting grassland and in highly fertilized swards in cold climates.

MAIN SHORTCOMINGS
Undesirable weed in arable land and in ley-crop rotations. Badly accepted by grazing animals probably because of the leaf roughness. Soluble sugar content low and fibre content high. Sometimes astonishingly low in production. Extremely variable species for a great number of characteristics.

FIGURE 5.94
Evolution of the digestibility of *Elymus repens* compared with *Lolium perenne* during an uninterrupted growth period in spring (100 kg N/ha) (linear regression equations and determination coefficients shown)

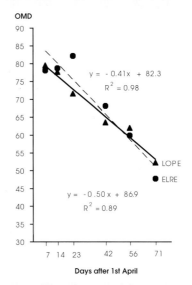

ELRE: *E. repens;* LOPE: *L. perenne*

Source: Peeters, 1992

FESTUCA ARUNDINACEA SCHREB.

SYNONYM
F. elatior L., *F. elatior* L. ssp. *arundinacea*
(Schreb.) Hack.

ETYMOLOGY
LATIN: *festuca* = straw, hay and wisp of straw,
derived from CELTIC: fest = pasture.
These species were often dominant in
pastures.
LATIN: *arundinaceus* = similar to reed (= *arundo*).

COMMON NAMES
ENGLISH: Tall fescue
FRENCH: Fétuque élevée, fétuque roseau
GERMAN: Rohr schwingel

DISTRIBUTION
Native to Europe. Has become
subcosmopolitan in temperate regions.

FIGURE 5.95
Inflorescence and part of the leaf of *Festuca
arundinacea*

PLATE 5.39
Festuca arundinacea

PLATE 5.40
Festuca arundinacea

DESCRIPTION

MORPHOLOGICAL DESCRIPTION
Perennial plant, very robust, hairless (except auricles), caespitose to shortly rhizomatous. Stems erect, 50–110 (–150) cm high. Leaf blade rolled when young, large (3–10 mm), flat, strongly veined, coarse, rough on the upper side, shiny below, dark green. Sheaths of the inferior leaves purplish red. Ligule short, greenish. Auricles strong, ciliate. Panicle-like inflorescence, spreading even after flowering, oblong, loose. Spikelets 4–7-flowered, briefly aristate, 10–15 mm long. Variable species (Figure 5.95) (Plates 5.39 and 5.40).

PHYSIOLOGICAL PECULIARITIES
Hybridizes with *Lolium perenne*: x *Festulolium holmbergii* and with *Festuca pratensis*: x *Festuca braunii*. Often invaded by endophytes (*Neotyphodium* spp.).

1000-SEED WEIGHT
1.8–2.5 g (average-sized seeds).

CHROMOSOME NUMBER
2n = 42 (hexaploid).

BIOLOGICAL FLORA
Gibson and Newman (2001).

ECOLOGICAL REQUIREMENTS

ALTITUDE – VEGETATION BELTS.
Mainly in the lowlands, for example, only below 300 m in the UK (Spedding and Diekmahns, 1972). From hill to mountain belts.

SOIL MOISTURE
Adapted to wide range of soil moisture conditions. Tolerates a sodden soil in winter remarkably well. Very resistant to summer drought. Thrives on wet to dry soils. Very well adapted to fluctuations in water supply. Able to absorb water far beneath the surface because of its deep root system (Figure 5.96).

FIGURE 5.96
Ecological optimum and range for soil pH and humidity of *Festuca arundinacea*

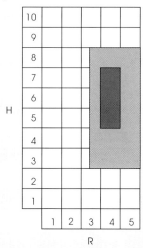

R: 1 = extremely acid and 5 = alkaline
H: 1 = extremely dry and 10 = permanently flooded

Complete key in Chapter 2, section 2.8.

FIGURE 5.97
Ecological optimum and range for nutrient availability and management of *Festuca arundinacea*

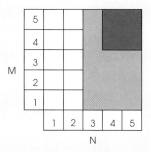

N: 1 = very low and 5 = very high
M: 1 = one defoliation per year and
5 = 5 defoliations per year or more

Complete key in Chapter 2, section 2.8.

SOIL FERTILITY
Large range for nutrient availability but excluded from very poor soils (mesotrophic to eutrophic species). Widely encountered on acid to alkaline soils (Figures 5.96 and 5.97). Salt-tolerant.

SOIL TYPE AND TOPOGRAPHY
Often found on heavy clay soils or on deep loam but is indifferent to soil texture.

CLIMATE REQUIREMENT
Very good resistance to climatic extremes of heat, drought and cold. However, the ecotypes and varieties native to the Mediterranean basin are sensitive to low temperatures. Infection by endophytes confers a higher resistance to climatic stress, especially to drought.

TOLERANCE OF CUTTING AND GRAZING
Largely indifferent to the management regime. Tolerates trampling. Sensitive to a low defoliation height, it disappears if defoliated too close. Infection by endophytes confers a higher tolerance of intensive grazing (Figure 5.97).

SOCIABILITY – COMPETITIVENESS – PLANT COMMUNITIES
Very competitive when established but very sensitive to competition during establishment phase. Rarely constitutes large spontaneous populations and is mainly distributed in small patches in small depressions.

Mostly frequent in *Cynosurion cristati*, *Bromion racemosi*, *Arrhenatherion elatioris*, intensive grasslands, *Agropyro-Rumicion*, *Filipendulion ulmariae* and *Molinion caeruleae*.

COMPATIBILITY WITH OTHER FORAGE GRASSES AND LEGUMES
Incompatible with other forage grasses.

INDICATOR VALUE
Good fertility, often deep soil.

AGRONOMIC CHARACTERISTICS

SOWING RATES
Pure: 20–30 kg/ha.
Mixed: not usually mixed with other grasses when used for forage production. It can be associated with *Trifolium pratense* or *T. repens* or *Medicago sativa* but its sowing rate stays the same. In Switzerland, it is included at 8 kg/ha

in a long-term mixture for dry areas unsuitable for *Lolium perenne* (Mosimann *et al.*, 1996a).

VARIETIES
A list of forage varieties is shown in Table 5.38.

FERTILIZATION
Needs high fertilization to express its full potential.

DRY MATTER YIELDS AND SEASONALITY OF PRODUCTION
One of the most productive forage grasses and under optimum conditions, its yield can reach or even exceed 15 tonnes DM/ha in Western Europe. Summer production is clearly higher than *Lolium* spp., *Festuca pratensis* or *Phleum pratense*. Tables 5.39 and 5.40 show the high yields obtained in the west of France.

These tables demonstrate the yield potential of *F. arundinacea* when it is grown in optimum conditions (deep and fertile soil, wet climate), a potential higher than *Dactylis glomerata* or *Lolium perenne*.

Slightly lower yields were cited for a less favourable region, the Ardennes (B) at 500 m (Table 5.41). However, *F. arundinacea* was still the most productive of all tested species.

In Tuscany (I), Talamucci (1975) obtained maximum annual yield (12 tonnes DM/ha) from cutting every 4 weeks. By cutting more often, every 3 weeks, the yield was clearly lower (almost 8 tonnes DM/ha) but by cutting every 5–6 weeks, the production ranged between 11 and 12 tonnes DM/ha. In another experiment, the yield reached 14 tonnes DM/ha with 300 kg N/ha.

In association with *Medicago sativa*, late types of *F. arundinacea* are more suitable than early types, the mixture being more productive and less dominated by the grass.

F. arundinacea has a very regular growth throughout the season. It continues to grow in autumn when the growth of all other forage grasses have stopped.

NUTRITIVE VALUE
The digestibility of *F. arundinacea* is much

TABLE 5.38

Examples of varieties of *Festuca arundinacea*

Variety	Seed breeder (country)
Advance	CEBECO (NL)
Barcel	Barenbrug (NL)
Dovey	WPBS (UK)
Kora	OSEVA (CZ)
Lutine	INRA (F)
Ondine	INRA (F)
Pastelle	RAGT (F)
Stef	ZNRO (PL)

TABLE 5.39

Annual yields (tonnes DM/ha) of *Festuca arundinacea* compared with *Dactylis glomerata* and *Lolium perenne* in a 4-year experiment in Normandy, France

	Species and varieties		
	F. arundinacea Manade	*D. glomerata* Floréal	*L. perenne* Vigor
Average yields (1964–67)	19.3	14.0	15.7
Minimum yields (1964–67)	15.5	12.5	12.5
Maximum yields (1964–67)	21.3	15.9	19.4

Source: Vivier, 1972

TABLE 5.40

Annual yields (tonnes DM/ha) of *Festuca arundinacea* with increasing nitrogen rates in Normandy, France 1967–71

	Annual N fertilization (kg/ha)			
Species (variety)	0	150	300	450
F. arundinacea (Manade)	8.0	12.7	15.9	17.4

Source: Vivier, 1972

TABLE 5.41

Comparison of annual yields (tonnes DM/ha) of varieties of *Festuca arundinacea*, *Lolium perenne*, *Festuca pratensis* and *Phleum pratense* in Ardennes, Belgium

	F. arundinacea	*L. perenne* early var.	*L. perenne* late var.	*F. pratensis*	*P. pratense*
Average yield	12.2	9.5	10.6	10.1	11.4
Yield of the best variety	13.4	11.2	12.1	10.5	12.4
Yield of the poorest variety	11.0	8.4	9.3	9.8	10.3

Source: Toussaint and Lambert, 1973

lower than that of the other main grasses. The new varieties have softer leaves and a higher digestibility than most of the wild ecotypes that thrive in permanent grasslands, ecotypes which always have much coarser leaves and significantly lower digestibilities than *Lolium perenne*. Protein and mineral contents are similar to those of the other main forage grasses.

ACCEPTABILITY

Badly accepted by animals even though the new varieties are better ingested than the old varieties or the wild types. Must be grazed early in spring and then the animals must be returned at short intervals to the paddocks, at least every 4 weeks, a management that requires a high rate of N application. It is especially important to avoid offering highly acceptable species before grazing cattle on *F. arundinacea*, especially if the latter is rather mature. The intake of *F. arundinacea* silage is lower than *Lolium perenne* silage.

TOXICITY

Ingestion of *F. arundinacea* can lead to diverse physiological troubles for grazing animals because of endophytes of *Neotyphodium* spp. In southeast USA, 'fescue foot' disease appears sporadically as lameness, a rough coat and a dry gangrene of the body extremities (tail, feet, ears). The infection of plant tissues by endophytes leads mainly to the production of alkaloids and perloline, with maximum concentrations in summer. Ingestion of infected *F. arundinacea* results in rumination and tissue metabolism problems (Sleper and Buckner, 1995).

ANIMAL PERFORMANCE

Animal performance on *F. arundinacea* swards can be high if the sward is well managed but infection by endophytes can considerably reduce animal production (Bacon and Siegel, 1988; Bush and Burrus, 1988; Stuedemann and Hoveland, 1988). The decrease is not linear and should only appear at 15–20 percent infection (Fribourg *et al.*, 1989).

DISEASES

Puccinia coronata can affect *F. arundinacea*, but less dramatically than *Lolium perenne*. *Puccinia graminis* can also be observed as well as a smut (*Urocystis agropyri*), *Drechslera* spp., *Spermospora* leaf spot (*Spermospora lolii*), *Rhynchosporium orthosporum*, *R. secalis*, *Blumeria* (= *Erysiphe*) *graminis* and *Claviceps purpurea* (Raynal *et al.*, 1989; Ellis and Ellis, 1997). *F. arundinacea* is moderately susceptible to snow mould, though more resistant than *Lolium multiflorum* and more susceptible than *Phleum pratense*. This disease, which appears under strong snow cover, is caused by a complex of fungi including *Microdochium* (= *Fusarium*) *nivale* and *Typhula incarnata* and leads to the disappearance of plants in winter.

Xanthomonas translucens pv. *graminis* can cause plant mortality as in other forage grasses.

Infection by endophytes of *Neotyphodium* spp. confers a higher resistance to diseases.

USE FOR PURPOSES OTHER THAN FORAGE PRODUCTION

Often used for the creation of lawns in dry areas. Also used in mixtures for sowing new ski pistes (at moderate altitude), on road and motorway sides (dry conditions) and in field margins.

MAIN QUALITIES

Very productive species. Adapted to extreme environmental conditions, e.g. summer drought, winter flooding, and is particularly well suited to warm and dry climates. Can be managed under cutting or grazing. Once established, its persistence is often notable (10 years or more).

MAIN SHORTCOMINGS

Develops slowly after sowing. Poorly accepted by grazing animals, especially if they have the choice of other species. For this reason it is often grazed only at the beginning of spring and then ensiled. Its endophyte content must be monitored in order to avoid toxicity levels that are detrimental to cattle.

FESTUCA OVINA L.

There are several subspecies in Europe. All grow on nutrient-poor, dry soils. The different taxa of *F.* group *ovina* are considered here as a whole.

ETYMOLOGY
LATIN: *festuca* = straw, hay and wisp of straw, derived from CELTIC: fest = pasture.
These species were often dominant in pastures.
LATIN: *ovis* = ewe.
The grass is grazed by sheep.

COMMON NAMES
ENGLISH: Sheep's fescue
FRENCH: Fétuque ovine
GERMAN: Schaf schwingel

DISTRIBUTION
Native to Europe, temperate Asia and North America.

FIGURE 5.98
Whole plant and part of the leaf of *Festuca ovina*

A. PEETERS

PLATE 5.41
Festuca ovina

DESCRIPTION

MORPHOLOGICAL DESCRIPTION
Perennial plant, small-sized, hairless, caespitose. Stems erect, 20–50 cm high. Needle-like leaf blade, narrow (about 1 mm), greyish green to bluish green, rough. Ligule very short. Auricles very reduced, sometimes distinct. Panicle-like inflorescence, short (3–10 mm), with erect or more or less spreading branches. Spikelets with 3–8 aristate flowers (Figure 5.98) (Plates 5.41 to 5.43).

PHYSIOLOGICAL PECULIARITIES
Maximum number of leaves per tiller: 6.2 (Sydes, 1984), which is higher than for most grasses.

1000-SEED WEIGHT
0.8–1.0 g (average-sized seeds).

CHROMOSOME NUMBER
2n = 14 (diploid) or 28 (tetraploid). Other chromosome numbers have also been observed: 2n = 35, 42 or 56.

ECOLOGICAL REQUIREMENTS

ALTITUDE – VEGETATION BELTS
From sea level up to 2 800 m in the Alps (Rameau *et al.*, 1993), but more abundant in the mountains; from hill to alpine belts.

SOIL MOISTURE
Optimum on dry to very dry soils (Figure 5.99).

SOIL FERTILITY
Only on soils very poor in nutrients, extremely acid to alkaline (Figures 5.99 and 5.100).

SOIL TYPE AND TOPOGRAPHY
Can be found on sandy soils or on very shallow stony soils originating from very diverse parental rocks.

CLIMATE REQUIREMENTS
Widespread in climates of temperate and Mediterranean areas. In high-rainfall areas, thrives in warm microclimates and/or excessively drained (very dry) soils. High

PLATE 5.42
Festuca ovina

PLATE 5.43
Festuca ovina

resistance to cold and drought. Does not tolerate soil wetness and shade.

TOLERANCE OF CUTTING AND GRAZING
Usually grazed extensively by sheep. Cutting is almost impossible because of its very low biomass production. Tolerates frequent mowings in lawns (almost 20 cuts/year) (Figure 5.100).

SOCIABILITY – COMPETITIVENESS – PLANT COMMUNITIES
Poorly competitive. Under difficult circumstances has adopted a strategy of survival by absorbing scarce water and nutrient resources and keeping these resources by a slow growth and a low tissue turnover. Low aerial part:below ground part ratio compared with fast-growing grasses.

Mostly frequent in *Mesobromion erecti, Koelerio-Phleion phleoidis, Festucion valesiacae, Agrostis-Festuca* grasslands, *Juncion squarrosi, Violo-Nardion strictae* and *Molinion caeruleae*.

COMPATIBILITY WITH OTHER FORAGE GRASSES AND LEGUMES
Incompatible with productive forage grasses.

INDICATOR VALUE
Dry to very dry, nutrient-poor soil.

AGRONOMIC CHARACTERISTICS

SOWING RATES
Pure: 30–40 kg/ha.
Mixed: 3–4 kg/ha. In Switzerland, *F. ovina* is included in complex mixtures adapted to mountain areas, particularly in dry situations. It is associated with *Festuca rubra, Agrostis capillaris, Poa pratensis, Cynosurus cristatus, Lotus corniculatus* and *Trifolium repens* (Mosimann *et al.*, 1996a).

VARIETIES
A list of varieties is shown in Table 5.42.

FIGURE 5.99
Ecological optimum and range for soil pH and humidity of *Festuca* group *ovina*

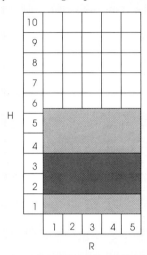

R: 1 = extremely acid and 5 = alkaline
H: 1 = extremely dry and
10 = permanently flooded

Complete key in Chapter 2, section 2.8.

FIGURE 5.100
Ecological optimum and range for nutrient availability and management of *Festuca* group *ovina*

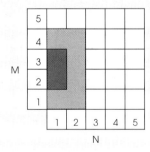

N: 1 = very low and 5 = very high
M: 1 = one defoliation per year and
5 = 5 defoliations per year or more

Complete key in Chapter 2, section 2.8.

Ornamental varieties: Amethystina, Blaufuchs, Grünling, Seeigel, Sohrewald, Soling, Superba.

FERTILIZATION
Low or nil.

DRY MATTER YIELDS AND SEASONALITY OF PRODUCTION
Very low productivity at the sites where it thrives but, cultivated in pure stand in favourable conditions, its yield can be surprisingly high.

In Switzerland, Lehmann *et al.* (1992) recorded a production higher than that from *Poa pratensis*. The average annual yield was 9.8 tonnes DM/ha with a maximum of 13.5 and a minimum of 8.1 tonnes DM/ha. In the Falkland Islands, in very difficult conditions, *F. ovina* (var. Novina) was the most productive of 13 tested grasses including *Lolium perenne, Dactylis glomerata, Phleum pratense, Holcus lanatus, Festuca rubra* and *F. arundinacea*. The average annual yield in a 3-year experiment was 4.1 tonnes DM/ha versus 3.6 tonnes DM/ha for *Festuca rubra* and 2.1 tonnes DM/ha for *Lolium perenne* (Davies and Riley, 1992). However, because of a low digestibility, the yield of DOM was one of the lowest and well below *Dactylis glomerata, Holcus lanatus, Festuca rubra* or *Agrostis capillaris*.

NUTRITIVE VALUE
Digestibility is particularly low, much lower than *Festuca rubra*. In the Falkland Islands, its digestibility (DOMD) was the lowest of 13 tested grasses: 54 percent versus 64 percent for *Festuca rubra* and 73.2 percent for *Lolium perenne* (Davies and Riley, 1992).

ACCEPTABILITY
Very low, although better accepted by sheep than by cattle.

ANIMAL PERFORMANCE
Very low.

DISEASES
In common with other *Festuca* spp., attacked by rusts: *Puccinia coronata, P. festucae* and *Uromyces dactylidis*. Also attacked by *Ustilago striiformis* (Ellis and Ellis, 1997).

USE FOR PURPOSES OTHER THAN FORAGE PRODUCTION
Included in the composition of fine lawns as a secondary component of the mixture with *Festuca rubra* and *Agrostis capillaris*. Can also contribute to stabilizing soils after building ski slopes in mountain areas. Diverse species of the *F. ovina* group are cultivated in gardens and parks, particularly bluish green to bluish forms that are used in rockeries or on the margin of mixed floral borders. Constituent of wild flora mixtures for dry, calcareous or acid soils (Crofts and Jefferson, 1999).

MAIN QUALITIES
Well adapted to dry rangelands, where it is grazed extensively by sheep.

MAIN SHORTCOMINGS
Extremely low nutritive value and even with a moderate yield under favourable conditions, only very low or insignificant animal performance is obtained.

TABLE 5.42
Examples of varieties of *Festuca ovina* used in lawn mixtures

Variety	Seed breeder (country)
Barfina	Barenbrug (NL)
Barok	Barenbrug (NL)
Barrespo	Barenbrug (NL)
Bornito	Barenbrug (NL)
Ridu	DLF (DK)
Spartan *	Pick Seed (USA)
Triana	DLF (DK)

* Spartan is the only recommended variety for forage production in Switzerland (Lehmann *et al.*, 1992)

FESTUCA PRATENSIS HUDS.

SYNONYM
F. elatior L. ssp. *pratensis* (Huds.) Hack, *F. elatior* auct. non L.

ETYMOLOGY
LATIN: *festuca* = straw, hay and wisp of straw, derived from CELTIC: fest = pasture. These species were often dominant in pastures.
LATIN: *pratensis* = from meadows.

COMMON NAMES
ENGLISH: Meadow fescue
FRENCH: Fétuque des prés
GERMAN: Wiesen schwingel

DISTRIBUTION
Native to Europe and western Asia.

FIGURE 5.101
Inflorescence and part of the leaf of *Festuca pratensis*

PLATE 5.44
Festuca pratensis

PLATE 5.45
Festuca pratensis

DESCRIPTION

MORPHOLOGICAL DESCRIPTION
Perennial plant, robust, hairless, caespitose.
Stems erect, 50–80 (–120) cm high. Leaf
blade rolled when young, large (2–5 mm)
(narrower than *Festuca arundinacea*, but
wider than *Lolium perenne*), with at least 16
distinct veins on the upper surface (less than
16 veins for *L. multiflorum*, which it
resembles morphologically at the vegetative
stage), shiny on the lower surface. Sheaths of
lower leaves purplish red. Ligule short.
Auricles well developed. Panicle-like
inflorescence, spreading during flowering,
contracted afterwards, long. Spikelets 5–11-
flowered (Figure 5.101) (Plates 5.44 and 5.45).

PHYSIOLOGICAL PECULIARITIES
Hybridizes with *Lolium perenne*: x *Festulolium
loliaceum* and *L. multiflorum*: x *Festulolium
braunii*.

1000-SEED WEIGHT
1.7–2.1 g (average-sized seeds).

CHROMOSOME NUMBER
2n = 14 (diploid).

ECOLOGICAL REQUIREMENTS

ALTITUDE – VEGETATION BELTS
From the lowlands to high altitudes in
mountain areas; from hill to alpine belts.

SOIL MOISTURE
Optimum on wet soils. Tolerates flooded
soils in winter and a certain degree of
drought in summer. Therefore is adapted to
water regime fluctuations. Often abundant
in wide alluvial valleys (Figure 5.102).

SOIL FERTILITY
Large range for nutrient availability,
but absent from very poor soils
(mesotrophic to eutrophic species).
Optimum on slightly acid to neutral soils,

FIGURE 5.102
Ecological optimum and range for soil pH and
humidity of *Festuca pratensis*

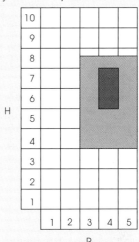

R: 1 = extremely acid and 5 = alkaline
H: 1 = extremely dry and 10 = permanently flooded

Complete key in Chapter 2, section 2.8.

FIGURE 5.103
Ecological optimum and range for nutrient
availability and management of *Festuca pratensis*

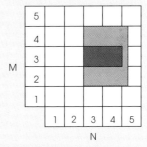

N: 1 = very low and 5 = very high
M: 1 = one defoliation per year and
5 = 5 defoliations per year or more

Complete key in Chapter 2, section 2.8.

but also frequent on alkaline soils (Figures
5.102 and 5.103).

SOIL TYPE AND TOPOGRAPHY
Broadly indifferent to soil texture. Well
represented on clay soils of large valleys.
Can be abundant on peat soils.

CLIMATE REQUIREMENTS
Large climate range. Well adapted to continental climates where it is an alternative to *Lolium perenne*. Resists cold very well. Tolerates temporary drought. Quite heat-resistant, though less than *Festuca arundinacea* or *Dactylis glomerata*.

TOLERANCE OF CUTTING AND GRAZING
Largely indifferent to the management regime but absent or very rare in intensively grazed grasslands. Optimum in alternate cutting-grazing systems. Resists trampling well. Poor persistence in exclusively cut temporary grasslands (Figure 5.103).

SOCIABILITY – COMPETITIVENESS – PLANT COMMUNITIES
Moderately competitive but is dominated by *Lolium perenne* and *Dactylis glomerata* in mixed sowings.
 Mostly frequent in *Cynosurion cristati, Arrhenatherion elatioris* and *Bromion racemosi*. Less frequent in *Mesobromion erecti*.

COMPATIBILITY WITH OTHER FORAGE GRASSES AND LEGUMES
Very compatible with many forage species because of its moderate competitiveness. Associates well with *Phleum pratense* in temporary cutting grassland in cold areas (mountains, Scandinavia). Can be incorporated in mixtures adapted to

moderately intensive systems, e.g. with *Festuca rubra, Agrostis capillaris, Poa pratensis*, as is carried out in Switzerland (Mosimann *et al.*, 1996a). In mixture with *Lolium perenne*, is strongly dominated by it, even though in practice this mixture is regularly used. Very compatible with legumes: *Trifolium repens, T. pratense, Medicago sativa*, sainfoin (*Onobrychis viciifolia*), but *Trifolium pratense* and *Medicago sativa* often dominate it in the mixtures. Ideal species with *Onobrychis viciifolia*.

INDICATOR VALUE
Often associated with an alternative water regime: rather dry in summer and wet in winter. Sward used for cutting-grazing or cut for hay.

AGRONOMIC CHARACTERISTICS

SOWING RATES
Pure: 20–30 kg/ha.
Mixed: 10–15 kg/ha.
• Simple mixtures with legumes: *Medicago sativa, Onobrychis viciifolia* or *Trifolium pratense*.
• Simple mixtures with *Phleum pratense* or *Lolium perenne*, or with these 2 species together, which can also include *Trifolium pratense* and *T. repens*.
• Complex mixtures. In Switzerland, *F. pratensis* is included in almost all mixtures except in long-term mixtures in areas favourable to *Lolium perenne*.

TABLE 5.43
Examples of varieties of *Festuca pratensis*

Variety	Seed breeder (country)
Bartran	Barenbrug (NL)
Cosmolit	Steinach (D)
Cosmos	Steinach (D)
Darimo	Mommersteeg (NL)
Merifest	DvP (B)
Pradel	RAC (CH)
Présent	RAC (CH)
Senu Pajbjerg	DLF (DK)
Stella	CEBECO (NL)

VARIETIES
A list of forage varieties is shown in Table 5.43.

FERTILIZATION
High levels needed to express its full potential, but yield is satisfactory even with moderate fertilization.

DRY MATTER YIELDS AND SEASONALITY OF PRODUCTION
Very productive though usually slightly less than other major grasses. In Western Europe, *F. pratensis* annual yield often ranges between 10–12 tonnes DM/ha in hay or silage production systems (2–4 cuts/year) with high N rates annually (200–300 kg N/ha). Production differences among varieties are lower than for other productive grasses (Toussaint and Lambert, 1973). *F. pratensis-Phleum pratense* associations constitute a balanced mixture, but the yield of the mixture is not necessarily higher than pure-sown *Phleum pratense* (Frame, Hunt and Harkess, 1971).

NUTRITIVE VALUE
Similar to *Lolium perenne*. The rate of digestibility reduction with advancing maturity is also similar, but its soluble carbohydrate content is slightly lower, though higher than for *Phleum pratense* or *Dactylis glomerata*. In mixture with *Phleum pratense* for hay or silage production, it is difficult to determine the optimum cutting period because the 2 species have very different heading dates, *F. pratensis* being much earlier than *Phleum pratense*. However, the *F. pratensis* heading date should be used as the guide, because the digestibility of *Phleum pratense* starts to fall even before heading. In mixture with early varieties of *Lolium perenne*, the choice of an ideal cutting date is easier because this type of *Lolium perenne* has heading dates similar to *F. pratensis*.

ACCEPTABILITY
Highly acceptable.

DISEASES
Puccinia coronata can affect *F. pratensis* though in

a less dramatic way than *Lolium perenne*. *Puccinia graminis* is also observed as well as *Drechslera* spp. which are frequently present, *Spermospora lolii, Rhynchosporium orthosporum, R. secalis, Blumeria (= Erysiphe) graminis* and *Claviceps purpurea* (Raynal *et al.*, 1989; Ellis and Ellis, 1997). Resistant to snow mould, more so than *Lolium perenne* and *Festuca arundinacea*, but less so than *Phleum pratense*. This disease, which appears under strong snow cover, is caused by a complex of fungi including *Microdochium (= Fusarium) nivale* and *Typhula incarnata*.

   Xanthomonas translucens pv. *graminis* can cause plant mortality.

USE FOR PURPOSES OTHER THAN FORAGE PRODUCTION
Constituent of wild flora mixtures for damp acid or neutral soils (Crofts and Jefferson, 1999). Sometimes used for sowing new ski pistes at moderate altitude. Can be sown in field margins.

MAIN QUALITIES
Productive species, well adapted to cold climates (mountain areas and northern regions), and to wet and cool soils, even flooded in winter. Better drought resistance than *Lolium perenne* and so better adapted to continental climates. Mainly used for hay and silage production and for mixed cutting-grazing management. Can be mixed with *Phleum pratense* or early *Lolium perenne* for silage, or with late *L. perenne* for cutting and grazing, being then considered as an insurance in case of failure of the main species. It is also a useful component of complex mixtures for rather extensive grazing in mountain areas. High acceptability and nutritive value.

MAIN SHORTCOMINGS
Lacks persistence. No specific quality that distinguishes it radically from *Phleum pratense* or *Lolium perenne*. Less persistent than *Phleum pratense* and *Lolium perenne* in cut mixtures. Less adapted to grazing than *Lolium perenne*. Plays a secondary role in mixtures.

FESTUCA RUBRA L.

There are many subspecies adapted to
several oligotrophic, mesotrophic and saline
environments. The subspecies *rubra* and
commutata are described here.

ETYMOLOGY
LATIN: *festuca* = straw, hay and wisp of straw,
derived from CELTIC: fest = pasture.
These species were often dominant in
pastures.
LATIN: *ruber* = red.
The panicle is reddish at a certain stage of
maturity.

COMMON NAMES
ENGLISH: Red fescue
FRENCH: Fétuque rouge
GERMAN: Roter schwingel

DISTRIBUTION
Native to Europe and/or temperate Asia

FIGURE 5.104
Inflorescence and part of the stem, rhizome
and leaf of *Festuca rubra*

PLATE 5.46
Festuca rubra

PLATE 5.47
Festuca rubra

according to the subspecies. Some subspecies are native to North America. Has become subcosmopolitan in temperate regions.

DESCRIPTION

MORPHOLOGICAL DESCRIPTION
Perennial plant, small-sized, hairless, caespitose or rhizomatous. Stems erect, 20–70 (–100) cm high. Needle-like leaf blade when young. Leaf blade long, narrow, dark green, in needle form at least for lower leaves. Upper leave blade flat, very veined. Ligule very short. Auricles reduced to minute excrescence. Panicle-like inflorescence, spreading, often reddish. Spikelets with 4–8 aristate flowers, 7–12 mm long. Upper glume pointed, lemma ciliate or aristate (Figure 5.104) (Plates 5.46 and 5.47).

PHYSIOLOGICAL PECULIARITIES
Maximum number of leaves per tiller: 5.0 (Sydes, 1984).

1000-SEED WEIGHT
0.9–1.1 g (average-sized seeds).

CHROMOSOME NUMBER
2n = 28 (tetraploid), 42 (hexaploid) or 56 (octoploid).
Other chromosome numbers have also been observed: 2n = 14 or 70.

ECOLOGICAL REQUIREMENTS

ALTITUDE – VEGETATION BELTS
From sea level (sand dunes, salt marshes, cliffs) up to almost 2 800 m in the Alps (Rameau *et al.*, 1993); from hill to alpine belts.

SOIL MOISTURE
Very large range (Figure 5.105).

SOIL FERTILITY
Optimum on moderately nutrient-rich soils (mesotrophic species). Disappears with high

FIGURE 5.105
Ecological optimum and range for soil pH and humidity of *Festuca* group *rubra*

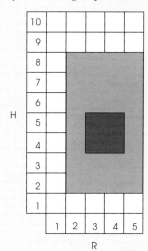

R: 1 = extremely acid and 5 = alkaline
H: 1 = extremely dry and
10 = permanently flooded

Complete key in Chapter 2, section 2.8.

FIGURE 5.106
Ecological optimum and range for nutrient availability and management of *Festuca* group *rubra*

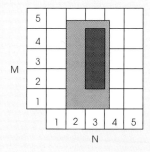

N: 1 = very low and 5 = very high
M: 1 = one defoliation per year and
5 = 5 defoliations per year or more

Complete key in Chapter 2, section 2.8.

fertilization (Figures 5.105 and 5.106). The subspecies *litoralis* is salt-tolerant.

SOIL TYPE AND TOPOGRAPHY
All soil types.

CLIMATE REQUIREMENTS
Large range of temperate climates. Good drought resistance. Tolerates cold very well.

TOLERANCE OF CUTTING AND GRAZING
Largely indifferent to the defoliation frequency. Can dominate the sward in hay meadows or in extensively grazed swards. Good resistance to trampling. Tolerates frequent mowings (almost 20 cuts/year) in lawns (Figure 5.106).

In Belgium (Figures 5.107 and 5.108), the % FM of *F. rubra* was measured in an experiment where it was mixed either with

Phleum pratense, or with *Lolium perenne*. *F. rubra* dominated *Lolium perenne* in all situations, except in the frequent cutting regime (5–6 cuts/year) and with the highest N rate (300 kg N/ha annually). In this cutting regime, its abundance decreased with increasing N rate. *F. rubra* dominance was the strongest in the 2 cuts/year regime. *Phleum pratense* was much more aggressive and was only dominated by *F. rubra* at nil N. In all other cases, the abundance of *F. rubra* was clearly reduced.

SOCIABILITY – COMPETITIVENESS – PLANT COMMUNITIES
Moderately competitive species. More aggressive than *Agrostis capillaris, Cynosurus*

FIGURES 5.107 AND 5.108
Proportion (% FM) of *Festuca rubra* in mixture with *Lolium perenne* or *Phleum pratense* in 3 cutting regimes and 4 levels of nitrogen fertilization, after 5 years of production

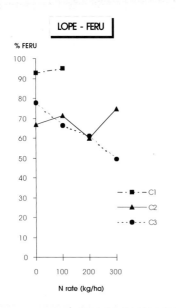 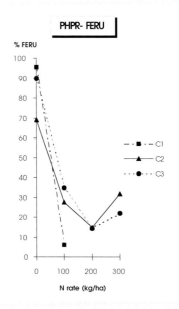

FERU: *F. rubra*; LOPE: *L. perenne*; PHPR: *P. pratense*. C1 = 2 cuts/year; C2 = 4 cuts/year; C3 = 5–6 cuts/year

Source: Peeters. Decamps and Janssens, 1999

cristatus or *Trisetum flavescens* when conditions are optimum for it. Can be dominant or abundant in extensively managed species-rich swards.

Mostly frequent in *Agrostis-Festuca* grasslands, *Cynosurion cristati*, *Arrhenatherion elatioris*, *Polygono-Trisetion flavescentis*, *Nardion strictae*, *Violo-Nardion strictae*, *Molinion caeruleae* and *Armerion maritimae* (*F. rubra* ssp. *litoralis*). Less frequent in intensive grasslands.

COMPATIBILITY WITH OTHER FORAGE GRASSES AND LEGUMES
In Switzerland, often included in mixtures for dry soils, for mountain areas or even for average soil fertility and extensive use

(Mosimann *et al.*, 1996a). For dry soils, it is mixed with *Dactylis glomerata*, *Poa pratensis*, *Lolium perenne* (early varieties) and *Trifolium repens*. In mountain areas, the mixtures are more complex and include the same species plus *Lotus corniculatus*, *Festuca pratensis*, *F. ovina* (for grazing), *Cynosurus cristatus* (for grazing), *Phleum pratense*, *Trisetum flavescens* (for cutting) and *Alopecurus pratensis* (for cutting). The mixtures suitable for extensive use include *Festuca pratensis*, *Lolium perenne*, *Phleum pratense*, *Trisetum flavescens*, *Trifolium repens* and *T. pratense*.

INDICATOR VALUE
Not very intensive management.

TABLE 5.44

Example of varieties of *Festuca rubra*

Variety	Seed breeder (country)
Barcrown	Barenbrug (NL)
Barpusta	Barenbrug (NL)
Boreal	Canada Dpt Agric. (CAN)
Cascade	AES, Oregon (USA)
Echo	Prodana (DK)
Reverent	KWS (D)
Roland	Steinach (D)
Rubina	DLF (DK)
Runo	ZNRO (PL)
S59	WPBS (UK)

TABLE 5.45

Annual yields (tonnes DM/ha) of *Festuca rubra* compared with *Lolium perenne* in Scotland, UK

Species/variety	Annual N fertilization (kg/ha)				
	0	120	240	360	480
6 cuts/year					
3-year average					
L. perenne	2.4	6.1	9.7	11.9	13.2
F. rubra					
S59	3.3	6.4	9.0	11.0	10.9
Boreal	2.7	6.3	9.1	10.4	11.4
3 cuts/year					
1-year experiment					
L. perenne	3.8	6.0	10.8	14.7	15.8
F. rubra	4.6	7.1	11.7	13.8	14.3

Source: Frame, 1991

AGRONOMIC CHARACTERISTICS

SOWING RATES

Pure: 30–40 kg/ha.

Mixed: 10–25 kg in simple mixture with *Trifolium repens* and one or several grasses including *Lolium perenne, Agrostis capillaris* and *Poa pratensis;* 3–6 kg/ha in complex mixtures adapted to areas unfavourable to *Lolium perenne* (Lehmann *et al.,* 1992).

VARIETIES

A list of varieties is shown in Table 5.44.

FERTILIZATION

Moderate requirements for fertilizers; good response to fertilizer N.

DRY MATTER YIELDS AND SEASONALITY OF PRODUCTION

High yields that reach those of *Lolium perenne* or *Holcus lanatus*. Its dry matter yield often exceeds those of *Lolium perenne* at low to moderate N rates.

In Scotland (UK), *F. rubra* yields were higher than for *Lolium perenne* up to about 150 kg N/ha. Above this rate, *Lolium perenne* was more productive (Table 5.45) in a 6 cuts/year regime. In a 3 cuts/year regime, the best *F. rubra* variety produced more than *Lolium perenne* up to 240 kg N/ha.

In an experiment in Belgium, *F. rubra* produced more than *Lolium perenne* in every cutting regime (2, 4 and 6 cuts/year) and at all N rates (0, 100, 200 and 300 kg N/ha). Its yield was often similar to that of *Phleum pratense* (Figures 5.109 to 5.112).

In association with *Trifolium repens*, without N application and in a 6 cuts/year regime, *F. rubra* was more productive than *Lolium perenne* and 5 other grasses, including *Holcus lanatus* (Frame, 1990). In a 4 cuts/year regime with 360 kg N/ha, *F. rubra* had the highest dry matter yields of 7 grasses in the trial, including *Lolium perenne* and *Holcus lanatus* (Frame, 1989). However, because of

the low OMD of *F. rubra*, the *Lolium perenne* DOM yields were higher.

In a frequent cutting regime (1 cut/month), *Lolium perenne* produced more (11.0 tonnes DM/ha) than *F. rubra* (10.0 tonnes DM/ha) (Haggar, 1976) because the spring growth of *F. rubra* was clearly lower than *L. perenne*, though the summer growth of *F. rubra* was slightly higher. In a less frequent cutting regime (3 cuts/year), *F. rubra* accumulated more biomass than *Lolium perenne* during an uninterrupted growth period in spring, and the summer yield was also higher (Table 5.46).

It can be concluded that *F. rubra* only expresses its full potential if it is not defoliated too early in spring. This probably explains its disappearance from intensively grazed grasslands.

In grazed swards, *F. rubra* gives a very stable productivity throughout the year (about 70 kg DM/ha.day with 400 kg N/ha) due to its good summer growth (Haggar, 1976).

NUTRITIVE VALUE

Not highly digestible and much less so than other productive grasses such as *Lolium perenne* (Table 5.47).

The rate of digestibility reduction with advancing maturity is faster than *Lolium perenne, Poa trivialis, Agrostis stolonifera* and *Holcus lanatus*, especially at the beginning of growth in spring (Haggar, 1976).

At the same N rate, *F. rubra* has a higher N content than *Lolium perenne* (Haggar, 1976; Frame, 1991) but is poorer in minerals, especially in Ca (Frame, 1991).

ACCEPTABILITY

Acknowledged to be poorly accepted by cattle, but better ingested by sheep. In mountain areas, it constitutes one of the best forage species. Its nutritive value, acceptability and production are much higher than *Nardus stricta* for instance. Well accepted by deer when sown in forest clearings.

FIGURES 5.109–5.112
Annual yields of *Festuca rubra* compared with *Lolium perenne* and *Phleum pratense* in 3 cutting
regimes and several nitrogen fertilization levels (average of 3 production years)

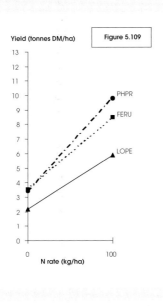

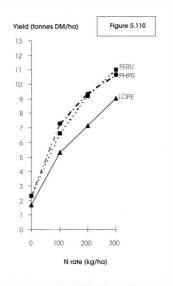

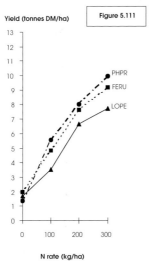

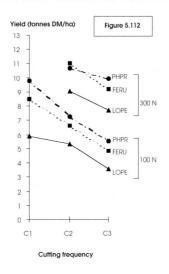

FERU: *F. rubra;* LOPE: *L. perenne;* PHPR: *P. pratense.* Fig. 5.109: 2 cuts/year (C1); Fig. 5.110: 4 cuts/year (C2);
Fig. 5.111: 5–6 cuts/year (C3); Fig. 5.112: 100 and 300 kg N/ha

Source: Peeters and Decamps, 1999

TABLE 5.46

Yields (kg DM/ha) during an uninterrupted growth period, 2 regrowths and total annual yield of *Festuca rubra* compared with *Lolium perenne*

	1st growth cycle						Regrowths		Total
	27 March	10 April	24 April	8 May	22 May	5 June	1 August	5 Sept	
L. perenne	20	60	220	608	1 413	5 148	1 950	695	7 693
F. rubra	59	276	651	2 150	3 814	6 863	2 189	1 390	10 442

Note: annual N fertilization: 120 kg/ha
Source: Haggar, 1976

TABLE 5.47

Comparison of the digestibility of *Festuca rubra* and *Lolium perenne* according to several authors

Type of measurement	Digestibility (%)		References
	L. perenne	*F. rubra*	
D-value	65.4	61.2	Haggar (1976), monthly cuts, 1-year average
OMD	76.5	67.5	Frame (1989), 4 cuts/year, 3-year average
OMD	79.6	72.4	Frame (1991), 6 cuts/year, 2-year average

ANIMAL PERFORMANCE

Animal performance from *F. rubra*-dominated swards is very variable. For example, in Scotland (UK), an *Agrostis/Festuca* community gave 2 250 grazing days/year for sheep of about 50 kg LW and a sheep LWG of about 54 g/day; this represented a surplus of 17 percent of sheep grazing days relative to a *Molinia* community and 33 percent compared with a *Nardus* community (Common *et al.*, 1991b).

DISEASES

Attacked by rusts: *Puccinia graminis, P. festucae* and *Uromyces dactylidis*. It is also attacked by *Ustilago striiformis* (Ellis and Ellis, 1997).

USE FOR PURPOSES OTHER THAN FORAGE PRODUCTION

Widely used for fine lawns in gardens and parks. It is not widely used in sport fields and well trampled lawns because it does not tolerate trampling as well as *Lolium perenne* turf varieties. Associated in lawn mixtures with *F. ovina, Agrostis capillaris* and *Poa pratensis*. Can also contribute to stabilizing soils after building ski slopes in mountain areas, on roadside or motorway verges and on industrial or polluted soils. Can be sown in set-asides and forest clearings. Constituent of wild flora mixtures for calcareous and damp acid or neutral soils (Crofts and Jefferson, 1999). In Scandinavia, covers wood roofs. Globally, Canada and Denmark are the 2 main producers of seed for amenity purposes.

MAIN QUALITIES

Productive, very persistent and tolerant of a wide range of ecological conditions. Compared with other species, it produces well on soils of poor to moderate fertility. Its summer growth ability means it can sustain production when many other species have low productivity. Compatible with *Trifolium repens*. Not very susceptible to disease. In the mountains, outside areas favourable to *Lolium perenne*, the *Festuca rubra-Agrostis capillaris* association constitutes the best vegetation in grazed swards.

MAIN SHORTCOMINGS

Low digestibility and lower intake by stock compared with *Lolium perenne*. Poorer in minerals than many other grasses. Intolerant of intensive grazing.

GLYCERIA FLUITANS (L.) R. BROWN

ETYMOLOGY
GREEK: γλυκερος = glyceros = sweet.
In Poland, the grain is used to make flour
and to prepare a milk soup (a kind of
porridge).
LATIN: *fluitans* = floating.
The leaves of this species can float on the
surface of ponds and rivers.

COMMON NAMES
ENGLISH: Floating sweet-grass
FRENCH: Glycérie flottante
GERMAN: Flutender schwaden

PLATE 5.48
Glyceria fluitans

FIGURE 5.113
Inflorescence and part of the stem, rhizome and leaf of *Glyceria fluitans* (a)
and *Glyceria maxima* (b)

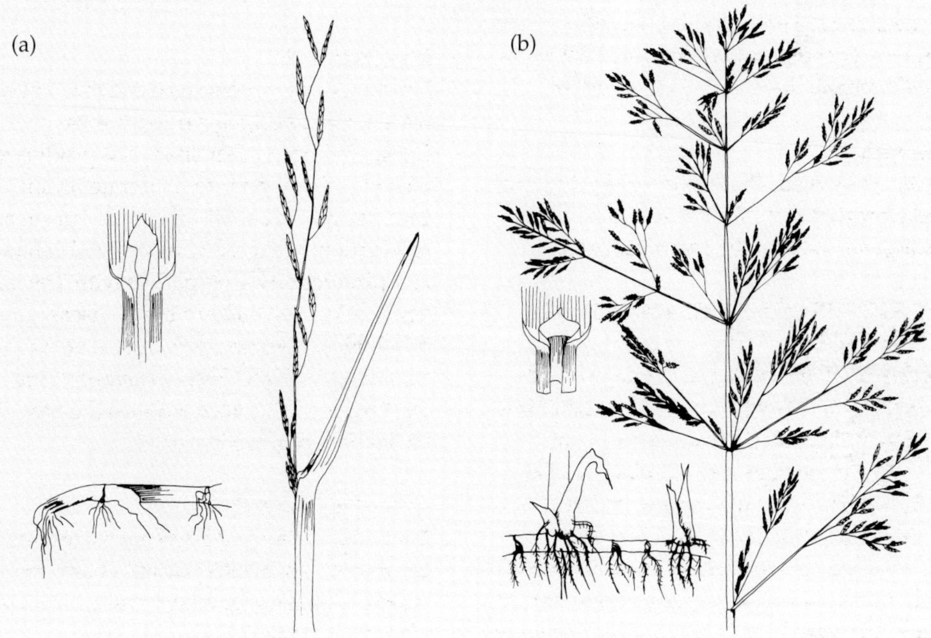

(a)

(b)

GLYCERIA MAXIMA (HARTM.) HOLMBERG

SYNONYM
G. spectabilis Mert et Koch, *G. aquatica* (L.) Wahlemb.

ETYMOLOGY
GREEK: γλυκερος = glyceros = sweet.
In Poland, the grain is used to make flour and to prepare a milk soup (a kind of porridge).
LATIN: *maximus* = the largest.
This species is very tall.

COMMON NAMES
ENGLISH: Reed sweet-grass
FRENCH: Glycérie aquatique
GERMAN: Wasser schwaden

DISTRIBUTION
Glyceria fluitans: Native to Europe, western Asia and eastern North America.

Glyceria maxima: Native to Europe, except the southwest and to median Asia.

DESCRIPTION

MORPHOLOGICAL DESCRIPTION
Perennial plants, medium (*G. fluitans*) or large-sized (*G. maxima*), hairless, caespitose (*G. maxima*), or stoloniferous (*G. fluitans*). Stems erect (*G. maxima*) or horizontal then ascendent (*G. fluitans*), 100–250 cm high (*G. maxima*) or 40–120 cm high (*G. fluitans*). Leaf blade folded when young, with reticulate veins (presence of short veins perpendicular to and joining the long main veins), large (5–10 mm in *G. fluitans*) (10–18 mm in *G. maxima*), flat, a little rough. Ligule short (3–6 mm), truncate and pointed in the middle (*G. maxima*), or long (5–15 mm) and acute or rounded on the top (*G. fluitans*). No auricles. Panicle-like inflorescence, loose, spreading, contracted after flowering (*G. fluitans*). Spikelets 4–9-flowered, 6–12 mm long, compressed on the side (*G. maxima*),

PLATE 5.49
Glyceria fluitans

PLATE 5.50
Glyceria maxima

PLATE 5.51
Glyceria maxima

or spikelets 5–14-flowered, 18–32 mm long, cylindrical or slightly compressed (*G. fluitans*) (Figure 5.113) (Plates 5.48 to 5.51).

1000-SEED WEIGHT
About 1.2 g (average-sized seeds).

CHROMOSOME NUMBER
G. fluitans: 2n = 40 (tetraploid).
G. maxima: 2n = 60 (hexaploid).

ECOLOGICAL REQUIREMENTS

ALTITUDE – VEGETATION BELTS
From the lowlands to high elevations in mountain areas; from hill to mountain belts.

SOIL MOISTURE
Optimum on very wet or boggy soils and on riversides. *G. fluitans* can cover the surface of ponds or rivers with its floating leaves (Figures 5.114 and 5.115).

SOIL FERTILITY
Optimum on nutrient-rich alluvial, or peat soils, but tolerates large range of soil pH including acid to very acid soils (Figures 5.114 to 5.117).

SOIL TYPE AND TOPOGRAPHY
Mainly associated with peat soils.

CLIMATE REQUIREMENTS
Large climate range.

TOLERANCE OF CUTTING AND GRAZING
G. fluitans tolerates grazing very well. *G. maxima* is adapted to cutting (Figures 5.116 and 5.117).

SOCIABILITY – COMPETITIVENESS – PLANT COMMUNITIES
Very competitive in optimum conditions. Both species can develop into large, almost pure populations.

FIGURES 5.114 AND 5.115
Ecological optimum and range for soil pH and humidity of *Glyceria fluitans* and *Glyceria maxima*

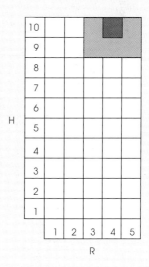

R: 1 = extremely acid and 5 = alkaline
H: 1 = extremely dry and 10 = permanently flooded

Complete key in Chapter 2, section 2.8.

FIGURES 5.116 AND 5.117
Ecological optimum and range for nutrient availability and management of *Glyceria fluitans* and
Glyceria maxima

Glyceria fluitans

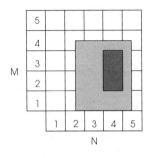

Glyceria maxima

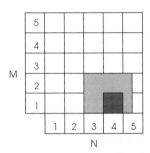

N: 1 = very low and 5 = very high
M: 1 = one defoliation per year and 5 = 5 defoliations per year or more

Complete key in Chapter 2, section 2.8.

Glyceria fluitans
Mostly frequent in *Filipendulion ulmariae,
Bromion racemosi,* humid improved
grasslands, *Phragmition australis* and
Sparganio-Glycerion maximae.

Glyceria maxima
Mostly frequent in *Filipendulion ulmariae* and
Phragmition australis.

*COMPATIBILITY WITH OTHER FORAGE
GRASSES AND LEGUMES*
G. fluitans is often associated with *Holcus
lanatus, Poa trivialis* and *Agrostis stolonifera.*
G. maxima is often found with *Phalaris
arundinacea, Phragmites australis* and
bulrushes (*Typha* spp.).

INDICATOR VALUE
Glyceria fluitans
Very wet soil, quite rich or rich
in nutrients.

Glyceria maxima
Very wet soil, quite rich or rich in nutrients.
Sward never grazed or grazed with a very
low stocking rate.

AGRONOMIC CHARACTERISTICS

VARIETIES
Ornamental variety: *G. maxima*: Variegata.

FERTILIZATION
The communities in which the 2 *Glyceria*
species grow are generally not fertilized,
though some *G. fluitans* grasslands do
receive fertilization.

*DRY MATTER YIELDS AND SEASONALITY OF
PRODUCTION*
Not usually measured but likely to be
moderate to high.

NUTRITIVE VALUE
G. fluitans is very digestible. *G. maxima* is
much higher in fibre.

ACCEPTABILITY
G. fluitans is well accepted by grazing animals
and the hay is also well ingested. *G. maxima*
hay is eaten if cut at a young stage of maturity.

TOXICITY
G. maxima is a cyanogenic species containing

high levels of hydrocyanic acid (HCN) (Faliu, 1981).

DISEASES
G. fluitans is attacked by *Puccinia brachypodii* var. *nemoralis*, while *G. maxima* is attacked by *P. coronata*. *Ustilago longissima* is very frequent on *G. maxima*, but can also be found on *G. fluitans* (Ellis and Ellis, 1997).

USE FOR PURPOSES OTHER THAN FORAGE PRODUCTION
G. maxima is regularly used for ornamental purposes on pond margins. It develops into large tufts topped by many branched panicles which contrast with the robust stem. The seeds of the 2 species were consumed in Poland as a human food.

MAIN QUALITIES
Both species are well adapted to boggy grasslands where they produce a yield that few other species could equal in this difficult environment. The forage nutritive value of *G. fluitans* is acceptable and it is also well ingested by grazing animals.

MAIN SHORTCOMINGS
G. maxima is very fibrous and can only produce acceptable hay if it is mown at a young growth stage. The sites of the 2 species correspond to the development area of a snail (*Lymnea* spp.) that is intermediary host of liver fluke (*Fasciola hepatica*) and so it is better to remove *Glyceria* spp. sites from the grazed area.

Holcus lanatus L.

FIGURE 5.118
Inflorescence and part of the stem and leaf
of *Holcus lanatus*

ETYMOLOGY

Greek: ολκος = holcos derived from ελκω = elco = I pull out, I take out.
The plant was used for binding fingers for taking out spines.
Latin: lanatus = covered by wool, woolly. This grass is covered by soft hairs.

COMMON NAMES

English: Yorkshire fog
French: Houlque laineuse, houlque velue
German: Wolliges honiggras

DISTRIBUTION

Native to Europe, western and eastern Asia, Northern Africa and North America. Has become subcosmopolitan especially in temperate regions.

PLATE 5.52
Holcus lanatus

PLATE 5.53
Holcus lanatus

PLATE 5.54
Holcus lanatus

DESCRIPTION

MORPHOLOGICAL DESCRIPTION
Perennial plant, medium-sized, densely and softly hairy, caespitose. Stems erect, 20–80 cm high. Leaf blade rolled when young, large, floppy with dense but short and soft hairs, yellowish green. Sheath of lower leaves striated by purplish red vertical lines. Ligule short, hairy, truncated. No auricles. Inflorescence in panicle, dense or spreading, of variable colour (whitish, pinkish white, reddish), 10–20 cm long. Spikelets 2-flowered, lower flower fertile, upper only male, 4–5 mm long (Figure 5.118) (Plates 5.52 to 5.54).

1000-SEED WEIGHT
0.3–0.4 g (small seeds).

CHROMOSOME NUMBER
2n = 14 (diploid).

BIOLOGICAL FLORA
Beddows (1961).

ECOLOGICAL REQUIREMENTS

ALTITUDE – VEGETATION BELTS
From the lowlands up to almost 1 500 m in the Alps (Rameau *et al.*, 1993); from hill to mountain belts and even to the base of the subalpine belt.

SOIL MOISTURE
Large range for soil moisture. Can develop important populations on dry sands, but optimum on cool and moist soils, present even on boggy soil (Figure 5.119).

SOIL FERTILITY
Large range for nutrient availability. Can grow on poor soils, but produces much more on nutrient-rich soils. Large range also for pH, but optimum on acid soils (Figures 5.119 and 5.120).

SOIL TYPE AND TOPOGRAPHY
Thrives on all textures. Frequent on peaty soils.

FIGURE 5.119
Ecological optimum and range for soil pH and humidity of *Holcus lanatus*

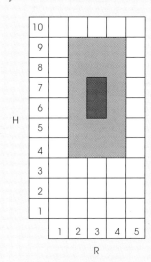

R: 1 = extremely acid and 5 = alkaline
H: 1 = extremely dry and 10 = permanently flooded

Complete key in Chapter 2, section 2.8.

FIGURE 5.120
Ecological optimum and range for nutrient availability and management of *Holcus lanatus*

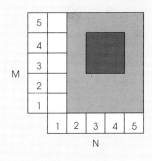

N: 1 = very low and 5 = very high
M: 1 = one defoliation per year and
5 = 5 defoliations per year or more

Complete key in Chapter 2, section 2.8.

CLIMATE REQUIREMENTS
Few specific climate requirements, but particularly abundant in oceanic climates. Can suffer from frost in winter.

TOLERANCE OF CUTTING AND GRAZING
Not very sensitive to management regimes (Figure 5.120). Tolerates trampling and frequent cuts, but a strong mortality is often observed after a late-season cut. However, it redevelops rapidly by producing new seedlings. Particularly abundant in mixed cutting-grazing systems. *H. lanatus* has an earlier growth than *Lolium perenne* in spring and thus *L. perenne* is often suppressed when the first defoliation is late, for instance in wet grasslands where grazing has to be delayed due to a lack of soil firmness. Close grazing and frequent grazing rotations (6–7 per year) or intensive continuous grazing encourage the growth of *Lolium perenne* at the expense of *H. lanatus.*

Watt and Haggar (1980) showed that *H. lanatus* dominated *Lolium perenne* in all their cutting treatments, except the frequent cutting regime with the low cutting height. When *H. lanatus* was mixed with *Phleum pratense* or with *Lolium perenne* in a Belgian experiment, it dominated *L. perenne* in all cutting regimes and at all N rates, though this dominance decreased when the cutting frequency increased. *H. lanatus* was dominated much more by *Phleum pratense*, especially with high N rates (Figures 5.121 and 5.122).

SOCIABILITY – COMPETITIVENESS – PLANT COMMUNITIES
Competitive. Can form dense populations.

FIGURES 5.121 AND 5.122
Proportion (% FM) of *Holcus lanatus* in mixture with *Lolium perenne* or *Phleum pratense* in 3 cutting regimes and 4 nitrogen fertilizer levels, after 5 years of production

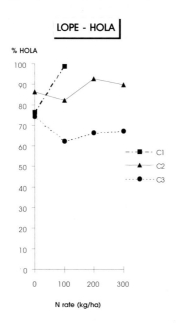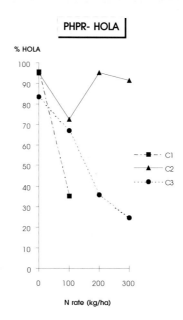

HOLA: *H. lanatus;* LOPE: *L. perenne;* PHPR: *P. pratense.* C1 = 2 cuts/year; C2 = 4 cuts/year; C3 = 5–6 cuts/year

Source: Peeters, Decamps and Janssens, 1999

One of the most widespread species in temperate grasslands.

Mostly frequent in *Cynosurion cristati*, intensive grasslands, humid improved grasslands, *Arrhenatherion elatioris*, *Bromion racemosi*, *Molinion caeruleae* and *Juncion acutiflori*.

COMPATIBILITY WITH OTHER FORAGE GRASSES AND LEGUMES
Under grazing, compatible with *Lolium perenne* in good grasslands and with *Festuca rubra* and *Agrostis* species in poorer grasslands.

AGRONOMIC CHARACTERISTICS

SOWING RATES
Pure: 20–30 kg/ha.
Mixed: 10–15 kg/ha with *Lolium perenne, Poa pratensis, Agrostis capillaris, Festuca rubra, Trifolium repens*, or only *Trifolium repens.*

H. lanatus is rarely sown in practice, partly because of a lack of commercial seed. Could be included in grazing mixtures, on poor quality soils, e.g. nutrient-poor, damp, as has already been done in New Zealand.

VARIETIES
Few selected varieties: Massey Basyn, Melita and Forester (NZ), Hola (CZ).

FERTILIZATION
Needs moderate to high fertilization to reach its full potential.

DRY MATTER YIELDS AND SEASONALITY OF PRODUCTION
Average to high production, often higher than for *Lolium perenne*, especially if fertilization is low.

In an experiment in Scotland (UK), *H. lanatus* yields were higher than *Lolium perenne* up to an N rate of about 150 kg N/ha. Above this rate, *Lolium perenne* was more productive (Table 5.48).

In an experiment in Belgium, *H. lanatus* produced more than *Lolium perenne* in every cutting regime (2, 4 and 6 cuts/year) and at all N rates (0, 100, 200 and 300 kg N/ha). Its yield was frequently on a par with that of *Phleum pratense* (Figures 5.123 to 5.126).

Conversely, in a cutting experiment (5–6 cuts/year) over 4 years and with 3 fertilizer levels (150, 300 and 400 kg N/ha), *Lolium perenne* had a 14 percent superiority in DM yield (Harvey *et al.*, 1984).

Several authors showed that *H. lanatus* produced as much or more than *Lolium perenne* at low N rates and infrequent cutting regimes (Haggar, 1976; Hayes, 1979; Haggar and Standell, 1982; Watt, 1987). Some experiments have shown that *H. lanatus* is even superior at a high rate of 400 kg N/ha (Haggar, 1976), or that the yields of the 2 species are equivalent at 130 or 300 kg N/ha (Anon., 1979; Smith and Allcock, 1985). In a grazing experiment with sheep, *Lolium perenne* production was higher than *H. lanatus* in the first year after sowing, but in the second year the difference was narrowed down.

It is possible that differences among the results of these trials are due to the effects of different cutting frequency and severity. Watt and Haggar (1980) showed that the DM yields of *H. lanatus* decrease faster than those of *Lolium perenne* when cutting frequency increases or when cutting height

TABLE 5.48

Annual yields (tonnes DM/ha) of *Holcus lanatus* compared with *Lolium perenne* in Scotland, UK

Species	Annual N fertilization (kg/ha)				
	0	120	240	360	480
L. perenne	2.4	6.1	9.7	11.9	13.2
H. lanatus	3.7	6.7	8.8	10.6	11.5

Note: 6 cuts/year regime
Source: Frame, 1991

FIGURES 5.123–5.126

Annual yields of *Holcus lanatus* compared with *Lolium perenne* and *Phleum pratense* in 3 cutting regimes and different nitrogen fertilizer levels (average of 3 production years)

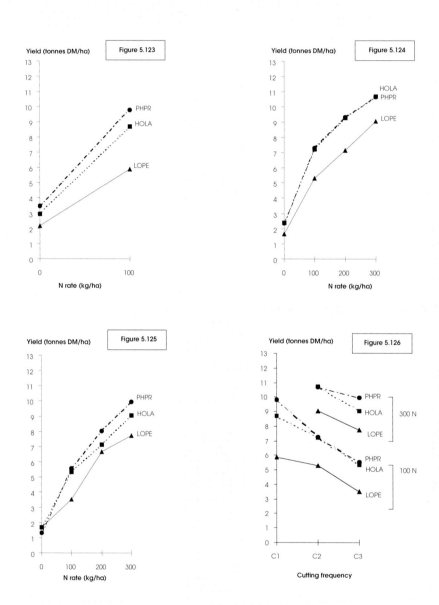

HOLA: *H. lanatus;* LOPE: *L. perenne;* PHPR: *P. pratense.* Fig. 5.123: 2 cuts/year (C1); Fig. 5.124: 4 cuts/year (C2); Fig. 5.125: 5–6 cuts/year (C3); Fig. 5.126: 100 and 300 kg N/ha

Source: Peeters and Decamps, 1999

decreases. While *H. lanatus* growth starts earlier than *Lolium perenne* in spring, the growth rate is then similar, but the summer growth of *H. lanatus* is greater and so *H. lanatus* has a less marked seasonality of annual production (Haggar, 1976).

The growth earliness of *H. lanatus* is illustrated in Table 5.49. A faster biomass accumulation in spring could explain its superiority over *Lolium perenne* in mixtures when the first defoliation is delayed.

NUTRITIVE VALUE

The pattern of digestibility (DOMD) during the first growth cycle is parallel to that of *Lolium perenne*, but constantly inferior by 3–4 percent since the growth of *H. lanatus* is faster than *L. perenne* in spring (Haggar, 1976). This implies that *H. lanatus* must be harvested at an earlier date than *Lolium perenne*. In frequent cutting systems (5–6 cuts/year), the digestibility of *H. lanatus* was not lower than *Lolium perenne* (Watt, 1987; Frame, 1991), and was even 4.6 percent higher in one trial (Haggar, 1976). In infrequent cutting systems (3 cuts/year), *H. lanatus* was less digestible, at least for certain cuts, by about 3–4 percentage units of digestibility (Harvey, Crothers and Hayes, 1984; Frame, 1991).

It is likely that a faster lignification rate of the stems of *H. lanatus* causes the digestibility differences. This could also explain why there is no digestibility difference between the 2 species with frequent defoliations because these prevent stem formation.

Nitrogen and mineral (P, K, Mg) contents are often higher in *H. lanatus* than in *Lolium perenne* (Frame, 1991; Harvey, Crothers and Hayes, 1984).

ACCEPTABILITY

Has the reputation of not being very acceptable due to the hairiness of the leaves and stems, but Watt (1978) attributes stem lignification as the reason. Several observations have shown that at a leafy stage, *H. lanatus* is well accepted by cattle and sheep, and this tends to support the contention that leaf hairiness is less important than stem quality.

Ingestion is adversely affected by the presence of rust disease, though frequent use of the sward decreases the rate of infection by this fungus. The use of the New Zealand cultivar, Massey Basyn, which is much more rust-resistant than most of the ecotypes available, is warranted when sowing out.

ANIMAL PERFORMANCE

Animal performances from *H. lanatus* and *Lolium perenne* are very close if the swards are frequently and closely grazed. Similar LWGs were obtained with sheep in New Zealand in a trial with var. Massey Basyn versus *Lolium perenne* (Watkin and Robinson, 1974). In Scotland (UK), Morton, Bolton and Hodgson (1992) obtained identical LWGs per animal with the 2 species, but the number of sheep grazing days was higher by 15 percent on *Lolium perenne* plots in the

TABLE 5.49

Yields (kg DM/ha) over an uninterrupted growth period, 2 regrowths, and total annual yield of *Holcus lanatus* compared to *Lolium perenne*

	1st growth cycle						Regrowths		Total
	27 March	10 April	24 April	8 May	22 May	5 June	1 August	5 Sept	
L. perenne	20	60	220	608	1 413	5 148	1 950	695	7 693
H. lanatus	245	480	941	2 047	3 386	7 049	2 858	1 004	10 911

Note: annual N fertilization: 120 kg/ha
Source: Haggar, 1976

first year of the experiment. In the second year, the superiority (7 percent) remained but it was no longer statistically significant.

DISEASES
H. lanatus is often heavily invaded by rusts: *Puccinia coronata* and *P. holcina*. A smut (*Tilletia holci*) is frequent as well as *Blumeria* (= *Erysiphe*) *graminis*, while *Drechslera* spp. are also observed. In wet sites, *Ramularia holci-lanati* can occur in spring. *H. lanatus* and *Holcus mollis* are hosts of *Colletotrichum holci* (Ellis and Ellis, 1997).

USE FOR PURPOSES OTHER THAN FORAGE PRODUCTION
Can be used to stabilize poor soils and north-facing slopes. However, should be mixed with *Agrostis capillaris* and *Festuca rubra* rather than sown pure.

MAIN QUALITIES
Has good production especially in summer. At low N fertilization, it is almost always superior to *Lolium perenne*. Suitable for semi-intensive systems for beef production. Adapted to a large range of soils, even rather poor and humid types.

MAIN SHORTCOMINGS
Badly accepted by animals if it is not frequently grazed and if attacked by rust. Frost-sensitive and the sward can be partially destroyed during cold winters, especially after late and high N application in autumn. Herbage difficult to wilt when cut for conservation.

HOLCUS MOLLIS L.

FIGURE 5.127
Inflorescence and part of the stem, rhizome
and leaf of *Holcus mollis*

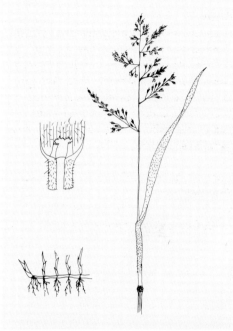

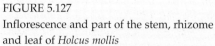

ETYMOLOGY
GREEK: ολκος = holcos derived from
ελκω = elco = I pull out, I take out.
The plant was used for binding fingers in
order to take out spines.
LATIN: *mollis* = flexible, soft.
This grass is covered by soft hairs.

COMMON NAMES
ENGLISH: Creeping soft-grass
FRENCH: Houlque molle
GERMAN: Weiches honiggras

DISTRIBUTION
Native to Western and median Europe.

DESCRIPTION
MORPHOLOGICAL DESCRIPTION
Perennial plant, medium-sized, hairy to
glabrescent, with thin, creeping rhizomes.

PLATE 5.55
Holcus mollis

PLATE 5.56
Holcus mollis

A. PEETERS

A. PEETERS

Stems erect, 20–80 cm high. Leaf blade rolled when young, wide, with a dense pilosity on the upper part, hairless below, greyish green. Nodes densely hairy, much more hairy than the sheath. Ligule long (up to 5 mm), oval-oblong. No auricles. Panicle-like inflorescence, contracted or spreading, reddish white to pink, 4–10 cm long. Spikelets 2-flowered, the lower flower fertile, the upper only male, 5–6 mm long (Figure 5.127) (Plates 5.55 and 5.56).

CHROMOSOME NUMBER
2n = 14 (diploid), 28 (tetraploid), 35 (pentaploid) or 42 (hexaploid).
Another chromosome number has also been observed: 2n = 49.

BIOLOGICAL FLORA
Ovington and Scurfield (1956).

ECOLOGICAL REQUIREMENTS

ALTITUDE – VEGETATION BELTS
From the lowlands up to 1 500 m in the Alps (Rameau *et al.*, 1993); from hill to mountain belts and even to the base of the subalpine belt.

SOIL MOISTURE
Optimum on well drained soils (Figure 5.128).

SOIL FERTILITY
Optimum on nutrient-poor soils and very to extremely acid soils (Figures 5.128 and 5.129).

SOIL TYPE AND TOPOGRAPHY
Optimum on siliceous soils.

CLIMATE REQUIREMENTS
Few specific requirements. Tolerates shade.

TOLERANCE OF CUTTING AND GRAZING
Thrives in cutting grassland and extensively grazed swards (Figure 5.129).

SOCIABILITY – COMPETITIVENESS – PLANT COMMUNITIES
Not very competitive, but can constitute

FIGURE 5.128
Ecological optimum and range for soil pH and humidity of *Holcus mollis*

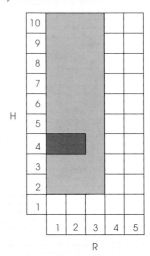

R: 1 = extremely acid and 5 = alkaline
H: 1 = extremely dry and
10 = permanently flooded

Complete key in Chapter 2, section 2.8.

FIGURE 5.129
Ecological optimum and range for nutrient availability and management of *Holcus mollis*

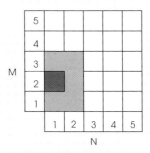

N: 1 = very low and 5 = very high
M: 1 = one defoliation per year and
5 = 5 defoliations per year or more

Complete key in Chapter 2, section 2.8.

large populations in abandoned grasslands.
     Mostly frequent in *Violo-Nardion strictae* and *Deschampsia flexuosa* grasslands.

COMPATIBILITY WITH OTHER FORAGE
GRASSES AND LEGUMES
Incompatible with productive forage species.

INDICATOR VALUE
Very poor acid soil. Extensive management.

AGRONOMIC CHARACTERISTICS

VARIETIES
Ornamental variety: Albovariegatus.

FERTILIZATION
The communities where *H. mollis* grows are
not fertilized.

*DRY MATTER YIELDS AND SEASONALITY OF
PRODUCTION*
Moderate to low production.

NUTRITIVE VALUE
Probably markedly lower than *Holcus*

lanatus, because of a considerably lower
leaf:stem ratio.

ACCEPTABILITY
Poorly accepted by grazing animals because
it produces stems quite rapidly and is very
hairy.

DISEASES
Rusts are frequent on *H. mollis*: *Puccinia
coronata* and *P. holcina*. A smut
(*Tilletia holci*) and *Blumeria* (= *Erysiphe*)
graminis are widespread, while
Colletotrichum holci can be present (Ellis and
Ellis, 1997).

MAIN QUALITIES
Moderate production on poor soils.

MAIN SHORTCOMINGS
Poorly accepted by grazing animals on
account of propensity for stemminess.

HORDEUM SECALINUM SCHREB.

FIGURE 5.130
Inflorescence and part of the leaf of *Hordeum secalinum*

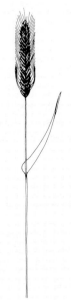

SYNONYM

H. pratense Huds., *H. nodosum* auct non L.

ETYMOLOGY

LATIN: *hordeum* = barley.
Hordeum could come itself from *hordus* = heavy; the bread made with barley is heavy.
LATIN: *secalinus* = like rye (= *secale*).

COMMON NAMES

ENGLISH: Meadow barley
FRENCH: Orge faux-seigle
GERMAN: Roggen gerste

DISTRIBUTION

Native to Western and southern Europe, to Asia Minor and Northern Africa.

R E N SMITH, IGER

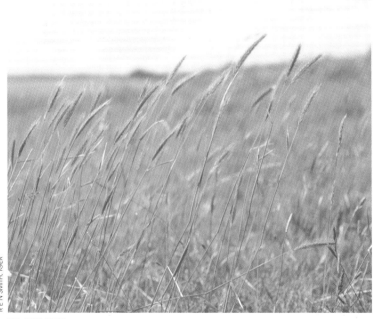

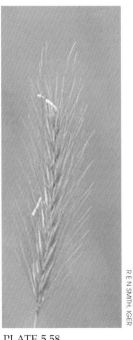

PLATE 5.57
Hordeum secalinum

R E N SMITH, IGER

PLATE 5.58
Hordeum secalinum

DESCRIPTION

MORPHOLOGICAL DESCRIPTION
Perennial plant, medium-sized, slightly
hairy, caespitose. Stems erect, 30–80 cm high.
Leaf blade rolled when young, flat, narrow,
sometimes hairy. Sheath always hairy. Ligule
short, truncate. Very short auricles. Spike-
like inflorescence, slender, cylindrical.
Spikelets with 1 fertile flower. Spikelets
clustered in threes, the laterals sterile.
Spikelets aristate, the centre spikelet with a
long awn (Figure 5.130) (Plates 5.57 and
5.58).

1000-SEED WEIGHT
4.0–5.0 g (large seeds).

CHROMOSOME NUMBER
2n = 28 (tetraploid).
Other chromosome numbers have also been
observed: 2n = 14 or 42.

ECOLOGICAL REQUIREMENTS

ALTITUDE – VEGETATION BELTS
Mostly in the lowlands, in the hill belt.

SOIL MOISTURE
Optimum on normally drained to slightly
wet soils (Figure 5.131).

SOIL FERTILITY
Large range of nutrient availability.
Optimum on neutral to alkaline soils
(Figures 5.131 and 5.132).

SOIL TYPE AND TOPOGRAPHY
Most frequent on clay soils.

CLIMATE REQUIREMENTS
Large climate range. Tolerates drought.

TOLERANCE OF CUTTING AND GRAZING
Much more abundant in grazing or in mixed
grazing-cutting grassland than in exclusively
cut grasslands (Figure 5.132).

FIGURE 5.131
Ecological optimum and range for soil pH and
humidity of *Hordeum secalinum*

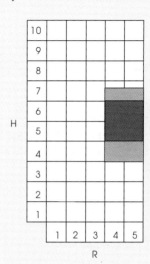

R: 1 = extremely acid and 5 = alkaline
H: 1 = extremely dry and
10 = permanently flooded

Complete key in Chapter 2, section 2.8.

FIGURE 5.132
Ecological optimum and range for nutrient
availability and management of *Hordeum
secalinum*

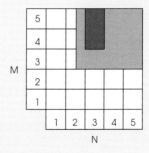

N: 1 = very low and 5 = very high
M: 1 = one defoliation per year and
5 = 5 defoliations per year or more

Complete key in Chapter 2, section 2.8.

SOCIABILITY – COMPETITIVENESS – PLANT COMMUNITIES
Poorly competitive.
    Mostly frequent in *Cynosurion cristati* and intensive grasslands.

COMPATIBILITY WITH OTHER FORAGE GRASSES AND LEGUMES
Rarely very invasive in *Lolium perenne* swards.

AGRONOMIC CHARACTERISTICS

FERTILIZATION
The communities in which *H. secalinum* grows are either not fertilized or lightly fertilized.

DRY MATTER YIELDS AND SEASONALITY OF PRODUCTION
Low production.

NUTRITIVE VALUE
Digestibility probably low to moderate.

ACCEPTABILITY
Refused by animals after heading.

MAIN QUALITIES
Not very invasive.

MAIN SHORTCOMINGS
Poorly productive. Heads rapidly and is then rejected by grazing animals.

KOELERIA PYRAMIDATA (Lam.) Beauv.

FIGURE 5.133
Inflorescence and part of the stem and leaf
of *Koeleria pyramidata*

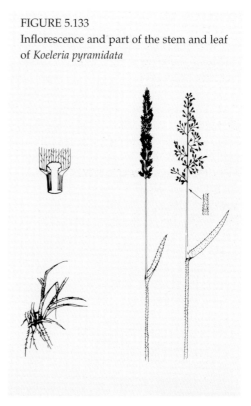

SYNONYM
K. cristata Pers.

ETYMOLOGY
Dedicated to Louis Koeler (1765–1807) from
Mainz, author of a book on grasses from
France and Germany (1802).
LATIN: *pyramidatus* = like a pyramid
(= *pyramis*).

COMMON NAMES
ENGLISH: Crested hair-grass
FRENCH: Koelérie pyramidale
GERMAN: Grosses schillergras, grosse
kammschmiele, pyramiden kammschmiele

DISTRIBUTION
Native to median Europe.

PLATE 5.59
Koeleria pyramidata

PLATE 5.60
Koeleria pyramidata

DESCRIPTION

MORPHOLOGICAL DESCRIPTION
Perennial plant, small-sized, slightly hairy, caespitose. Stems erect, 20–110 cm high. Leaf blade rolled when young, narrow (2–4 mm), short, erect, rough on the upper side, smooth on the under side, hairless or hairy on both sides, with a ciliate margin, glaucescent. Ligule short (1 mm), truncate. No auricles. Spike-like panicle, compact, 4–15 cm long, whitish green to purple. Spikelets 2–4-flowered, almost 6–7 mm long (Figure 5.133) (Plates 5.59 and 5.60).

PHYSIOLOGICAL PECULIARITIES
Maximum number of leaves per tiller: 5.4 (Sydes, 1984), which is higher than for most grasses.

1000-SEED WEIGHT
0.2–0.3 g (small seeds).

CHROMOSOME NUMBER
2n = 14 (diploid), 28 (tetraploid), 42 (hexaploid), 70 (decaploid) or 84 (dodecaploid).

ECOLOGICAL REQUIREMENTS

ALTITUDE – VEGETATION BELTS
From the lowlands to high altitudes in mountain areas; from hill to alpine belts.

SOIL MOISTURE
Restricted to dry and to extremely dry soils (Figure 5.134).

SOIL FERTILITY
Only on soils very poor in nutrients, slightly acid to alkaline (Figures 5.134 and 5.135).

SOIL TYPE AND TOPOGRAPHY
Shallow, stony, calcareous soils, but also on sand.

CLIMATE REQUIREMENTS
Large climate range. Very resistant to drought.

FIGURE 5.134
Ecological optimum and range for soil pH and humidity of *Koeleria pyramidata*

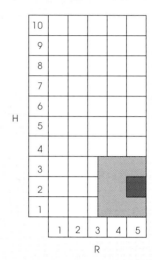

R: 1 = extremely acid and 5 = alkaline
H: 1 = extremely dry and 10 = permanently flooded

Complete key in Chapter 2, section 2.8.

FIGURE 5.135
Ecological optimum and range for nutrient availability and management of *Koeleria pyramidata*

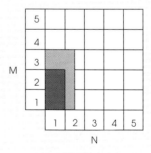

N: 1 = very low and 5 = very high
M: 1 = one defoliation per year and
5 = 5 defoliations per year or more

Complete key in Chapter 2, section 2.8.

TOLERANCE OF CUTTING AND GRAZING
Thrives on poor grasslands grazed by sheep (Figure 5.135).

SOCIABILITY – COMPETITIVENESS – PLANT COMMUNITIES
Poorly competitive.
   Mostly frequent in *Mesobromion erecti, Festucion valesiacae* and *Molinion caeruleae*.

COMPATIBILITY WITH OTHER FORAGE GRASSES AND LEGUMES
Thrives in spontaneous mixtures with *Bromus erectus, Festuca ovina, Trisetum flavescens, Avenula* spp. and *Brachypodium pinnatum*.

INDICATOR VALUE
Very poor, dry soil. Extensive management.

AGRONOMIC CHARACTERISTICS

VARIETIES
Variety for low-maintenance green areas: Barkoel (Barenbrug, NL).

FERTILIZATION
The communities where *K. pyramidata* grows are not fertilized.

DRY MATTER YIELDS AND SEASONALITY OF PRODUCTION
Very low production in spontaneous growths.

NUTRITIVE VALUE
A great part of the aerial biomass is constituted by stems and so the quality is certainly poor.

ACCEPTABILITY
Poorly accepted.

DISEASES
The rust *Puccinia longissima* has been sometimes observed (Ellis and Ellis, 1997).

USE FOR PURPOSES OTHER THAN FORAGE PRODUCTION
Can be incorporated in species-rich mixtures suited to dry, calcareous conditions.

MAIN QUALITIES
Constituent of poor grasslands grazed by sheep.

MAIN SHORTCOMINGS
Low production, poor quality and low acceptability.

LOLIUM X *HYBRIDUM* HAUSSKN.

Hybrid between *Lolium perenne* L. and *L. multiflorum* Lam. However, many varieties are morphologically closer to *Lolium multiflorum* than to *L. perenne*.

ETYMOLOGY
CELTIC: loloa = darnel and LATIN: *lolium* = darnel.
LATIN: *hybridus* = hybrid.

COMMON NAMES
ENGLISH: Hybrid ryegrass
FRENCH: Ray-grass hybride
GERMAN: Bastard-Raigras

DISTRIBUTION
In the areas common to the 2 parents, e.g. Europe and other temperate areas.

DESCRIPTION

MORPHOLOGICAL DESCRIPTION
Depending upon the breeding, some hybrids are close to *Lolium perenne,* while others are close to *L. multiflorum*, and so may be indistinguishable from their parents; still others are intermediate between the parents.

1000-SEED WEIGHT
Almost 2 g (average-sized seeds).

CHROMOSOME NUMBER
2n = 14 (diploid).
2n = 28 (tetraploid).

ECOLOGICAL REQUIREMENTS
Intermediate between the requirements of the parents.

Usually good winter hardiness. Higher tiller density and thus better ground cover than *Lolium multiflorum*.

AGRONOMIC CHARACTERISTICS

SOWING RATES
Pure: diploid varieties: 25–30 kg/ha;
        tetraploid varieties: 40–45 kg/ha.
Mixed: with *Trifolium pratense* or *T. repens*:
        diploid varieties: 20–25 kg/ha;
        tetraploid varieties: 30–40 kg/ha.

VARIETIES
The forage varieties of *Lolium x hybridum* can be clustered according to their ploidy level (Table 5.50).

FERTILIZATION
High levels needed to express its full potential.

DRY MATTER YIELDS AND SEASONALITY OF PRODUCTION
High to very high production. Most have the early-season and late-season growth of *Lolium multiflorum*, particularly the hybrids

TABLE 5.50
Examples of varieties of *Lolium* x *hybridum*

Variety	Seed breeder (country)
Diploids	
Barcolte	Barenbrug (NL)
Barsilo	Barenbrug (NL)
Pirol	Steinach (D)
Tetraploids	
AberExcel	IGER (UK)
AberLinnet	IGER (UK)
Antilope	FAL (CH)
Barvisto	Barenbrug (NL)
Dalita	DLF (DK)
Merini	DvP (B)
Siriol	PBI (UK)

close to *L. multiflorum*, combined with some of the sward density of *L. perenne*. Yields lower than *Lolium multiflorum* in the first harvest year, but third year yields can be even better. Spring growth can be lower than Italian types. Good growth in the autumn thus allowing an extension of the grazing period.

NUTRITIVE VALUE
The digestibility of the hybrids declines more slowly with advancing maturity than *Lolium multiflorum*. Summer digestibility better than *Lolium multiflorum* but lower than *L. perenne*. High soluble carbohydrate content.

ACCEPTABILITY
Highly acceptable.

DISEASES
Good disease resistance.

USE FOR PURPOSES OTHER THAN FORAGE PRODUCTION
Can be sown as a green manure or a nitrate catch crop, also in field margins to limit runoff and erosion.

MAIN QUALITIES
Usually more persistent than *Lolium multiflorum* though less than *L. perenne*. Highly productive. Persistence for 4–5 years in ideal conditions, but 3–4 years in more usual conditions. Combines the spring-growth type of *Lolium multiflorum* with the leafy growth of *L. perenne* in summer. New varieties are more intermediate between the 2 parents, allowing a better flexibility in the cutting time and a good grazing potential. More compatible with *Trifolium repens* than *Lolium multiflorum*.

MAIN SHORTCOMINGS
Cannot be incorporated in long-term mixtures for sowing permanent grassland.

LOLIUM MULTIFLORUM LAM.

FIGURE 5.136
Inflorescence and part of the leaf of *Lolium multiflorum*

There are 2 subspecies of *L. multiflorum*:
- *westerwoldicum* , Westerwold ryegrass, an annual plant that dies after seed formation.
- *multiflorum* , Italian ryegrass (*Italicum* type) that is a short-lived perennial and usually persists for 2–3 years.

SYNONYM
L. italicum A. Braun, *L. perenne* L. ssp. *multiflorum* (Lam.) Husnot

ETYMOLOGY
CELTIC: loloa = darnel and LATIN: *lolium* = darnel.
LATIN: *multus* = in great number, numerous and LATIN: *flos* = flower.
This grass has a great number of flowers in each spikelet.

A. PEETERS

PLATE 5.61
Lolium multiflorum

P. ALLARD

PLATE 5.62
Lolium multiflorum

COMMON NAMES
ENGLISH: Italian ryegrass
FRENCH: Ray-grass d'Italie
GERMAN: Italienisches Raigras

DISTRIBUTION
Native to Central and southern Europe,
Northern Africa and southwest Asia.
Introduced to most temperate regions.

DESCRIPTION

MORPHOLOGICAL DESCRIPTION
Annual or perennial plant, very robust, rather
large, hairless, caespitose. Stems erect,
20–100 cm high. Leaf blade rolled when
young, large (up to 10 mm), long (6–25 cm),
flexible, pale green, shiny on the lower side,
less than 16 well marked veins on the upper
side (more than 16 veins in *Festuca pratensis*,
which it resembles vegetatively). Sheaths of
the lower leaves purplish red. Ligule quite
short, membranous. Auricles long, well
developed. Spike-like inflorescence. Spikelets
10–14-flowered (up to 20–25) attached to the
axis by one of their sides. One glume per
spikelet but 2 on the top spikelet. Glume
shorter than the spikelet. Lemma exceeds the
glume in length and is usually aristate
(Figure 5.136) (Plates 5.61 and 5.62).

PHYSIOLOGICAL PECULIARITIES
Hybridizes with *Lolium perenne*: *L.* x
hybridum Hausskn.
  Hybridizes very rarely with *Festuca
pratensis* and *F. arundinacea*: x *Festulolium*.
This hybridization is used in plant breeding
to develop hybrids with resistance to
drought and better persistence.
  The roots of seedlings of *L. multiflorum*
and its hybrids fluoresce under ultraviolet
light whereas roots of *Lolium perenne* do not.
This technique is used in germination tests
to distinguish both species.

1000-SEED WEIGHT
Multiflorum diploid varieties: 2.0–2.5 g

FIGURE 5.137
Ecological optimum and range for soil pH and
humidity of *Lolium multiflorum*

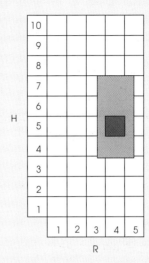

R: 1 = extremely acid and 5 = alkaline
H: 1 = extremely dry and 10 = permanently flooded

Complete key in Chapter 2, section 2.8.

FIGURE 5.138
Ecological optimum and range for nutrient
availability and management of *Lolium
multiflorum*

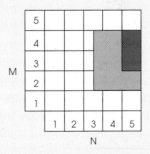

N: 1 = very low and 5 = very high
M: 1 = one defoliation per year and 5 = 5
defoliations per year or more

Complete key in Chapter 2, section 2.8.

(average-sized seeds), tetraploid varieties:
3.0–4.6 g (average-sized to large seeds),
Westerwoldicum diploid varieties: 2.5–3.0 g

(average-sized seeds), tetraploid varieties: 3.7–5.1 g (large seeds).

CHROMOSOME NUMBER
2n = 14 (diploid).
2n = 28 (tetraploid).

BIOLOGICAL FLORA
Beddows (1973).

ECOLOGICAL REQUIREMENTS

ALTITUDE – VEGETATION BELTS
From the lowlands to about 800 m in the Alps (Caputa, 1967); restricted to the hill belt, sometimes sown in the mountain belt.

SOIL MOISTURE
Optimum on normally drained soils. Dry or very wet soils are not suitable. Very sensitive to winter flooding (Figure 5.137).

SOIL FERTILITY
Needs nutrient-rich soils, slightly acid to alkaline (Figures 5.137 and 5.138).

SOIL TYPE AND TOPOGRAPHY
Suited to a range of soil textures.

CLIMATE REQUIREMENTS
Requires mild or warm climates. Very sensitive to winter cold and to summer drought. Tolerates heat, provided the water supply is sufficient. Can be cultivated in rather cold climates, but its persistence is then reduced.

TOLERANCE OF CUTTING AND GRAZING
Very well adapted to cutting, even frequently (4 cuts/year). Very sensitive to trampling and so does not tolerate exclusive grazing, but can be grazed occasionally, particularly in a rotational system (Figure 5.138).

SOCIABILITY – COMPETITIVENESS – PLANT COMMUNITIES
The most competitive of all grasses of temperate regions, from the time of germination. It is not highly persistent and so its competitive ability does not last very long, unless it has the opportunity to reseed naturally. If fertilization is sufficient, it suppresses all other species. This characteristic can be used in ley-crop rotations to control weeds, especially *Elymus repens*.

COMPATIBILITY WITH OTHER FORAGE GRASSES AND LEGUMES
Compatible with crimson clover (*Trifolium incarnatum*) or Egyptian clover (*T. alexandrinum*) in mixtures for several months or with *T. pratense* for 2–3 years. Except for these legumes, it is better to cultivate *L. multiflorum* in pure stand.

INDICATOR VALUE
Good fertility soil.

AGRONOMIC CHARACTERISTICS

SOWING RATES
Pure: diploid variety: 25–30 kg/ha;
        tetraploid variety: 40–45 kg/ha.
Mixed: with *Trifolium pratense*:
        diploid variety: 20–25 kg/ha;
        tetraploid variety: 30–40 kg/ha.

VARIETIES
The forage varieties of *L. multiflorum* can be grouped according to their persistence (1 or 2–3 years) and their ploidy level (Table 5.51).

FERTILIZATION
High levels needed to express its full potential.

DRY MATTER YIELDS AND SEASONALITY OF PRODUCTION
Very high production, being one of the most productive grasses of temperate regions. Yield regularly reaches 15 tonnes DM/ha in favourable conditions in Western Europe. In southern Europe, under irrigation, its yield

TABLE 5.51
Examples of varieties of *Lolium multiflorum*

	Varieties	Seed breeder (country)	Varieties	Seed breeder (country)
		Diploid		Tetraploid
Westerwold type (annual varieties: they head in the year of sowing)	Barcomet Ducado Limella Mendoza Merwester Shoot Topspeed	Barenbrug (NL) Zelder (NL) DSV (D) DvP (B) DvP (B) Limagrain Genetics (F) Advanta (NL)	Barspectra Biliken Caramba Clipper Lemnos Libonus Missyl Torero	Barenbrug (NL) Petersen Saatzucht (D) Mommersteeg (NL) Limagrain Genetics (F) DvP (B) DSV (D) RAGT (F) Advanta (NL)
Italicum type (varieties persisting 2–3 years: they do not head in the year of sowing)	AberComo Axis Barprisma Jericho Licarno Meribel Merode Meryl Prestyl Podium Zarastro	IGER (UK) FAL (CH) Barenbrug (NL) Verneuil Semunion (F) DSV (D) DvP (B) DvP (B) DvP (B) RAGT (F) Limagrain Genetics (F) DLF (DK)	Caballo Gemini Lipurus Meroa Montblanc Racine Tonyl Wilo Zorro	Limagrain Genetics (F) DvP (B) FAL (CH) DvP (B) Advanta (NL) DvP (B) RAGT (F) Prodana (DK) DLF (DK)

can reach 20–25 tonnes DM/ha. However, summer drought considerably decreases its production. Mainly cultivated for hay (2–3 cuts/year) or silage (4 cuts/year). Growth starts very early in spring, well before even early varieties of *Lolium perenne*. Growth continues in autumn later than that of *Lolium perenne*.

The year after sowing, *L. multiflorum* starts growth very early in spring and can often be ensiled around mid-April in the mild Atlantic climates of Western Europe, while *Lolium perenne* is only harvestable 15 days later. Alternatively, the stock can be turned out very early in spring. The grazing season can also be extended in autumn by prolonged growth.

NUTRITIVE VALUE
The digestibility of *L. multiflorum* is high and similar to *Lolium perenne* up to the pre-heading stage, but thereafter it declines more rapidly. Regrowths have lower digestibility than *Lolium perenne* regrowths, because *L. multiflorum* quickly produces new ears after a cut. It must thus be cut at a

young growth stage, otherwise, while production is abundant, the quality of the forage is rather low. Autumn regrowths are leafy and the digestibility is excellent. Its N content tends to be low if N application is limited. Soluble carbohydrate content is high and this aids a good fermentation in silage-making. The mineral content is normal except for Na, which is higher than that of other grasses. *L. multiflorum* wilts easily and so is suitable for hay-making, the quality being dependent on the stage of growth at cutting.

ACCEPTABILITY
Excellent at a leafy stage, particularly at the beginning of spring and in autumn. In the year of sowing, *L. multiflorum* ssp. does not produce stems and is therefore highly acceptable. The presence of stems in regrowths the following year can markedly decrease intakes at grazing. Unless used at a leafy stage, *L. multiflorum* ssp. *westerwoldicum*, which heads in the year of sowing, has reduced acceptability because of the stems.

TOXICITY

No toxicity unless ergot (*Claviceps purpurea*) or some other fungi develop.

DISEASES

The main leaf diseases of *L. multiflorum* are *Puccinia coronata, Drechslera dictyoides, D. siccans* and *Rhynchosporium orthosporum* (Raynal *et al*., 1989; Ellis and Ellis, 1997). The attacks of *Puccinia coronata* can be spectacular. It thrives mainly at the end of summer up to October–November, but it can appear as early as the month of July. The parasitism leads to a drying of infected areas and DM production can be reduced by almost 30 percent (Lancashire and Latch, 1966; O'Rourke, 1967). Digestibility is also reduced (Potter, 1987). There are big differences between varieties for susceptibility to this disease. *Puccinia graminis* is less important, but it is also observed as well as *P. hordei, P. striiformis* and *P. brachypodii* var. *poae-nemoralis*.

Other diseases are less detrimental, though better studied than for most forage grasses. The extensive list below does not mean that *L. multiflorum* is susceptible to a larger number of diseases than other forage grasses.

Rhynchosporium spp. and *Drechslera* spp. also cause leaf necrosis and decrease nutritive value (digestibility, soluble carbohydrates and even amino acid contents) (Lam, 1985). *Drechslera* spp. lead to a large decline in quality in comparison with the small leaf area attacked. *Rhynchosporium* spp. have a smaller effect per unit of leaf area. *Drechslera* spp. are transmitted by seeds and can cause seedling blight or even inhibit germination. Severe attacks of *Drechslera* spp. are more common on *L. multiflorum* than on *Lolium perenne* (Lam, 1984).

These foliar diseases predispose grasses in general to attacks by opportunistic fungi causing snow mould. This disease, which appears in case of heavy snow cover is caused by a complex of fungi including *Microdochium* (= *Fusarium*) *nivale* and *Typhula incarnata*, and leads to the disappearance of plants in winter. *L. multiflorum* is very susceptible to this fungus, less resistant than *Lolium perenne* and much more susceptible than *Phleum pratense*.

The bacteria, *Xanthomonas translucens* pv. *graminis*, causes significant damage and can also lead to plant death. It thrives in spring, but more so in summer and is favoured by dry and warm weather. It seems to be one of the main causes of mortality in *L. multiflorum*.

Barley Yellow Dwarf Virus (BYDV) and Ryegrass Mosaic Virus (RMV) can develop on *Lolium* spp. BYDV often exists in a latent stage, but when it develops in *Lolium perenne*, yield losses of about 20 percent have been noted (Catherall, 1987). Yield losses caused by this virus are more important on *Lolium perenne* than on *L. multiflorum* while the reverse is observed for RMV. BYDV is transmitted only by aphids while RMV is transmitted by a dust mite and can affect the yield by more than 20 percent (Catherall, 1987). Effects of RMV are more important in spring and when the fertilization level is high (A'Brook and Heard, 1975). Yield loss is also accompanied by a decrease in forage quality. Other fungal species are observed, but they are less important: 3 smut species (*Urocystis agropyri, Ustilago striiformis* and *Tilletia lolii*), *Claviceps purpurea* and *Spermospora lolii. Blumeria* (= *Erysiphe*) *graminis* is very widespread but causes less damage than *Puccinia coronata. Septoria tritici* var. *lolicola* is also observed on *Lolium* spp., though it differs from strains which develop on wheat (*Triticum* spp.).

ANIMAL PERFORMANCE

Mainly grazed in early spring and in autumn, and at these times can provide good LWGs. Rarely grazed throughout the grazing season.

Performances of animals fed with silage of *L. multiflorum* are comparable with those of animals fed with *Lolium perenne* silage.

In commercial farms, animal performance based on *L. multiflorum* hay is often disappointing because the hay has been cut too late. Farmers tend to harvest *L. multiflorum* at the same time as other grasses when the physiological stage of this species is too advanced.

USE FOR PURPOSES OTHER THAN FORAGE PRODUCTION
Good green manure and nitrate catch crop.

MAIN QUALITIES
Very productive with a good early-season and late-season growth. Good nutritive quality if well managed but moderate quality if used at too late a growth stage, especially in hay production. Well suited to silage because of its high soluble carbohydrate content. It establishes very rapidly and is very competitive against weeds including *Elymus repens*. Sown in spring, it is very productive during the establishment year compared with other forage grasses. This is particularly true for *L. multiflorum* ssp. *westerwoldicum*, which can be used in arable rotations in intercropping. It mixes very well with *Trifolium pratense*. Can be easily incorporated in crop rotations.

MAIN SHORTCOMINGS
Not very persistent, i.e. lasts 2–3 years. Some varieties are quite sensitive to rusts that decrease forage quality. Quickly produces stems in spring and during regrowths, which also reduces quality. Produces forage with a rather low leaf:stem ratio. Forage quality rapidly declines once the optimum growth stage is past. Difficult to control its rapid maturing and seed production. Seed may spread around and that is the reason why it can behave as a weed in arable land (in cereal crops mainly). In grazed swards, it can be rejected by cattle because of fast stem production compared with the other grasses of the mixture. Sensitive to cold, waterlogging and drought.

LOLIUM PERENNE L.

FIGURE 5.139
Inflorescence and part of the leaf of *Lolium perenne*

ETYMOLOGY

CELTIC: loloa = darnel and *LATIN*: *lolium* = darnel.
LATIN: *perennis* = perennial.
This species is persistent compared with other species of this genera.

COMMON NAMES

ENGLISH: Perennial ryegrass
FRENCH: Ray-grass anglais
GERMAN: Englisches Raigras

DISTRIBUTION

Native to Europe, temperate Asia and Northern Africa. Has become subcosmopolitan in temperate regions: introduced to North and Latin America, Australia and New Zealand.

A. PEETERS

PLATE 5.63
Lolium perenne

F. QUERTAINMONT

PLATE 5.64
Lolium perenne

A. PEETERS

PLATE 5.65
Lolium perenne

DESCRIPTION

MORPHOLOGICAL DESCRIPTION
Perennial plant, robust, medium-sized, hairless, caespitose. Stems erect, 10–60 (–90) cm high. Leaf blade folded when young, narrow (2–6 mm), rather long (3–20 cm), dark green, shiny on the lower side, veins well marked on the upper side. Ligule short, greenish to transparent. Auricles narrow. Sheaths of the lower leaves red to purplish red. Spike-like inflorescence. Spikelets attached to the axis by one of their sides. One glume per spikelet except on the top spikelet where there are 2. Glume shorter than the spikelet. Spikelets 6–10-flowered (up to 14). Glume(s) usually exceed(s) the lower lemma. Lemma without awn (Figure 5.139) (Plates 5.63 to 5.65).

PHYSIOLOGICAL PECULIARITIES
The most studied forage grass in the world. It has some physiological pecularities in common with other grasses, while others are specific and so a short summary of these characteristics is presented below.

The tiller density is very high under grazing (20 000–50 000 tillers/m^2 and even up to 70 000 tillers/m^2 under sheep grazing) but lower in a cutting regime (6 000–15 000 tillers/m^2). The maximum number of leaves per tiller is 3. Almost 100 positive day.degrees are required to produce a new leaf and each leaf has a lifetime of 300 day.degrees. Thus, the leaf turnover is very fast, which is a competitive advantage in frequently defoliated swards. In spring, the first leaves that appear are relatively small but the next leaves increase in size until maximum size is reached. In autumn, the opposite evolution is observed.

The roots are densely fasciculate. They can reach 1–1.5 m deep though the great majority of roots can be found in the first 15 cm layer of the soil. Root growth starts early in spring, almost 1–2 months before the leaves grow, slows down in summer and accelerates again in autumn. It is thus parallel to the aerial part growth except in spring. The roots have a lifetime of 2–3 months during the growing season. Roots of seedlings of *L. perenne* do not fluoresce under ultraviolet light while those of *Lolium multiflorum* and its hybrids do. This technique is used in germination tests to distinguish between the species.

L. perenne does not head during the year of sowing. During the following years, there is almost always a degree of heading in the regrowth after the first cut. The interval between the spring heading date of the earliest and the latest varieties is noteworthy and important for cutting management in this species. It can reach 6 weeks at low altitudes in oceanic climates.

Hybridizes with *Festuca pratensis*: x *Festulolium loliaceum* (Huds.) P. Fourn. and more rarely with *Festuca arundinacea*: x *Festulolium holmberii* (Dörfl.) P. Fourn., also with *Lolium multiflorum*: *Lolium* x *hybridum* Hausskn. These opportunities of hybridization are used in plant breeding to obtain hybrids that are more resistant to drought (x *Festulolium*), more persistent (x *Festulolium holmberii*) or more productive (*Lolium* x *hybridum*).

1000-SEED WEIGHT
Diploids: 1.3–2.7 g (average-sized seeds).
Tetraploids: 2.0–4.0 g (average-sized to large seeds).

CHROMOSOME NUMBER
2n = 14 (diploid).
2n = 28 (tetraploid).

BIOLOGICAL FLORA
Beddows (1967).

ECOLOGICAL REQUIREMENTS

ALTITUDE – VEGETATION BELTS
From the lowlands to almost 1 200 m in the Alps (Caputa, 1967); from hill to mountain belts.

FIGURE 5.140
Ecological optimum and range for soil pH and humidity of *Lolium perenne*

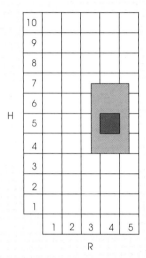

R: 1 = extremely acid and 5 = alkaline
H: 1 = extremely dry and 10 = permanently flooded

Complete key in Chapter 2, section 2.8.

FIGURE 5.141
Ecological optimum and range for nutrient availability and management of *Lolium perenne*

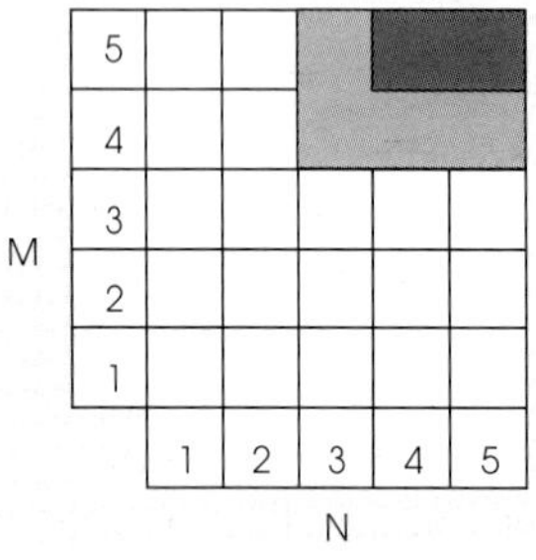

N: 1 = very low and 5 = very high
M: 1 = one defoliation per year and
5 = 5 defoliations per year or more

Complete key in Chapter 2, section 2.8.

SOIL MOISTURE
Optimum on normally drained soils. Dry or wet soils are not suitable. Very sensitive to winter flooding (Figure 5.140).

SOIL FERTILITY
Optimum on nutrient-rich or highly fertile soils, slightly acid to neutral (Figures 5.140 and 5.141).

SOIL TYPE AND TOPOGRAPHY
Indifferent to soil texture though loams and clays are the most suitable. Sands can be suitable if the water supply (irrigation or water table close to the surface) and the nutrient availability are sufficient. Rarer on peat soils.

CLIMATE REQUIREMENTS
Adapted to oceanic, mild and wet climates. Sensitive to intense frost, drought and high temperatures. Less cold-sensitive and more heat-sensitive than *Lolium multiflorum*. Intolerant of shade.

TOLERANCE OF CUTTING AND GRAZING
Adapted to high defoliation frequency (continuous grazing or 5–7 grazing periods/year) (Figure 5.141). Very resistant to trampling. Disappears quite quickly (after 3–4 years) in exclusive cutting systems (2–4 cuts/year) though it tolerates frequent mowings (almost 20 cuts/year) in lawns.

SOCIABILITY – COMPETITIVENESS – PLANT COMMUNITIES
Very competitive from germination, though less so than *Lolium multiflorum*. Heavily fertilized and frequently defoliated (1 cut per 3–4 weeks), it can suppress almost all species, particularly under grazing. Tetraploid varieties are more aggressive than diploid varieties, but because of the openness of a tetraploid sward, it is more compatible with *Trifolium repens* than a diploid sward.

In a Belgian trial, *L. perenne* was dominated by *Phleum pratense* when mixed together, especially in the infrequent cutting regime, though in other regimes, the mixture was better balanced (Figure 5.142).

Mostly frequent in *Cynosurion cristati*, intensive grasslands, humid improved

FIGURE 5.142
Proportion (% FM) of *Phleum pratense* in mixture with *Lolium perenne* in 3 cutting regimes and 4 levels of nitrogen fertilization, after 5 years of production

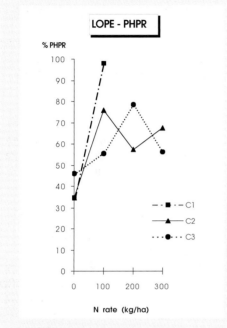

PHPR: *P. pratense;* LOPE: *L. perenne*
C1 = 2 cuts/year; C2 = 4 cuts/year; C3 = 5–6 cuts/year

Source: Peeters, Decamps and Janssens, 1999

grasslands and lawns. Less frequent in *Polygonion avicularis*.

COMPATIBILITY WITH OTHER FORAGE GRASSES AND LEGUMES
Compatible with *Poa trivialis, P. pratensis, Holcus lanatus, Phleum pratense* (mainly in cutting), *Trifolium pratense* (in cutting) and *T. repens* (in grazing).

INDICATOR VALUE
Good fertility soil. Mainly grazed sward.

AGRONOMIC CHARACTERISTICS

SOWING RATES
Pure: diploid varieties: 25–30 kg/ha;

tetraploid varieties: 40–50 kg/ha.
In simple mixture with *Trifolium repens*:
    diploid varieties: 20–25 kg/ha;
    tetraploid varieties: 30–40 kg/ha.
More complex mixtures mainly with *Festuca pratensis, Phleum pratense* and *Trifolium pratense*:
    diploid varieties: 10–15 kg/ha;
    tetraploid varieties: 15–20 kg/ha.
Complex mixture: 5–10 kg/ha with many other grasses and legumes in mixtures adapted to mountain areas (Mosimann *et al.*, 1996a).

VARIETIES
Forage varieties can be grouped in 6 groups according to their heading dates and ploidy level (Table 5.52).

FERTILIZATION
High levels needed to express its full potential.

DRY MATTER YIELDS AND SEASONALITY OF PRODUCTION
High production of about 12–15 tonnes DM/ha in optimum conditions, though not the most productive species. In cutting trials, *Lolium multiflorum, Festuca arundinacea, Dactylis glomerata* and *Arrhenatherum elatius* almost always have higher yields when similarly fertilized, while *Phleum pratense* has higher or similar yields. Other species, such as *Festuca rubra* and *Holcus lanatus* can equal or exceed the production of *L. perenne* under certain conditions (see previous profiles). However, the digestible organic matter yield of *L. perenne* often exceeds those of the low-digestibility species (*Festuca rubra, F. arundinacea*) or the moderate-digestibility species (*Arrhenatherum elatius, Dactylis glomerata*). A vast collaborative network, including 16 countries, 32 sites and lasting 5 years, organized within the framework of the FAO network for grassland and fodder crops in Europe, showed the high yield potential of *L. perenne* (Table 5.53).

TABLE 5.52

Examples of varieties of *Lolium perenne*

	Varieties Diploid	Seed breeder (country)	Varieties Tetraploid	Seed breeder (country)
Early	Barmore	Barenbrug (NL)	Abereclair	IGER (UK)
	Indiana	Limagrain Genetics (F)	Anaconda	Mommersteeg (NL)
	Lacerta	FAL (CH)	Aubisque	Mommersteeg (NL)
	Livorno	DSV (D)	Impresario	Limagrain Genetics (F)
	Merbo	DvP (B)	Labrador	DLF - Trifolium (DK)
	Rebecca	DvP (B)	Merlinda	DvP (B)
Intermediate	Abermont	IGER (UK)	Barmedia	Barenbrug (NL)
	Foxtrot	DLF (DK)	Calibra	Limagrain Genetics (F)
	Premium	CEBECO (NL)	Missouri	DLF - Trifolium (DK)
	Recolta	CEBECO (NL)	Pandora	DvP (B)
	Ritz	DvP (B)	Proton	Mommersteeg (NL)
Late	Cancan	Limagrain Genetics (F)	Bocage	Carneau (F)
	Herbie	Advanta (NL)	Ernesto	DvP (B)
	Jumbo	DLF (DK)	Montagne	Mommersteeg (NL)
	Liparis	DSV (D)	Pastoral	RAGT (F)
	Norton	Limagrain Genetics (F)	Pomerol	DvP (B)
	Ohio	Limagrain Genetics (F)	Tivoli	DLF (DK)
	Sponsor	CEBECO (NL)	Ventoux	Mommersteeg (NL)
	Veritas	Advanta (NL)		

TABLE 5.53

Annual yields (tonnes DM/ha) and growth rates (kg DM/ha.day) of *Lolium perenne*

	Average	Standard deviation	Number of data (site x year)
Not irrigated			
Average annual yields	11.8	3.0	79
Maximum annual yields	19.7	-	1
Minimum annual yields	2.2	-	1
Maximum yields after 4 weeks of growth	4.0	1.0	79
Growth rate at the peak of production	121	28	79
Irrigated			
Average annual yields	13.8	3.1	36
Maximum annual yields	18.9	-	1
Minimum annual yields	5.8	-	1
Maximum yields after 4 weeks of growth	4.0	0.9	36
Growth rate at the peak of production	126	28	36

Note: monthly cutting regime; annual N fertilization: 600 kg/ha
Source: Peeters and Kopec, 1996

In Belgium, on a soil low in organic matter and with low N mineralization, *L. perenne* was always less productive than *Phleum pratense* in all cutting regimes (2, 4 and 6 cuts/year) and at all N rates (0, 100, 200 and 300 kg N/ha) (Figures 5.143 to 5.146). Although *L. perenne* is adapted to frequent defoliation and even requires frequent defoliation to persist (almost 1 defoliation per 3–4 weeks), its maximum yield is reached with 3 cuts/year. Above this cutting frequency, the yield decreases as for other grasses. Early varieties start their growth relatively early in the spring. Late varieties start at the beginning of April or in mid-April in Western Europe at low altitude and in early May at moderate altitude. A maximum growth rate is reached at the end

FIGURES 5.143–5.146

Annual yields of *Lolium perenne* compared with *Phleum pratense* in 3 cutting regimes and several nitrogen fertilization levels (average of 3 production years)

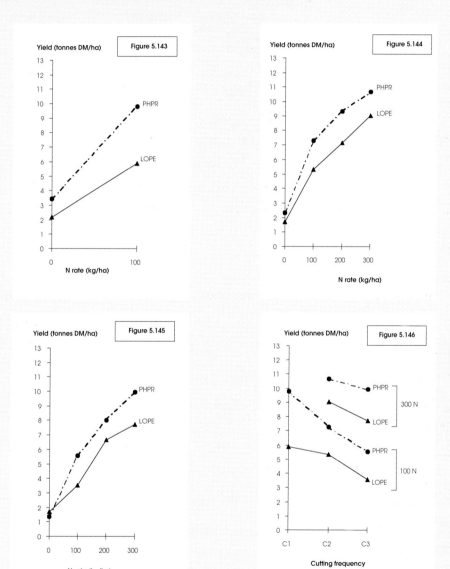

PHPR: *P. pratense;* LOPE: *L. perenne.* Fig. 5.143: 2 cuts/year (C1); Fig. 5.144: 4 cuts/year (C2); Fig. 5.145: 5–6 cuts/year (C3); Fig. 5.146: 100 and 300 kg N/ha

Source: Peeters and Decamps, 1999

of May or beginning of June, i.e. the first growth peak. By mid-June, the growth rate slows down sharply and reaches a summer minimum. At about the end of summer, the growth recovers and reaches a second peak, though it is clearly lower than the first. Thereafter, the growth rate decreases once more and becomes virtually nil by the end of October or in November. This growth pattern is similar for other grasses but the seasonality of production can be more pronounced, e.g. *Alopecurus pratensis, Poa trivialis*, or less pronounced, e.g. *Poa pratensis, Festuca rubra, Dactylis glomerata*, than for *L. perenne*.

NUTRITIVE VALUE

High digestibility (one of the best among all grass species) and rich in protein and soluble carbohydrate. Mineral contents are similar to other productive grasses but Na content is particularly high. The tetraploid varieties have slightly higher water and soluble carbohydrate contents than for diploid varieties.

The digestibility of early varieties declines faster than late varieties in spring and so must be defoliated earlier. Their digestibility often declines even faster during the second period of growth. However, the decline of digestibility of *L. perenne* is slower than that of other grasses, even *Phleum pratense* (-0.41 versus -0.48 percent per day) (Figure 5.147). When N supply is not a limiting factor, the N content is related to the accumulation of harvestable biomass according to the following equation:

$$N\ (\%\ \text{in DM}) = 5.7\ Y^{-0.50}$$

where Y = yield in tonnes DM/ha (Greenwood *et al.*, 1990). This equation is valid for other C3 grasses too.

ACCEPTABILITY

Highly acceptable, on a par with *Phleum pratense, Festuca pratensis* and *Lolium multiflorum*. The ingested amounts can reach 15–18 kg DM/cow.day, for a cow of 600 kg LW. There are small intake differences

FIGURE 5.147
Evolution of the digestibility of *Lolium perenne* compared with *Phleum pratense* during an uninterrupted growth period in spring (100 kg N/ha) (linear regression equations and determination coefficients shown)

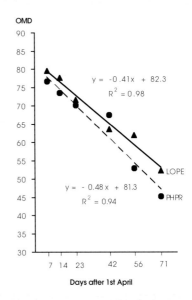

LOPE: *L. perenne;* PHPR: *P. pratense*

Source: Peeters, 1992

among varieties, but intakes are greater for tetraploid varieties.

TOXICITY

No toxicity except when attacked by endophytic fungi, ergot (*Claviceps purpurea*) or other fungal diseases.

DISEASES

The main leaf diseases of *L. perenne* are *Puccinia coronata, Drechslera dictyoides, D. siccans* and *Rhynchosporium orthosporum* (Raynal *et al.*, 1989; Ellis and Ellis, 1997). The attacks of *Puccinia coronata* can be spectacular. It thrives mainly at the end of summer up to October–November, but it can appear as early as the month of July. The parasitism leads to a drying of infected areas and DM production can be reduced by almost 30 percent (Lancashire and Latch, 1966; O'Rourke, 1967). Digestibility is also reduced (Potter, 1987). There are big differences among varieties for susceptibility to this disease. *Puccinia graminis* is less important, but it is also observed as well as *P. hordei*, *P. striiformis* and *P. brachypodii* var. *poae-nemoralis*.

Other diseases are less detrimental, though better studied than for most forage grasses. The extensive list below does not mean that *L. perenne* is susceptible to a larger number of diseases than other forage grasses. *Rhynchosporium* spp. and *Drechslera* spp. also cause leaf necrosis and decrease nutritive value (digestibility, soluble sugars and even amino acid contents) (Lam, 1985). *Drechslera* spp. leads to a large decline in quality in comparison with the small leaf area attacked. *Rhynchosporium* spp. have a smaller effect per unit of leaf area. *Drechslera* spp. are transmitted by seeds and can cause seedling blight or even inhibit germination. Severe attacks of *Drechslera* spp. are less common on *L. perenne* than on *Lolium multiflorum* (Lam, 1984).

L. perenne is moderately susceptible to snow mould, though more resistant than *Lolium multiflorum* and more susceptible than *Phleum pratense*. This disease, which appears in case of heavy snow cover, is caused by a complex of fungi including *Microdochium* (= *Fusarium*) *nivale* and *Typhula incarnata*, and leads to plant disappearance in winter. *Microdochium nivale* can also provoke leaf spots during the other seasons.

Neotyphodium lolii is sometimes observed as endophyte (Baert, Verbruggen and Carlier, 1994). It is likely to produce alkaloids toxic for cattle, but also beneficial effects on plant growth and on their resistance to some insects and nematodes.

The bacteria, *Xanthomonas translucens* pv. *graminis*, can cause significant damage and lead to plant death. It thrives in spring, but more so in summer and is favoured by dry and warm weather.

The viruses BYDV and RMV can develop on *Lolium* spp. BYDV often exists in a latent stage, but when it develops, yield losses of about 20 percent have been noted (Catherall, 1987). The yield losses caused by this virus are more important on *L. perenne* than on *L. multiflorum*, while the reverse is observed for RMV. The BYDV is transmitted only by aphids, while RMV is transmitted by a dust mite and its effects are mostly important in spring and when the fertilization level is high (A'Brook and Heard, 1975). Yield loss is also accompanied by a decrease in forage quality. Other fungal species are observed but they are less important: 3 smut species (*Urocystis agropyri*, *Ustilago striiformis* and *Tilletia lolii*), *Claviceps purpurea* and *Spermospora lolii*. *Blumeria* (= *Erysiphe*) *graminis* is very widespread, but causes less damage than *Puccinia coronata*. *Ascochyta* leaf blight (*Ascochita* spp.) is not very frequent, but in case of attack it can cause huge damage by provoking leaf drying. *Septoria tritici* var. *lolicola* is also observed on *Lolium* spp. though it differs from strains which develop on *Triticum* spp.

ANIMAL PERFORMANCE
Animal performance is high, particularly on well fertilized *L. perenne* (Table 5.54).

USE FOR PURPOSES OTHER THAN FORAGE PRODUCTION
Most widely used species in Europe for the creation of sports fields, lawns and for the greening of verges and roadsides, because of high resistance to trampling and frequent mowing. There are many varieties bred for this type of use. However, finer lawns are obtained with mixtures of *Festuca rubra*, *F. ovina* and *Agrostis capillaris*. Good green manure and nitrate catch crop, though not as good as *Lolium multiflorum* for these kinds of uses because of a slower establishment. Sometimes sown in field margins.

TABLE 5.54

Typical animal performance data from a pure-sown *Lolium perenne* sward in Western Europe

Stocking rate in spring (number of Holstein cows/ha)	5–6
Stocking rate in summer (number of Holstein cows/ha)	3–4
Number of cattle grazing days (day/ha.year)	800–900
Daily calf liveweight gains (g/day)	1 000

Milk production per cow (litre/cow.day)
        Spring
        Summer
        Autumn
	
	18–22
	14–18
	10–14

Milk production during the grazing period (litre/ha)	15 000–20 000

MAIN QUALITIES

Productive grass of excellent quality and with good intake characteristics. Very well adapted to oceanic, mild and wet climates. Realizes its full potential on highly N-fertilized, P- and K-rich soils, slightly acid to neutral. Compatible with *Trifolium repens*. Establishes rapidly and is very competitive against weeds. Persistent under grazing. Its high soluble carbohydrate content, one of the highest among grasses, facilitates a desirable lactic acid fermentation during silage-making.

MAIN SHORTCOMINGS

Lacks persistence in an exclusive cutting regime (3–4 years if it is cut 2–4 times per year). Sensitive to disease, especially to rust at the end of summer and this strongly decreases intake, though some varieties are more resistant than others. Sensitive to cold and so not well suited to mountain areas and to northern and continental countries. Summer production is rather low, especially during periods of drought. It wilts slowly, which creates difficulties during hay production.

MOLINIA CAERULEA (L.) MOENCH

There are 2 subspecies with particular ecological niches: *M. caerulea* ssp. *caerulea* and *M. caerulea* ssp. *arundinacea* (Schrank) K. Richt.

ETYMOLOGY
Dedicated to Molina (1740–1829), a Spanish botanist.
LATIN: *caeruleus* = dark blue.
The panicle can be purple.

COMMON NAMES
ENGLISH: Purple moor-grass, flying bent
FRENCH: Molinie, molinie bleue
GERMAN: Pfeifengras

DISTRIBUTION
Native to Europe, western Asia, Northern Africa and eastern North America.

FIGURE 5.148
Inflorescence and part of the leaf of *Molinia caerulea*

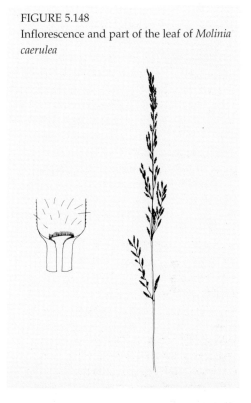

PLATE 5.66
Molinia caerulea

PLATE 5.67
Molinia caerulea

PLATE 5.68
Molinia caerulea

A. PEETERS

PH. LEVÊQUE, BIOS

A. PEETERS

DESCRIPTION

MORPHOLOGICAL DESCRIPTION
Perennial plant, glabrescent, caespitose.
Stems erect, 10–100 (–150) cm high
(in some subspecies up to 200–250 cm).
Long stems without leaves, with no visible
node (all nodes are located close to the
tillering level). Leaf blade rolled when
young, wide, rough. Ligule reduced to
a row of hairs. No auricles. Panicle-like
elongated inflorescence, erect, contracted
after flowering. Spikelets with 2–3 spaced
flowers (Figure 5.148) (Plates 5.66 to 5.68).

1000-SEED WEIGHT
0.50–0.70 g (small seeds).

CHROMOSOME NUMBER
M. caerulea ssp. *caerulea:* 2 n = 36 (tetraploid).
M. caerulea ssp. *arundinacea:* 2 n = 90 (decaploid).
Another chromosome number has also been
observed for *M. caerulea* ssp. *arundinacea:* 2n = 18.

BIOLOGICAL FLORA
Taylor, Rowland and Jones (2001).

ECOLOGICAL REQUIREMENTS

ALTITUDE – VEGETATION BELTS
From the lowlands up to 2 300 m in the
Alps (Caputa, 1967); from hill to alpine belts.

SOIL MOISTURE
Optimum on normally drained to boggy
soils, but very large range. However, the
subspecies *arundinacea* has its optimum on
drier soils. Tolerates large fluctuations in
water supply (Figure 5.149).

SOIL FERTILITY
Restricted to soils poor to very
poor in nutrients (oligotrophic species)
and usually very acid. However, the
subspecies *arundinacea* grows on calcareous
or marl soils. Thrives only on soils with low
P content (Figures 5.149 and 5.150).

FIGURE 5.149
Ecological optimum and range for soil pH and
humidity of *Molinia caerulea* (*M. caerulea* ssp.
caerulea on the left; *M. caerulea* ssp.
arundinacea on the right)

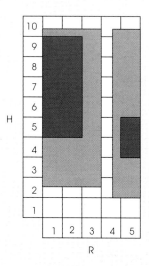

R: 1 = extremely acid and 5 = alkaline
H: 1 = extremely dry and 10 = permanently flooded

Complete key in Chapter 2, section 2.8.

FIGURE 5.150
Ecological optimum and range for nutrient
availability and management of *Molinia caerulea*

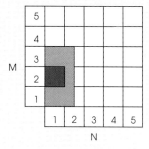

N: 1 = very low and 5 = very high
M: 1 = one defoliation per year and
5 = 5 defoliations per year or more

Complete key in Chapter 2, section 2.8.

SOIL TYPE AND TOPOGRAPHY
Mainly on peat soils, but also on humus-rich
sands (podzols).

Large climate range. Tolerates shade.

TOLERANCE OF CUTTING AND GRAZING
Well adapted to the infrequent cutting regime of a hay meadow. Much less abundant in grazed grasslands, though it can become dominant on some extensively grazed moors (Figure 5.150).

SOCIABILITY – COMPETITIVENESS – PLANT COMMUNITIES
Competitive in optimum conditions. Can form pure populations in abandoned grassland or rangeland. Highly favoured by fire (pyrophytic species).

Mostly frequent in *Molinion caeruleae, Violo-Nardion strictae, Juncion squarrosi* and *Mesobromion erecti (M. caerulea* ssp. *arundinacea).*

COMPATIBILITY WITH OTHER FORAGE GRASSES AND LEGUMES
Incompatible with productive forage species.

INDICATOR VALUE
Very poor soil, especially in P. Extensive management.

AGRONOMIC CHARACTERISTICS

VARIETIES
Ornamental varieties: *M. caerulea* ssp. *caerulea*: Dauerstrahl, Edith Dudszus, Heidebraut, Moorflamme, Moorhexe, Strahlenquelle, Rotschopf, Variegata; *M. caerulea* ssp. *arundinacea*: Bergfreund, Karl Foerster, Fontäne, Sky Racer, Transparent, Windspiel, Zuneigung.

FERTILIZATION
The communities in which *M. caerulea* grows are never fertilized in practice.

DRY MATTER YIELDS AND SEASONALITY OF PRODUCTION
The production of *Molinia* communities is very variable and large differences exist especially in relation to altitude. In Wales, a *Molinia* sward produced annually 1.8–2.2 tonnes DM/ha from unfertilized to highly N-fertilized treatments (Davies and Munro, 1974). In other locations, the annual DM yield has reached 4 tonnes/ha.

NUTRITIVE VALUE
The quality of *M. caerulea* grasslands is very low.

Common *et al.* (1991a) reported an average digestibility of 63 percent compared with 64 percent for *Nardus* and 66 percent for *Agrostis/Festuca* communities.

Several of the quality parameters are shown in Table 5.55. It appears that *M.*

TABLE 5.55
Chemical composition (g/kg DM) of *Molinia caerulea* compared with *Nardus stricta*, *Agrostis/Festuca* and *Lolium perenne* harvested at 3 dates in Scotland, UK

	Cut	N	Soluble carbohydrates	NDF	ADF	Lignin
M. caerulea	1	23.8	70	628	327	52.3
	2	15.5	51	665	414	94.5
	3	9.5	31	699	435	94.8
N. stricta	1	17.4	79	597	322	36.0
	2	13.7	52	620	356	53.1
	3	12.3	42	670	381	63.1
Agrostis/Festuca	1	32.2	50	467	265	26.4
	2	15.8	110	487	280	53.2
	3	13.7	82	624	334	53.3
L. perenne	1	22.8	189	394	227	16.1
	2	11.0	174	532	316	40.8
	3	7.6	147	578	346	48.5

Source: Armstrong, Common and Smith, 1986

caerulea is of course much poorer in soluble carbohydrates than *Lolium perenne*, but not poorer than the other species. It has acid detergent fibre (ADF) and neutral detergent fibre (NDF) contents comparable with *Nardus stricta* but higher than the other species. The lignin content is markedly higher than *Nardus stricta* and, of course, also higher than the other species. In this experiment, the N contents of all these grazed species are not very different.

ACCEPTABILITY

Poorly accepted by grazing animals, though it is preferred to *Nardus stricta*. It must be grazed at a very young stage. The accumulation of senescent material strongly decreases intake by stock.

ANIMAL PERFORMANCE

Low animal performance from *Molinia* communities. The *Festuca/Agrostis* community gave a surplus of 17 percent of sheep grazing days compared with the *Molinia* community in the study of Common *et al.* (1991b) in Scotland (UK).

DISEASES

Two particular rusts appear on *M. caerulea*: *Puccinia brunellarum-moliniae*, for which the alternative host is *Prunella vulgaris*, and *P. nemoralis*, for which the alternative host is *Melampyrum pratense* (Ellis and Ellis, 1997).

USE FOR PURPOSES OTHER THAN FORAGE PRODUCTION

Ornamental varieties used in mixed borders and in the understorey of trees. In Normandy, stems were used for binding cereal sheaves, for making straw hats in place of rice straw (Anon., 1881), for stuffing mattresses for enuretic children because the stems do not rot easily, for making small brooms (Bonnier, 1897) and woven baskets, and also as trays and filters in cider apple presses (M. Vivier, personal communication, 1992). Roots were used for making brushes (Bonnier, 1897).

MAIN QUALITIES

Dominant grass in wet or damp moors and in some poor grasslands. Acceptable forage when young.

MAIN SHORTCOMINGS

Very low nutritive value, including low mineral content. High in lignin and of low digestiblility. Animal performance very low.

NARDUS STRICTA L.

ETYMOLOGY
LATIN: *nardus* can mean perfume.
This grass has fragrant roots.
LATIN: *strictus* = narrow.
The inflorescence is very narrow.

COMMON NAMES
ENGLISH: Mat-grass
FRENCH: Nard, nard raide
GERMAN: Borstgras

DISTRIBUTION
Native to Europe, western Asia, the
mountains of Northern Africa, Greenland
and eastern North America.

FIGURE 5.151
Inflorescence, base of the plant and part of
the leaf of *Nardus stricta*

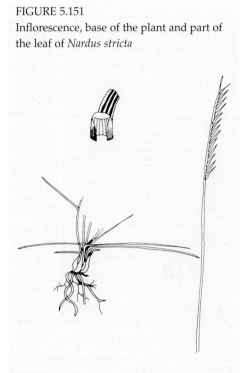

A. PEETERS

PLATE 5.69
Nardus stricta

PLATE 5.70
Nardus stricta

PLATE 5.71
Nardus stricta

DESCRIPTION

MORPHOLOGICAL DESCRIPTION
Perennial plant, small-sized, hairless, caespitose, with very short rhizomes. Stems erect, 10–40 cm high, stiff. Leaf blade needle, very narrow, hard, in a spiral, rough, bluish green. Many blades set out at right angles to the sheaths and stems. Ligule short but visible (up to 2 mm). No auricles. Spike-like, stiff, 1-sided inflorescence. Spikelets 1-flowered (Figure 5.151) (Plates 5.69 to 5.71).

1000-SEED WEIGHT
0.4–0.5 g (small seeds).

CHROMOSOME NUMBER
2n = 26 (diploid).
Another chromosome number has also been observed: 2n = 30.

BIOLOGICAL FLORA
Chadwick (1960).

ECOLOGICAL REQUIREMENTS

ALTITUDE – VEGETATION BELTS
From the lowlands up to 3 000 m in the Alps, often dominating grazed swards between 900 and 2 200 m in the Alps (Caputa, 1967); from hill to alpine belts.

SOIL MOISTURE
Optimum on dry soils but also sometimes on wet soils (Figure 5.152).

SOIL FERTILITY
Restricted to soils that are very poor in nutrients, especially P (oligotrophic species), and to acid and extremely acid soils (Figures 5.152 and 5.153).

SOIL TYPE AND TOPOGRAPHY
Thrives on a large range of soil textures, usually on a thick layer of badly rotted organic matter (namely, ranker).

FIGURE 5.152
Ecological optimum and range for soil pH and humidity of *Nardus stricta*

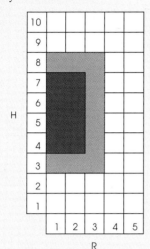

R: 1 = extremely acid and 5 = alkaline
H: 1 = extremely dry and 10 = permanently flooded

Complete key in Chapter 2, section 2.8.

FIGURE 5.153
Ecological optimum and range for nutrient availability and management of *Nardus stricta*

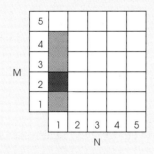

N: 1 = very low and 5 = very high
M: 1 = one defoliation per year and 5 = 5 defoliations per year or more

Complete key in Chapter 2, section 2.8.

CLIMATE REQUIREMENTS
Large climate range. Very resistant to cold and long periods of snow cover.

TOLERANCE OF CUTTING AND GRAZING
Species typical of poor, extensively grazed grasslands (Figure 5.153). Much rarer in cutting grassland where it is only encountered in isolated tufts. Very resistant to trampling. Decreases with high stocking rates.

SOCIABILITY – COMPETITIVENESS – PLANT COMMUNITIES
Very competitive in optimum conditions. Constitutes dense populations over huge areas in mountains. However, some *Nardus* communities can be species-rich.

Mostly frequent in *Violo-Nardion strictae, Nardion strictae, Juncion squarrosi* and *Polygono-Trisetion flavescentis*.

COMPATIBILITY WITH OTHER FORAGE GRASSES AND LEGUMES
Compatible with Alpine clover (*Trifolium alpinum*), a good spontaneous forage legume in the highlands. Never sown in forage mixtures.

INDICATOR VALUE
Acid soil, very poor in nutrients, especially in P. Extensive management, very low stocking rate.

AGRONOMIC CHARACTERISTICS

FERTILIZATION
N. stricta grasslands are never fertilized or only exceptionally.

DRY MATTER YIELDS AND SEASONALITY OF PRODUCTION
Low productivity, usually 1–4 tonnes DM/ha annually (Jarrige, 1979; Jeangros and Schmid, 1991; Common *et al.*, 1991a). Swards are almost always managed by grazing and better used by cattle than by sheep. The latter, being much more selective, avoid *N. stricta* thus allowing its spread to the detriment of other species (Jarrige, 1979; Common *et al.*, 1991a; Rodwell, 1992; Grant *et al.*, 1996).

NUTRITIVE VALUE
Produces a very poor quality forage, its digestibility being inferior to most other

TABLE 5.56

Nitrogen, phosphorus and potassium contents (g/kg DM) of the last completely spread leaves of *Nardus stricta* and *Agrostis capillaris* in a grazed *N. stricta* community

	N. stricta	*A. capillaris*
N	15.84	21.09
P	1.14	1.38
K	11.49	19.40

Source: Armstrong *et al.*, 1997

grasses, even those that are suited to poor soils (Armstrong, Common and Smith, 1986). Common, Wright and Grant (1998) obtained OMD values for swards dominated by *N. stricta* and maintained either short (4–5 cm), or tall (6–7 cm) of 64.8–67.6 percent (short) and 66.5–70.2 percent (tall). Armstrong *et al.* (1997) obtained the following values for OMD (average of 4 years and 3 treatments): 69.0 percent in May–June, 64.3 percent in July–August, 63.2 percent in September, with extreme values of 73.6 percent in May–June to 58.3 percent in September. The N, P and K contents of *N. stricta* leaves were consistently lower than *Agrostis capillaris* growing in the same community (Table 5.56).

ACCEPTABILITY
Intake low, even compared with other grasses typical of poor soils (Armstrong, Common and Smith, 1986). It has a preference rank lower than these grasses for grazing animals (Grant *et al.*, 1985). Cattle ingest more *N. stricta* than sheep.

ANIMAL PERFORMANCE
Animal performance from *N. stricta* communities is very low as shown below:
Common, Wright and Grant (1998):
- Number of grazing days of suckler cow/ha.year (1988–1992): 90–230,

- Milk production (litre/cow.day): 4.7–7.9,
- Daily LWG (g/day) of calves: 470–920.
Common *et al.* (1991b):
- Stocking rate at the beginning of the grazing season (kg LW of sheep/ha): 46–65,
- Daily LWG (g/day): –9,
- Number of grazing days of sheep/ha.year: 2 680,
- Intake (g OM/sheep.day): 1 370.
Jarrige (1979):
Stocking rate (animals/ha):
- ewes of 60 kg: 5–10,
- suckler cows (Aubrac breed): 0.80,
- heifers 13–15 months old: 1.1,
- fillies 13–15 months old: 0.70,
- suckler mares: 0.50.

USE FOR PURPOSES OTHER THAN FORAGE PRODUCTION
Can be used in mixtures for stabilizing soils after building ski slopes in mountain areas.

MAIN QUALITIES
Some value for grazing when young.

MAIN SHORTCOMINGS
Low productivity. Low nutritive value especially with a very low digestibility. Low acceptability by animals. Does not allow better species to develop in its optimal habitat.

PHALARIS ARUNDINACEA L.

FIGURE 5.154
Inflorescence and part of the stem, rhizome and leaf of *Phalaris arundinacea*

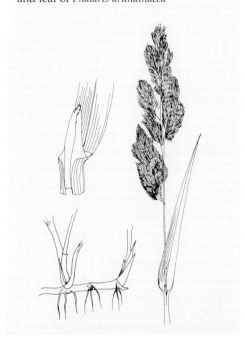

SYNONYM
Phalaroides arundinacea (L.) Rauschert,
Typhoides arundinacea (L.) Moench,
Baldingera arundinacea (L.) Dum.,
Digraphis arundinacea (L.) Trin.

ETYMOLOGY
GREEK: øαλαρις = phalaris derived from
øαλαρος = phalaros = white spotted,
shining.
The spikelets are silvery and shining.
LATIN: *arundinaceus* = similar to reed (=
arundo).

COMMON NAMES
ENGLISH: Reed canary-grass
FRENCH: Baldingère, alpiste roseau
GERMAN: Rohr glanzgrass

PLATE 5.72
Phalaris arundinacea

PLATE 5.73
Phalaris arundinacea

DISTRIBUTION

Native to temperate and cold areas of the northern hemisphere, and to mountains of eastern and southern Africa.

DESCRIPTION

MORPHOLOGICAL DESCRIPTION
Perennial plant, very robust, hairless, rhizomatous. Stems erect, 50–150 (–200) cm high. Leaf blade rolled when young, wide (8–16 mm), long, finely striated, almost smooth, pale green. Ligule long (almost 5 mm), oval-obtuse. No auricles. Panicle-like inflorescence, elongated, spreading at flowering then contracted, whitish green to purple. Spikelets 1-flowered (Figure 5.154) (Plates 5.72 and 5.73).

1000-SEED WEIGHT
0.60–0.80 g (small seeds).

CHROMOSOME NUMBER
2n = 28 (tetraploid) or 42 (hexaploid). Another chromosome number has also been observed: 2n = 14.

ECOLOGICAL REQUIREMENTS

ALTITUDE – VEGETATION BELTS
From the lowlands up to 1 500 m in the Alps (Rameau *et al.*, 1993); from hill to mountain belts and even to the base of the subalpine belt.

SOIL MOISTURE
Restricted to wet and boggy soils. Resistant to flooding and also to temporary drought (Figure 5.155).

SOIL FERTILITY
Large range of nutrient availability but favoured by fertile soils. Large range also for pH, thrives on acid as well as on alkaline soils (Figures 5.155 and 5.156).

SOIL TYPE AND TOPOGRAPHY
Mostly on peat soils but frequent also on clay soils. Less abundant on coarse-textured soils.

FIGURE 5.155
Ecological optimum and range for soil pH and humidity of *Phalaris arundinacea*

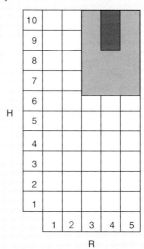

R: 1 = extremely acid and 5 = alkaline
H: 1 = extremely dry and
10 = permanently flooded

Complete key in Chapter 2, section 2.8.

FIGURE 5.156
Ecological optimum and range for nutrient availability and management of *Phalaris arundinacea*

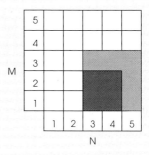

N: 1 = very low and 5 = very high
M: 1 = one defoliation per year and
5 = 5 defoliations per year or more

Complete key in Chapter 2, section 2.8.

CLIMATE REQUIREMENTS
Large climate range. Tolerant of winter cold.

TOLERANCE OF CUTTING AND GRAZING
Adapted to infrequent cutting regimes (2–3
cuts/year). Disappears when grazed.
Sensitive to trampling (Figure 5.156).

*SOCIABILITY – COMPETITIVENESS – PLANT
COMMUNITIES*
Very competitive in optimum conditions.
Can form vast dense populations on
nutrient-rich peats.

Mostly frequent in *Filipendulion ulmariae,
Magnocaricion* and *Phragmition australis*.

*COMPATIBILITY WITH OTHER FORAGE
GRASSES AND LEGUMES*
Compatible with *Phleum pratense* on wet but
not sodden soils.

INDICATOR VALUE
Soil nutrient-rich and often wet. High content
in soil organic matter. Sward cut 2 or 3 times
a year, or grazed with a low stocking rate.

AGRONOMIC CHARACTERISTICS

SOWING RATES
Pure: 10 kg/ha.

VARIETIES
No forage varieties in Europe. In the USA,
the 2 best varieties are Venture and Palaton.
They have a low content of the alkaloid,
gramine, and do not contain alkaloids of the
tryptamine and carboline types as other
varieties do (Sheaffer and Marten, 1995).
Ornamental varieties: Feesey = Mervy
Feesey, Luteo Picta, Picta, Strawberries 'n'
Cream, Streamlined.

FERTILIZATION
High levels needed to express its full potential.

*DRY MATTER YIELDS AND SEASONALITY OF
PRODUCTION*
Very high production, especially on wet soil,
its annual yield reaching 17 tonnes DM/ha.
Nevertheless, with low water supply, it is as

productive as *Dactylis glomerata* and *Bromus
inermis*, and even more productive than
Phleum pratense in North America (Sheaffer
and Marten, 1995).

NUTRITIVE VALUE
Digestibility moderate, probably similar to
Bromus inermis or *Dactylis glomerata*, but its
N content tends to be higher than other
productive grasses. However, its value falls
because of a high alkaloid content, mainly of
tryptamine-carboline, gramine and other
indol-alkaloids (Sheaffer and Marten, 1995).

ACCEPTABILITY
Poorly accepted by grazing animals,
especially in mixed swards, because of its
alkaloid content. The lack of intake results
from an indol-alkaloid content of 0.6 percent
in the DM, sometimes even 0.4 percent, but
there are large intake differences between
genotypes. Silage and hay are well accepted
(Sheaffer and Marten, 1995).

TOXICITY
The presence of alkaloids in variable
concentrations results in a variable degree of
toxicity. This toxicity is evidenced in lambs by
diarrhoea and growth reduction. However,
the varieties that are low in alkaloids allow
markedly higher LWGs than common
ecotypes. The toxicity appears at a content of
0.2 percent indol-alkaloids in the DM, the
diarrhoea being due to tryptamine-carboline
alkaloids (Sheaffer and Marten, 1995).

ANIMAL PERFORMANCE
LWGs of 0.6–0.9 kg/day for cattle are
achievable. The varieties low in alkaloids
can produce lamb growth of 0.1 kg/day, i.e.
as much as *Bromus inermis* (Sheaffer and
Marten, 1995).

*USE FOR PURPOSES OTHER THAN FORAGE
PRODUCTION*
Used for bedding in marshy regions of
Central Europe.

Ornamental varieties including variegated leaf varieties are planted in garden and parks in floral borders and in ponds.

DISEASES
Puccinia coronata appears on *P. arundinacea* as well as another rust (*P. sessilis*). A smut (*Ustilago striiformis*) develops on the leaves, as does another smut (*Tilletia menieri*) in the flowers. *Rhynchosporium secalis* is widespread on this grass (Ellis and Ellis, 1997).

USE FOR PURPOSES OTHER THAN FORAGE PRODUCTION
Wild type or ornamental varieties used beside water features or in mixed borders. Has hallucinogenic properties (Schultes, Hofmann and Rätsch, 2000).

MAIN QUALITIES
Very productive and of good digestibility when it is utilized at the right growth stage. High N content. Well adapted to wet soils and to cold climates, though its foliage is quickly destroyed by frost. Mostly suitable for silage production in peaty areas. Can also be grazed, especially if varieties low in alkaloids are used.

MAIN SHORTCOMINGS
Poorly accepted by grazing animals because of toxic alkaloid content. Growth of grazing animals is low and diarrhoea occurs in lambs. However, varieties bred with lower alkaloid content are much better accepted and less toxic.

PHLEUM ALPINUM L.

ETYMOLOGY

GREEK: øλεως = phleos = reed mace (=*Typha* spp.).
The inflorescence of this grass is similar to the poker-like inflorescence of *Typha* spp.
LATIN: *alpinus* = from the Alps.

COMMON NAMES

ENGLISH: Alpine cat's-tail
FRENCH: Fléole des Alpes
GERMAN: Alpen timothe

DISTRIBUTION

Mountain areas of Europe, Asia, North and South America, as well as cold areas of the northern hemisphere at low altitudes.

DESCRIPTION

MORPHOLOGICAL DESCRIPTION
Perennial plant, small-sized, hairless,

FIGURE 5.157
Inflorescence and part of the stem, leaf and glumes of *Phleum alpinum*

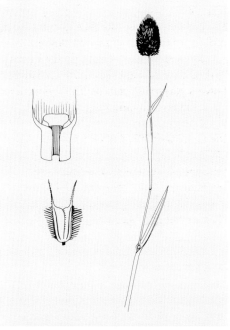

A. PEETERS

PLATE 5.74
Phleum alpinum

A. PEETERS

PLATE 5.75
Phleum alpinum

P. ALLARD

PLATE 5.76
Phleum alpinum

caespitose. Stems erect, 20–50 cm high, without bulb at the base. Leaf blade rolled when young, short, 3–5 mm wide. Ligule short (–2 mm), truncate. No auricles. Spike-like panicle, cylindrical, 1–3 cm long (much shorter than *P. pratense*). Spikelets 1-flowered (Figure 5.157) (Plates 5.74 to 5.76).

CHROMOSOME NUMBER
2n = 28 (tetraploid).

ECOLOGICAL REQUIREMENTS

ALTITUDE – VEGETATION BELTS
Mostly frequent in mountain and northern areas, even at low altitude. Very common between 1 400 and 2 400 m in the Alps (Caputa, 1967). From mountain to alpine belts; sometimes in the hill belt (Rameau *et al.*, 1993).

SOIL MOISTURE
Optimum on cool soils (Figure 5.158).

SOIL FERTILITY
Optimum on soils with moderate nutrient availability (mesotrophic species), slightly acid to neutral (Figures 5.158 and 5.159).

SOIL TYPE AND TOPOGRAPHY
Large range of soil textures.

CLIMATE REQUIREMENTS
Adapted to northern and mountain climates.

TOLERANCE OF CUTTING AND GRAZING
Thrives in grazed and cut mountain grasslands (Figure 5.159).

SOCIABILITY – COMPETITIVENESS – PLANT COMMUNITIES
Poorly competitive.
    Mostly frequent in *Poion alpinae* and *Polygono-Trisetion flavescentis*.

COMPATIBILITY WITH OTHER FORAGE GRASSES AND LEGUMES
Compatible with many other forage species

FIGURE 5.158
Ecological optimum and range for soil pH and humidity of *Phleum alpinum*

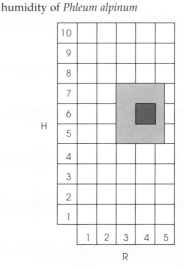

R: 1 = extremely acid and 5 = alkaline
H: 1 = extremely dry and
10 = permanently flooded

Complete key in Chapter 2, section 2.8.

FIGURE 5.159
Ecological optimum and range for nutrient availability and management of *Phleum alpinum*

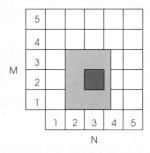

N: 1 = very low and 5 = very high
M: 1 = one defoliation per year and
5 = 5 defoliations per year or more

Complete key in Chapter 2, section 2.8.

adapted to mountain areas: *Trisetum flavescens, Festuca rubra, Agrostis capillaris,* Alpine meadow grass (*Poa alpina*), *Trifolium* spp.

AGRONOMIC CHARACTERISTICS

FERTILIZATION
The communities in which *P. alpinum* grows are only slightly fertilized or else not fertilized at all.

DRY MATTER YIELDS AND SEASONALITY OF PRODUCTION
Low to average production from sites where it appears spontaneously.

NUTRITIVE VALUE
Not well known, but probably rather good.

ACCEPTABILITY
Very well ingested at a leafy stage.

DISEASES
The diseases of *P. alpinum* are not well known, but they are probably similar to *Phleum pratense*. In particular, timothy eyespot (*Cladosporium phlei*) can thrive on both species (Ellis and Ellis, 1997).

USE FOR PURPOSES OTHER THAN FORAGE PRODUCTION
Can be incorporated in mixtures for sowing new ski pistes in mountain areas.

MAIN QUALITIES
Adapted to extensive systems in mountain areas where it is one of the best forage grasses.

MAIN SHORTCOMINGS
Low productivity.

PHLEUM PRATENSE L.

ETYMOLOGY
GREEK: øλεως = phleos = reed mace (= *Typha* spp.). The inflorescence of this grass is similar to the poker-like inflorescence of *Typha* spp.
LATIN: *pratensis* = from meadows.

COMMON NAMES
ENGLISH: Timothy
FRENCH: Fléole des prés
GERMAN: Wiesen lieschgras

DISTRIBUTION
Native to Europe, temperate Asia and Northern Africa. Has become subcosmopolitan in temperate regions.

DESCRIPTION
MORPHOLOGICAL DESCRIPTION
Perennial plant, robust, hairless, caespitose.

FIGURE 5.160
Inflorescence and part of the stem, leaf and glumes of *Phleum pratense*

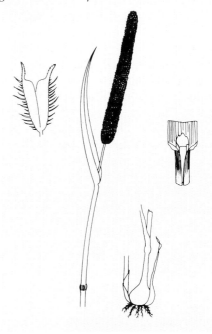

PLATE 5.77
Phleum pratense

PLATE 5.78
Phleum pratense

PLATE 5.79
Phleum pratense

Stems erect, 20–100 (–130) cm high, often bulging at the base and forming a small bulb. Leaf blade rolled when young, large (3–10 mm), flat, slightly rough on the margin, rather long (reaching 45 cm), pale green to greyish green. Ligule strong (reaching 6 mm long), obtuse, white. No auricles. Spike-like panicle, cylindrical, 6–20 (–30) cm long. Spikelets 1-flowered (Figure 5.160) (Plates 5.77 to 5.79).

PHYSIOLOGICAL PECULIARITIES
Tall grass, with erect stems that produce fewer tillers than *Lolium perenne*: 4 000–10 000 tillers/m². Leaves produced at a slightly faster rate than the leaves of *Lolium perenne*, the interval between the emergence of 2 successive leaves being about 60–70 positive day.degrees versus 100 day.degrees for *L. perenne*. The maximum number of leaves per tiller amounts to 6–7 (rarely 8) versus only 3 for *Lolium perenne* and the leaf lifetime is about 360 day.degrees (300 for *L. perenne*). Thus, *P. pratense* is able to accumulate a high level of standing biomass before senescence starts, giving it a competitive advantage over *Lolium perenne* in infrequent cutting regimes. *P. pratense* can head during the year of sowing while, the following year, some stem formation is observed after the first cut in addition to the main stem formation before a first cut of silage or hay.

1000-SEED WEIGHT
0.3–0.7 g (small seeds).

CHROMOSOME NUMBER
2n = 42 (hexaploid).

ECOLOGICAL REQUIREMENTS

ALTITUDE – VEGETATION BELTS
From lowlands to high altitudes in mountain areas; from hill to subalpine belts.

SOIL MOISTURE
Optimum on slightly wet soils (Figure 5.161).

FIGURE 5.161
Ecological optimum and range for soil pH and humidity of *Phleum pratense*

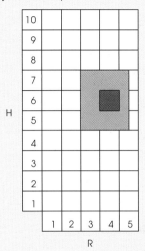

R: 1 = extremely acid and 5 = alkaline
H: 1 = extremely dry and
10 = permanently flooded

Complete key in Chapter 2, section 2.8.

FIGURE 5.162
Ecological optimum and range for nutrient availability and management of *Phleum pratense*

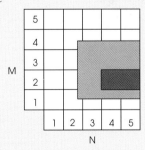

N: 1 = very low and 5 = very high
M: 1 = one defoliation per year and
5 = 5 defoliations per year or more

Complete key in Chapter 2, section 2.8.

SOIL FERTILITY
Thrives on fertile to very fertile soils, acid to alkaline. However, has lower requirements than *Lolium perenne*. Often grows on soils

rich in organic matter (Figures 5.161 and 5.162).

SOIL TYPE AND TOPOGRAPHY
Largely indifferent to soil texture. Less frequent on dry sands. Thrives on peaty soils if they are not too wet (boggy).

CLIMATE REQUIREMENTS
Large climate range but its good resistance to cold makes it a very useful forage plant in mountain areas and northern countries, e.g. in Scandinavia. Adapted to cool and wet climates. Sensitive to drought.

TOLERANCE OF CUTTING AND GRAZING
Species typical of cutting grassland. Optimum cutting frequency: 2 cuts/year. Does not resist trampling very well. In grazing, persists only sparsely in mixtures. Sensitive to frequent defoliations (Figure 5.162).

In a Belgian cutting experiment where it was mixed with *Lolium perenne, P. pratense* dominated *L. perenne* in almost every treatment, but especially in the infrequent cutting regime of 2 cuts/year (Figure 5.163).

SOCIABILITY – COMPETITIVENESS – PLANT COMMUNITIES
Grows slowly after sowing. Moderately competitive, and much less aggressive than the *Lolium* spp. Rarely forms vast spontaneous populations.

During the year of sowing, *P. pratense* is invaded by weeds more so than other productive grasses. For instance, Toussaint and Lambert (1973) noted an average of 80 percent (by fresh weight) weeds at the first cut after sowing, compared with 40 percent for *Lolium perenne* and 33 percent for *L. multiflorum*.

Mostly frequent in *Cynosurion cristati, Arrhenatherion elatioris, Polygono-Trisetion flavescentis* and intensive grasslands.

COMPATIBILITY WITH OTHER FORAGE GRASSES AND LEGUMES
Compatible with *Festuca pratensis, Lolium*

FIGURE 5.163
Proportion (% FM) of *Phleum pratense* in mixture with *Lolium perenne* in 3 cutting regimes and 4 levels of nitrogen fertilization, after 5 years of production

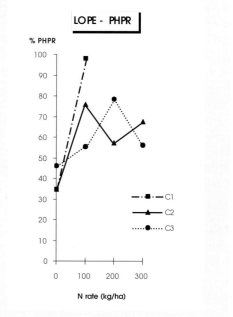

PHPR: *P. pratense;* LOPE: *L. perenne.* C1 = 2 cuts/year, C2 = 4 cuts/year, C3 = 5–6 cuts/year

Source: Peeters, Decamps and Janssens, 1999

perenne, Arrhenatherum elatius, Elymus repens (in cutting grassland), *Phalaris arundinacea* (on peat soil), *Poa trivialis, Trifolium pratense* and *T. repens,* alsike clover (*T. hybridum*), *Onobrychis viciifolia*.

INDICATOR VALUE
Good fertility soil, usually high organic matter content. Sward cut 2 or 3 times a year.

AGRONOMIC CHARACTERISTICS

SOWING RATES
Pure: 10–15 kg/ha, mainly for cutting.
Mixed: usually 6–10 kg/ha but 3 kg/ha in complex mixtures.
Simple cutting mixture with *Trifolium pratense*.

Simple grazing mixtures with *Lolium perenne* and *Trifolium repens*.

More complex mixtures, mainly for cutting, with *Festuca pratensis*, *Lolium perenne*, *Trifolium pratense* and sometimes *T. repens*. Complex mixtures for grazing or mixed use in mountain areas with *Poa pratensis*, *Festuca rubra*, *F. pratense*, *Lolium perenne* and *Trifolium repens* (Mosimann *et al.*, 1996a).

VARIETIES
A list of forage varieties is shown in Table 5.57.

FERTILIZATION
High levels needed to express its full potential, but yields are satisfactory with moderate fertilization levels.

DRY MATTER YIELDS AND SEASONALITY OF PRODUCTION
Very productive except for the year of sowing, when the yield is particularly low because of slow establishment. In cutting regimes, *P. pratense* can be more productive than *Lolium perenne*, especially if the cutting frequency is low (2–3 cuts/year) and N rates are nil to moderate (0–150 kg N/ha).

In an experiment carried out in Belgium, *P. pratense* was slightly more productive than *Lolium perenne* in all cutting regimes (2, 4 and 6 cuts/year) and at all N rates (0, 100, 200 and 300 kg N/ha) (Figures 5.164 to 5.167). Compared with 10 other grasses, only *Arrhenatherum elatius* yielded more in a 2 cuts/year regime (9.8 and 11.2 tonnes DM/ha respectively with 100 kg N/ha).

In a large research programme including 16 countries, 32 sites and 5 years of experiments organized within the framework of the FAO network for grassland and fodder crops in Europe, *P. pratense* produced the results shown in Table 5.58.

Lolium perenne was more productive than *P. pratense* in these trials but the 2 species were not always harvested together. At the sites where they were, *Lolium perenne* was superior to *P. pratense* in 94 percent of the cases without irrigation and in 89 percent of the cases with irrigation. The average difference in production was, respectively, 2 ± 1.3 tonnes DM /ha (n = 66) and 1.4 ± 1.2 tonnes DM/ha (n = 27). The maximum yields harvested after 4 weeks of growth without irrigation were identical for the 2 species: 4 tonnes DM/ha, and the maximum growth rates were also very similar (Table 5.59).

In Western Europe, the yield of *P. pratense* often reaches 12–15 tonnes DM/ha in a silage regime (3–4 cuts/year) with a high N fertilization (300–350 kg N/ha).

The growth of *P. pratense* is similar to

TABLE 5.57
Examples of varieties of *Phleum pratense*

Variety	Seed breeder (country)
Barliza	Barenbrug (NL)
Comer	DvP (B)
Erecta	DvP (B)
Lirocco	DSV (D)
Odenwälder	ZG (D)
Promesse	CEBECO (NL)
Rasant	ZG (D)
Richmond	Pickseed (CAN)
Tiller	Advanta (NL)
Toro	ISCF – Lodi (I)

FIGURES 5.164–5.167

Annual yields of *Phleum pratense* compared with *Lolium perenne* in 3 cutting regimes and several nitrogen fertilization levels (average of 3 production years)

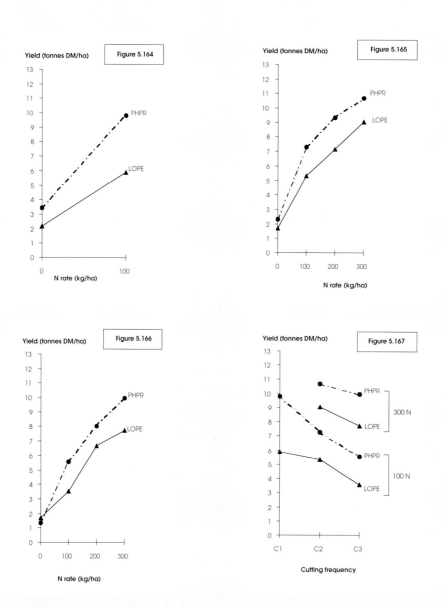

PHPR: *P. pratense;* LOPE: *L. perenne.* Fig. 5.164: 2 cuts/year (C1); Fig. 5.165: 4 cuts/year (C2); Fig. 5.166: 5–6 cuts/year (C3); Fig. 5.167: 100 and 300 kg N/ha

Source: Peeters and Decamps, 1999

TABLE 5.58

Annual yields (tonnes DM/ha) of *Phleum pratense* compared with *Lolium perenne*
at numerous sites in Europe

	P. pratense	Number of data (site x year)	*L. perenne*	Number of data (site x year)
Not irrigated				
Average yields	10.0	87	11.8	79
Maximum yields	15.4	1	19.7	1
Minimum yields	4.8	1	2.2	1
Irrigated				
Average yields	11.4	46	13.8	36
Maximum yields	17.2	1	18.9	1
Minimum yields	6.0	1	5.8	1

Note: monthly cutting regime; annual N fertilization: 600 kg/ha
Source: Peeters and Kopec, 1996

TABLE 5.59

Maximum growth rates (kg DM/ha.day) of *Phleum pratense* and *Lolium perenne*
(average of maximum rates per site, standard deviation and maximum individual values)

	P. pratense	*L. perenne*
Not irrigated	121 ± 25	121 ± 28
Irrigated	131 ± 24	136 ± 28
Maximum of individual values	194	227

Source: Peeters and Kopec, 1996

TABLE 5.60

Evolution of yields (tonnes DM/ha) of *Phleum pratense* compared with *Lolium perenne*
during the first growth cycle in the springs of 1991–93 in Belgium

	Dates					
1991	8 April	15 April	24 April	13 May	27 May	11 June
L. perenne	1.0	1.8	2.8	6.0	8.7	12.9
P. pratense	1.2	2.1	3.2	5.7	8.3	10.4
1992	15 April	29 April	7 May	19 May	27 May	9 June
L. perenne	0.9	2.8	4.7	7.8	10.1	9.9
P. pratense	1.7	3.5	5.2	8.1	10.1	12.0
1993	16 April	26 April	4 May	17 May	25 May	1 June
L. perenne	1.7	3.8	5.4	7.7	9.2	9.7
P. pratense	1.6	3.9	5.9	8.1	9.9	11.0

Note: N fertilization during the first growth cycle in spring: 100 kg/ha
Source: Peeters and Decamps, 1994

Lolium perenne at the beginning of the spring, but at the end of the first growth cycle (beginning of June), it often accumulates more standing biomass (10–12 tonnes DM/ha) (Table 5.60).

NUTRITIVE VALUE
It is noteworthy that the digestibility of *P. pratense* starts to decline prior to ear emergence in contrast to most grasses, which start their most rapid decline after ear

emergence. Nevertheless, the digestibility of *P. pratense* declines more slowly than *Dactylis glomerata, Festuca pratensis, Lolium multiflorum* or early *L. perenne* because of its markedly later heading. Therefore, high yields can be harvested at the first cut in spring without its digestibility being severely reduced.

Compared with an intermediate *Lolium perenne*, the rate of digestibility decline of *P. pratense* was a little faster (-0.48 versus -0.41 percent per day), though nevertheless, up to mid-May, the digestibilities of the 2 species were very similar (Figure 5.168).

The most compatible grass with *P. pratense* is *Festuca pratensis*, which has a much earlier heading date. It is thus difficult to determine the ideal cutting date of the mixture. However, the *Festuca pratensis* heading date should be used as the guide since the digestibility of *P. pratense* starts to fall even before heading. Some late *Lolium perenne* varieties head at the same time as *P. pratense* varieties, but these 2 species are not very compatible. At a leafy stage, the digestibility of *P. pratense* is equal or superior to *Lolium perenne*. However, the lignin content of *P. pratense* is a little higher than that of *Lolium perenne*. That is certainly due to the erect, tall stems of *P. pratense*, stems that need a sufficient support structure. Its content of soluble carbohydrates is inferior to *Lolium perenne*, making it less suitable for silage. The nutritive quality of *P. pratense* hay can be higher than *Lolium perenne* because of better air circulation in the stemmy *P. pratense* swaths, which allows faster wilting and a lower risk of rotting during hay making, but it can also be lower if the stand is cut too late.

The mineral contents are similar to other productive grasses except for Na, which is particularly low in *P. pratense*.

ACCEPTABILITY
Highly acceptable by grazing animals and at a leafy stage it has a higher acceptability

FIGURE 5.168
Evolution of the digestibility of *Phleum pratense* compared with *Lolium perenne* during an uninterrupted growth period in spring (100 kg N/ha) (linear regression equations and determination coefficients shown)

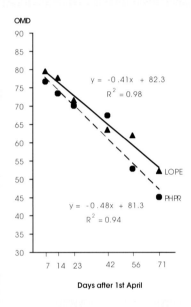

PHPR: *P. pratense;* LOPE: *L. perenne*

Source: Peeters, 1992

than *Lolium perenne*. Positive selection in mixed-species swards leads to it being overgrazed, thus decreasing its abundance in relation to other companion grasses such as *Lolium perenne*.

ANIMAL PERFORMANCE
Being mainly used for hay and silage production, animal performance depends on the physiological stage at cutting. *P. pratense* hay usually allows good animal performance provided it is cut at a satisfactory growth stage and is well made.

DISEASES
The main fungus that attacks the leaves is *Cladosporium phlei*, a disease specific to

Phleum spp. that is favoured by wet and cool weather.

Xanthomonas translucens pv. *phlei* can cause spectacular damage even in the spring. Plants cut at a young stage and heavily fertilized are particularly susceptible and there is high mortality among infected plants.

Very resistant to snow mould, unlike *Lolium* spp., *Festuca* spp. or *Dactylis glomerata*. Other observed diseases are: *Puccinia graminis, P. striiformis, Blumeria (= Erysiphe) graminis, Drechslera* spp., *Rhynchosporium secalis, Mastigosporium kitzebergense, Claviceps purpurea* and *Epichloë typhina* (Raynal *et al.*, 1989; Ellis and Ellis, 1997).

USE FOR PURPOSES OTHER THAN FORAGE PRODUCTION
Can be used in mixture for sowing new ski pistes in mountain areas.

In set-asides, provides a tall and permeable cover that is favourable to wildlife reproduction and shelter. Used in field margins for limiting runoff and erosion.

MAIN QUALITIES
Highly productive with very good quality and intake characteristics. Compatible with *Trifolium pratense* and *T. repens*. Well adapted to hay production and to cool, wet climates. Mainly suited to cold areas and meat production, but can also be used in dairy farms. Resistant to even exceptionally hard winters, making it an insurance species for grazing mixtures used in cold climates. Can be conserved as silage though less suitable than *Lolium perenne*.

MAIN SHORTCOMINGS
Slow growth after sowing and low yield during the establishment year. Sensitive to grazing and when added to mixtures for grazing, its abundance in the sward rapidly decreases. Not suited to favourable lowland climates and intensive systems.

POA ALPINA L.

ETYMOLOGY
GREEK: πoα = poa = sward.
LATIN: *alpinus* = from the Alps.

COMMON NAMES
ENGLISH: Alpine meadow grass
FRENCH: Pâturin des Alpes
GERMAN: Alpen rispengras

DISTRIBUTION
Native to cold areas and to mountains of the northern hemisphere.

DESCRIPTION

MORPHOLOGICAL DESCRIPTION
Perennial plant, small-sized, hairless, caespitose. Stems erect, 8–40 cm high, not very leafy. Leaf blade folded when young, short, firm, 2–5 mm wide. Ligule of the

FIGURE 5.169
Inflorescence and part of the stem and the leaf of *Poa alpina*

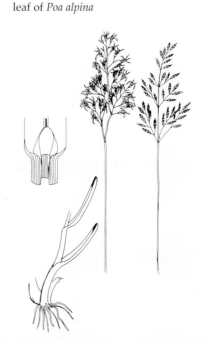

A. PEETERS

PLATE 5.80
Poa alpina

PLATE 5.81
Poa alpina

PLATE 5.82
Poa alpina

upper leaves long (up to 6 mm), oblong-acute. Ligule of the lower leaves shorter. No auricles. Panicle-like inflorescence, spreading, short, ovate or pyramidal, contracted after flowering. Spikelets 4–8-flowered, big, without woolly hairs at the base, often affected by vivipary (miniature plants developing in the flowers) (Figure 5.169) (Plates 5.80 to 5.82).

PHYSIOLOGICAL PECULIARITIES
Frequency of vivipary.

CHROMOSOME NUMBER
2n = 14, 28, 30–34 or 33–46–74.

ECOLOGICAL REQUIREMENTS

ALTITUDE – VEGETATION BELTS
From 1 500 m up to 3 600 m in the Alps (Rameau *et al.,* 1993); from the mountain belt to the top of the alpine belt.

SOIL MOISTURE
Optimum on normally drained soils (Figure 5.170).

SOIL FERTILITY
Optimum on soils moderately rich in nutrients (mesotrophic species) (Figures 5.170 and 5.171).

CLIMATE REQUIREMENTS
Adapted to northern and mountain climates.

TOLERANCE OF CUTTING AND GRAZING
Thrives mainly in grazed pastures and to a lesser extent in cutting grassland (Figure 5.171).

SOCIABILITY – COMPETITIVENESS – PLANT COMMUNITIES
Poorly competitive.
    Mostly frequent in *Poion alpinae* and *Polygono-Trisetion flavescentis.*

COMPATIBILITY WITH OTHER FORAGE GRASSES AND LEGUMES
Compatible with many other species such as *Festuca rubra, Agrostis capillaris, Phleum pratense, Trifolium* spp. that are adapted to mountain areas.

AGRONOMIC CHARACTERISTICS

VARIETIES
Ornamental variety: Vivipara.

FERTILIZATION
The communities in which *P. alpina* grows are slightly fertilized or else not fertilized at all.

DRY MATTER YIELDS AND SEASONALITY OF PRODUCTION
Low or moderate production in the sites where it appears spontaneously.

NUTRITIVE VALUE
Not known but probably good at a leafy stage.

ACCEPTABILITY
Very well accepted at a leafy stage.

DISEASES
The diseases of *P. alpina* are probably similar to other *Poa* spp. (Ellis and Ellis, 1997).

USE FOR PURPOSES OTHER THAN FORAGE PRODUCTION
Can be incorporated in mixtures for sowing new ski pistes in mountain areas.

MAIN QUALITIES
Well adapted to extensive systems in mountain areas.

MAIN SHORTCOMINGS
Not productive enough to be used in intensive systems.

FIGURE 5.170
Ecological optimum and range for soil pH and humidity of *Poa alpina*

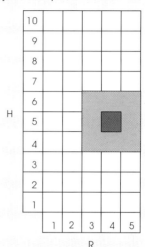

R: 1 = extremely acid and 5 = alkaline
H: 1 = extremely dry and
10 = permanently flooded

Complete key in Chapter 2, section 2.8.

FIGURE 5.171
Ecological optimum and range for nutrient availability and management of *Poa alpina*

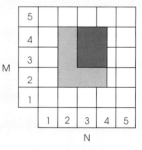

N: 1 = very low and 5 = very high
M: 1 = one defoliation per year and
5 = 5 defoliations per year or more

Complete key in Chapter 2, section 2.8.

POA ANNUA L.

ETYMOLOGY
GREEK: πoα = poa = sward.
LATIN: *annuus* = annual.

COMMON NAMES
ENGLISH: Annual meadow grass
FRENCH: Pâturin annuel
GERMAN: Einjähriges rispengras

DISTRIBUTION
Subcosmopolitan but only at altitude in tropical areas.

DESCRIPTION

MORPHOLOGICAL DESCRIPTION
Annual plant, small-sized, hairless, caespitose. Stems creeping to creeping-ascendent, sometimes rooted at the nodes, or erect, 5–25 cm high. Leaf blade folded when

FIGURE 5.172
Inflorescence and part of the leaf of *Poa annua*

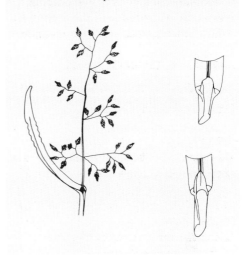

young, soft, 2–5 mm wide, flat, yellowish green, often accordion-pleated in the top third, with 2 highly visible veins on each side of the central nerve ('ski tracks') and with a hooded tip. Ligule visible, 2–5 mm

A. PEETERS

PLATE 5.83
Poa annua

P. ALLARD

PLATE 5.84
Poa annua

long, white. No auricles. Panicle-like inflorescence, pyramidal, few branches but all oriented to one side of the axis (180°). Spikelets 3–6-flowered, without woolly hairs at the base (Figure 5.172) (Plates 5.83 and 5.84).

1000-SEED WEIGHT
0.2–0.3 g (small seeds).

CHROMOSOME NUMBER
2n = 28 (tetraploid).

BIOLOGICAL FLORA
Hutchinson and Seymour (1982).

ECOLOGICAL REQUIREMENTS

ALTITUDE – VEGETATION BELTS
From the lowlands to the highlands; from hill to subalpine belts.

SOIL MOISTURE
Optimum on normally drained to cool soils (Figure 5.173).

SOIL FERTILITY
Optimum on nutrient-rich soils, neutral to slightly acid, but large range of distribution (Figures 5.173 and 5.174).

SOIL TYPE AND TOPOGRAPHY
All soil types.

CLIMATE REQUIREMENTS
Large climate range. Recovers from extreme weather periods through its seed bank in the soil.

TOLERANCE OF CUTTING AND GRAZING
Species typical of over-trampled places: paddock entrances, paths, damaged swards of pastures. Colonizes empty places, even gaps in the centre of paddocks created by animal poaching. Rare in cutting grassland (Figure 5.174).

FIGURE 5.173
Ecological optimum and range for soil pH and humidity of *Poa annua*

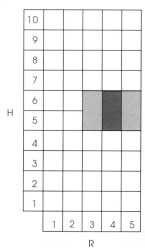

R: 1 = extremely acid and 5 = alkaline
H: 1 = extremely dry and 10 = permanently flooded

Complete key in Chapter 2, section 2.8.

FIGURE 5.174
Ecological optimum and range for nutrient availability and management of *Poa annua*

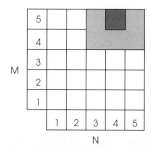

N: 1 = very low and 5 = very high
M: 1 = one defoliation per year and
5 = 5 defoliations per year or more

Complete key in Chapter 2, section 2.8.

SOCIABILITY – COMPETITIVENESS – PLANT COMMUNITIES
Uncompetitive. Often associated with other species such as greater plantain (*Plantago major*), knot grass (*Polygonum aviculare*), pineappleweed

(*Matricaria discoidea*) and common chickweed (*Stellaria media*), which colonize bare ground, e.g. at entrances of paddocks.

Mostly frequent in *Cynosurion cristati*, intensive grasslands, improved grasslands and *Polygonion avicularis*.

COMPATIBILITY WITH OTHER FORAGE GRASSES AND LEGUMES
Often associated with *Lolium perenne* in poached paddocks.

INDICATOR VALUE
Sward degraded by excessive trampling or grazing during wet periods that leads to uprooting of perennial grasses.

AGRONOMIC CHARACTERISTICS

FERTILIZATION
The communities in which *P. annua* grows are usually fertilized, even intensively.

DRY MATTER YIELDS AND SEASONALITY OF PRODUCTION
Low to very low production.

NUTRITIVE VALUE
Good quality.

ACCEPTABILITY
Very well accepted at a leafy stage.

DISEASES
Susceptible to rust and can be attacked by *Puccinia graminis*, which provides a continuous source of inoculum (Smith, Jackson and Woolhouse, 1989). Can also be invaded by *Puccinia poae-nemoralis, P. poarum* (alternative host: *Tussilago farfara*) and *Uromyces dactylidis* (Ellis and Ellis, 1997). It is also particularly susceptible to snow mould (*Microdochium* [= *Fusarium*] *nivale*), in contrast to *Poa pratensis*, and can also be attacked by *Colletotrichum graminicola*.

Xanthomonas translucens pv. *poae* causes a bacterial wilt and could be used as a biological means to control *P. annua* (Imaizumi *et al.*, 1997).

MAIN QUALITIES
Good nutritive value.

MAIN SHORTCOMINGS
Not very productive. Colonizes gaps in swards and prevents ingress and establishment of more productive species.

POA PRATENSIS L.

FIGURE 5.175
Inflorescence and part of the stem, rhizome
and leaf of *Poa pratensis*

Very variable species which includes several
subspecies. The subspecies described below
is *pratensis*.

ETYMOLOGY
GREEK: ποα = poa = sward.
LATIN: *pratensis* = from meadows.

COMMON NAMES
ENGLISH: Smooth meadow grass, smooth-
stalked meadow grass, meadow grass;
Kentucky bluegrass in USA
FRENCH: Pâturin des prés
GERMAN: Wiesen rispengras

DISTRIBUTION
Subcosmopolitan mainly in temperate and
cold areas.

PLATE 5.85
Poa pratensis

PLATE 5.86
Poa pratensis

DESCRIPTION

MORPHOLOGICAL DESCRIPTION
Perennial plant, medium-sized, usually hairless, rhizomatous. Stems erect or ascendent, 15–80 cm high. Leaf blade folded when young, long (up to 30 cm), rather narrow (1–5 mm), quite stiff, dark green to greyish green, with 2 visible veins on each side of the central nerve ('ski tracks') and with a hooded tip (by smoothing down, the top of the blade tears and looks like a cloven hoof). Ligule of the lower leaves very short, truncate. Ligule of the upper leaves longer (1–3 mm), but always rather short and truncate. No auricles. Panicle-like inflorescence, spreading, oblong-pyramidal, with branches clustered per 2 to 5 and tendency to form 3 short branches and 3 long branches at each level, though this feature is very irregular. Spikelets 3–5-flowered, with woolly hairs at the base. Spikelets more robust than *Poa trivialis* (Figure 5.175) (Plates 5.85 and 5.86).

PHYSIOLOGICAL PECULIARITIES
Apomictic reproduction (ovule producing a seed without pollen fertilization).

1000-SEED WEIGHT
0.3–0.5 g (small seeds).

CHROMOSOME NUMBER
2n = (28, 42), 50 to 124.

ECOLOGICAL REQUIREMENTS

ALTITUDE – VEGETATION BELTS
From the lowlands to high altitudes in mountain areas; from hill to alpine belts.

SOIL MOISTURE
Optimum on slightly dry (sand and shallow soils, namely) to slightly wet soils, but large range of distribution from wet to very dry soils (Figure 5.176).

SOIL FERTILITY
Large range for nutrient availability. Can be

FIGURE 5.176
Ecological optimum and range for soil pH and humidity of *Poa pratensis*

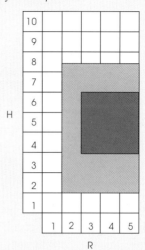

R: 1 = extremely acid and 5 = alkaline
H: 1 = extremely dry and 10 = permanently flooded

Complete key in Chapter 2, section 2.8.

FIGURE 5.177
Ecological optimum and range for nutrient availability and management of *Poa pratensis*

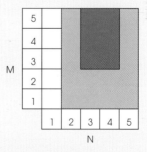

N: 1 = very low and 5 = very high
M: 1 = one defoliation per year and 5 = 5 defoliations per year or more

Complete key in Chapter 2, section 2.8.

abundant on soils moderately to very rich in nutrients and on acid to alkaline soils (mesotrophic to eutrophic species). However, can also survive on infertile soils (Figures 5.176 and 5.177). Salt-tolerant.

SOIL TYPE AND TOPOGRAPHY
Preference for light, sandy soils but large range.

CLIMATE REQUIREMENTS
Large climate range. Very resistant to cold, drought and heat. Marked thermophilous tendency. Can be a suitable alternative to *Lolium perenne* in grazed pastures in dry continental climates or on dry sands.

TOLERANCE OF CUTTING AND GRAZING
Very flexible for the management regime but more abundant in grazed pastures. Very resistant to trampling and to frequent defoliation (up to almost 20 mowings/year in lawns). In cutting grassland, it must be cut rather higher than the average for grasses to ensure good regrowth and persistence (Figure 5.177).

SOCIABILITY – COMPETITIVENESS – PLANT COMMUNITIES
Extremely slow growth after sowing, but becomes competitive. Once established in mixtures, develops patches in the sward that increase progressively with rhizome development. In favourable conditions, can constitute large almost monospecies populations, but is often associated with diversified communities, including species-rich grasslands.

Mostly frequent in *Cynosurion cristati*, intensive grasslands, *Arrhenatherion elatioris*, *Mesobromion erecti* and lawns.

COMPATIBILITY WITH OTHER FORAGE GRASSES AND LEGUMES
Often associated with *Lolium perenne*, *Poa trivialis*, *Holcus lanatus*, *Festuca rubra*, *Agrostis capillaris*, *Dactylis glomerata* and *Trifolium repens*, especially in grazed pastures.

AGRONOMIC CHARACTERISTICS

SOWING RATES
Pure: 15–25 kg/ha.
Mixed: 5–10 kg/ha in simple mixtures for grazing, with *Lolium perenne* and *Trifolium repens* mainly, though *Phleum pratense* can be added.

TABLE 5.61
Examples of varieties of *Poa pratensis*

Variety	Seed breeder (country)	Resistance to rust
Compact	Limagrain genetics (F)	Resistant
Lato	Steinach (D)	Resistant
Leikra*	Hellerud (N)	Very resistant
Licox	DSV (D)	Resistant
Monopoly	Mommersteeg (NL)	Very resistant
Tommy	Limagrain Genetics (F)	Resistant

*Adapted to high altitudes

TABLE 5.62
Relative yield (%) of *Poa pratensis* compared with *Lolium perenne* during the 3 first years of production (yield of L. perenne = 100)

	Year of production		
	1	2	3
P. pratensis	69	91	102

Source: recalculated from Frame, 1990

In dry areas, can be sown with *Dactylis glomerata* and *Festuca rubra*. In Switzerland, *P. pratensis* is included at 5–10 kg/ha in mixtures adapted to several altitudes.

A shallow sowing depth is necessary because of its small seed size. In mixture with other species, it is often sown too deep and this militates against a good establishment.

P. pratensis develops very slowly after sowing, even in pure stand. In mixture, it often appears only after 2–3 years, but then it spreads in the sward through its rhizomes, giving full production after 3–4 years.

VARIETIES
A list of varieties is shown in Table 5.61. It is important to choose varieties resistant to rusts.

DRY MATTER YIELDS AND SEASONALITY OF PRODUCTION
Moderately or even barely productive the first 2 years after the establishment year, *P. pratensis* becomes fully productive from the third to the fourth harvest year (Table 5.62).

Over 3 years, in a 6 cuts/year regime, *P. pratensis* was less productive than *Lolium perenne* regardless of N rate (Table 5.63).

Its digestible organic matter yield is even less in relation to that of *Lolium perenne* because of a lower digestibility.

It is particularly compatible with *Trifolium repens* in spite of its high sward density (Table 5.64).

A *P. pratensis-Trifolium repens* mixture can be as productive as a *Lolium perenne-T. repens* mixture from the first year of production, because of a greater contribution of *T. repens* in the total annual yield.

The growth of *P. pratensis* is slower than *Lolium perenne* in spring. It also accumulates less standing biomass at the end of the first cycle, i.e., by the beginning of June (Table 5.65).

NUTRITIVE VALUE
The digestibility of *P. pratensis* is appreciably lower than *Lolium perenne* (Table 5.66) and is often similar to that of *Festuca rubra*.

At the same levels of N application, its N content is higher than *Lolium perenne*. Its mineral content is similar to productive

TABLE 5.63
Annual yields (tonnes DM/ha) of *Poa pratensis* compared to *Lolium perenne* in Scotland, UK

Species	Annual N fertilization (kg/ha)				
	0	120	240	360	480
L. perenne	2.4	6.1	9.7	11.9	13.2
P. pratensis	1.5	4.9	8.0	10.2	11.8

Note: 6 cuts/year regime
Source: Frame, 1991

TABLE 5.64
Comparison of the abundance of *Trifolium repens* (% on a DM basis) in *Poa pratensis* and *Lolium perenne* swards for 3 years

	Year of production		
	1	2	3
L. perenne	23	34	42
P. pratensis	44	45	54

Note: 6 cuts/year regime with an annual fertilization of 50 kg N/ha applied in spring
Source: Frame, 1990

TABLE 5.65

Development of yields (tonnes DM/ha) of *Poa pratensis* compared to *Lolium perenne* during the first growth cycle in the springs of 1991–93 in Belgium

				Dates		
1991	8 April	15 April	24 April	13 May	27 May	11 June
L. perenne	1.0	1.8	2.8	6.0	8.7	12.9
P. pratensis	0.5	1.2	2.0	4.8	7.0	8.8
1992	15 April	29 April	7 May	19 May	27 May	9 June
L. perenne	0.9	2.8	4.7	7.8	10.1	9.9
P. pratensis	1.3	2.5	4.0	6.2	7.2	7.7
1993	16 April	26 April	4 May	17 May	25 May	1 June
L. perenne	1.7	3.8	5.4	7.7	9.2	9.7
P. pratensis	0.8	2.3	4.1	6.3	7.4	8.0

Note: N fertilization during the first growth cycle in spring: 100 kg/ha
Source: Peeters and Decamps, 1994

TABLE 5.66

Organic matter digestibility (% OMD) of *Poa pratensis* compared to *Lolium perenne* in several management systems

	L. perenne	*P. pratensis*
6 cuts/year regime 2-year average	79.6	70.7
3 cuts/year regime 1-year data	72.5	60.7
4 cuts/year regime 3-year average	76.5	67.4

Source: Frame, 1989; 1991

grasses, though slightly lower than *Lolium perenne*.

Figure 5.178 shows that the decline in the rate of digestibility for *P. pratensis* is much faster than for *Lolium perenne* (-0.63 versus -0.41 percent per day on average), especially after mid-May. This decline was the fastest of the 9 tested grasses and was rather close to that of *Arrhenatherum elatius*.

ACCEPTABILITY
Well accepted except when it is invaded by *Puccinia* spp., and then intake is very low.

ANIMAL PERFORMANCE
Animal performance when fed *P. pratensis* can be high, but especially if it is mixed with more digestible species like *Lolium perenne* or *Trifolium repens*.

DISEASES
Particularly susceptible to *Puccinia* spp., since it can be attacked by 6 different species. *Puccinia poarum* (alternative host: *Tussilago farfara*) and *P. poae-nemoralis* seem to be the most frequent in Western Europe. Can also be attacked by *Puccinia coronata*, *P. graminis*, *P. striiformis* ssp. *poae* and *Uromyces dactylidis* (Ellis and Ellis, 1997).

There are large variety differences in susceptibility to *P. poarum*, *P. poae-nemoralis* and *P. graminis*, but it is important to emphasize that the differences in susceptibility to these 3 fungi are not necessarily correlated (Smith, Jackson and Woolhouse, 1989).

This *Poa* is also attacked by *Drechslera poae* and by *Spermospora poagena*. In contrast to *Poa annua*, *P. pratensis* is rather resistant to *Microdochium* (= *Fusarium*) *nivale*.

USE FOR PURPOSES OTHER THAN FORAGE PRODUCTION

Often used in mixture with *Festuca rubra, F. ovina* and *Agrostis capillaris* for the sowing of fine lawns. Resistant to uprooting in sport grounds. Its good drought resistance enables lawns to keep a green appearance in summer.

Constituent of wild flora mixtures for neutral soils (Crofts and Jefferson, 1999). Regularly incorporated in seed mixtures for sowing new ski pistes in mountain areas. Good for stabilizing soils on road and motorway verges, as well as in disused industrial sites. Sown in forest clearings.

MAIN QUALITIES

Productive, very persistent grass. A good complement to *Lolium perenne* because it increases its yield year after year following sowing, while *L. perenne* production decreases. It can colonize the gaps in the sward and so prevent invasion by less productive species. Very compatible with *Trifolium repens*. Good drought resistance and contributes relatively more than *Lolium perenne* to summer yield. In continental climates and on dry soils, constitutes an alternative to *Lolium perenne*. In mountain areas, it is one of the best grasses along with *Festuca rubra* in grazed swards.

MAIN SHORTCOMINGS

Markedly lower digestibility than *Lolium perenne*. More susceptible to *Puccinia* spp.

FIGURE 5.178
Evolution of the digestibility of *Poa pratensis* compared with *Lolium perenne* during an uninterrupted growth period in spring (100 kg N/ha) (linear regression equations and determination coefficients shown)

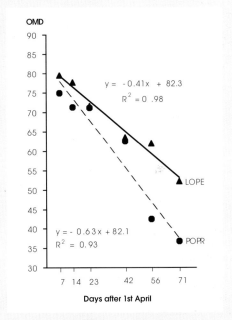

$$y = -0.41x + 82.3$$
$$R^2 = 0.98$$

$$y = -0.63x + 82.1$$
$$R^2 = 0.93$$

POPR: *P. pratensis;* LOPE: *L. perenne*

Source: Peeters, 1992

than *Lolium perenne* and when it is invaded by this disease in summer, it is generally rejected by cattle. Grows very slowly after sowing and only becomes productive after 2–3 years.

POA TRIVIALIS L.

FIGURE 5.179
Inflorescence and part of the leaf of *Poa trivialis*

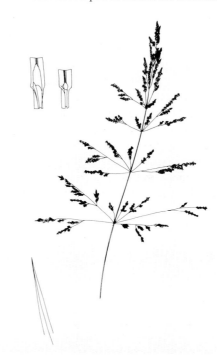

ETYMOLOGY
GREEK: ποα = poa = sward.
LATIN: *trivialis* = common.

COMMON NAMES
ENGLISH: Rough meadow grass, rough
stalked meadow grass; rough bluegrass in
USA
FRENCH: Pâturin commun
GERMAN: Gemeines rispengras

DISTRIBUTION
Native to temperate and cold areas of the
northern hemisphere.

DESCRIPTION

MORPHOLOGICAL DESCRIPTION
Perennial plant but short-lived, average-
sized, hairless, caespitose. Stems erect or

PLATE 5.87
Poa trivialis

PLATE 5.88
Poa trivialis

ascendent and in the latter case often rooted on the nodes, 20–100 cm high. Leaf blade folded when young, long (up to 20 cm), rather narrow (1.5–6 mm), rather soft, pale green to yellowish green, with 2 visible veins on each side of the central nerve ('ski tracks'), tip not or slightly hooded. Ligule of the lower leaves very short, acute. Ligule of the upper leaves, long (4–8 or even 10 mm), acute. No auricles. Panicle-like inflorescence, spreading, pyramidal, with branches clustered per 4–6 and tendency to form 3 short branches and 3 long branches at each level, but this feature is very irregular. Spikelets 2–4-flowered, with woolly hairs at the base. Spikelets thinner than *Poa pratensis* (Figure 5.179) (Plates 5.87 and 5.88). Very variable species.

1000-SEED WEIGHT
0.2–0.3 g (small seeds).

CHROMOSOME NUMBER
2n = 14 (diploid) or 28 (tetraploid).

ECOLOGICAL REQUIREMENTS

ALTITUDE – VEGETATION BELTS
From the lowlands up to 2 000 m in the Alps (Rameau *et al.*, 1993); from hill to subalpine belts.

SOIL MOISTURE
Large range of soil moisture. Optimum on slightly to rather wet soils. Much rarer on dry soils in contrast to *Poa pratensis* (Figure 5.180).

SOIL FERTILITY
Optimum on nutrient-rich soils (eutrophic species) and slightly acid to neutral soils (Figures 5.180 and 5.181).

SOIL TYPE AND TOPOGRAPHY
Adapted to a range of soil textures.

CLIMATE REQUIREMENTS
Large climate range. Tolerates shade, e.g. in old orchards. Drought-sensitive.

FIGURE 5.180
Ecological optimum and range for soil pH and humidity of *Poa trivialis*

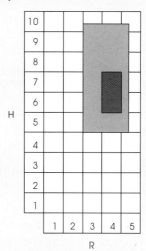

R: 1 = extremely acid and 5 = alkaline
H: 1 = extremely dry and 10 = permanently flooded

Complete key in Chapter 2, section 2.8.

FIGURE 5.181
Ecological optimum and range for nutrient availability and management of *Poa trivialis*

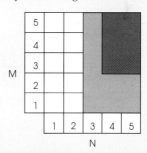

N: 1 = very low and 5 = very high
M: 1 = one defoliation per year and
5 = 5 defoliations per year or more

Complete key in Chapter 2, section 2.8.

TOLERANCE OF CUTTING AND GRAZING
Widespread in grazed pastures and in cutting grassland, so long as the level of soil fertility is sufficiently high. Abundant in cutting leys (Figure 5.181).

SOCIABILITY – COMPETITIVENESS – PLANT COMMUNITIES

Competitive. Opportunistic species with a short lifetime that quickly produces seeds and establishes in gaps in the swards of more productive species like *Lolium perenne*, *Phleum pratense* and *Festuca pratensis*.

Mostly frequent in intensive grasslands, humid improved grasslands, *Cynosurion cristati, Arrhenatherion elatioris, Polygono-Trisetion flavescentis, Bromion racemosi, Deschampsion caespitosae, Agropyro-Rumicion crispi* and lawns.

COMPATIBILITY WITH OTHER FORAGE GRASSES AND LEGUMES

Compatible with other productive forage species including *Trifolium repens* and *T. pratense*.

INDICATOR VALUE

Soil never dry or nutrient-poor.

AGRONOMIC CHARACTERISTICS

SOWING RATES

Pure: 15–25 kg/ha.
Rarely used in forage mixtures, but it appears spontaneously and often massively in suitable sites.

VARIETIES

Few bred varieties: Dasas and Omega ϕtofte (Dansk Planteforædling, DK), Sabre (Nungesser KG, D).

FERTILIZATION

The grasslands where it is found are usually fertilized, even heavily fertilized.

DRY MATTER YIELDS AND SEASONALITY OF PRODUCTION

Low to moderate production. In pure-sown stands, production only lasts 1 or 2 years, then the sward becomes very thin. Frame (1991) recorded an 'intermediate' production (8.8 tonnes DM/ha), lower than *Lolium perenne*, *Holcus lanatus* and *Festuca rubra*. In a trial carried out in Belgium (Peeters and Decamps, 1999), which included 3 cutting regimes (2, 4 or 6 cuts/year) and 4 N rates (0, 100, 200 and 300 kg N/ha), *P. trivialis* had the lowest production among the 12 tested grasses. Maximum annual yields ranged between 4.4 and 5.6 tonnes DM/ha, while *Lolium perenne* yields ranged between 5.9 and 9.1 tonnes DM/ha (Figures 5.182 to 5.185). *P. trivialis* was also the least productive of the 4 grasses tested by Haggar (1976): 7.8 tonnes DM/ha against 11.0 tonnes DM/ha for *Lolium perenne*, and also the least productive of the 7 grasses tested by Frame (1989): 8.4 tonnes DM/ha versus 15.3 tonnes DM/ha for *L. perenne*.

The production of *P. trivialis* is proportionately much more important in spring compared to most productive grasses. Summer and autumn regrowths are low (Table 5.67).

At the beginning of spring, the growth rate is similar to *Lolium perenne* and can reach 110 kg DM/ha.day, but from the end

TABLE 5.67

Yields (kg DM/ha) of the first cycle of uninterrupted growth, of 2 regrowths and annual production of *Poa trivialis* compared with *Lolium perenne*

	1st growth cycle						Regrowths		Total
	27 March	10 April	24 April	8 May	22 May	5 June	1 August	5 Sept	
L. perenne	20	60	220	608	1 413	5 148	1 950	695	7 693
P. trivialis	6	54	117	479	1 449	3 869	200	44	4 113

Note: annual N fertilization: 120 kg/ha
Source: Haggar, 1976

FIGURES 5.182–5.185

Annual yields of *Poa trivialis* compared with *Lolium perenne* and *Phleum pratense* in 3 cutting regimes and several nitrogen fertilization levels (average of 3 production years)

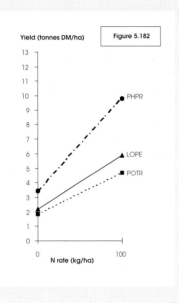

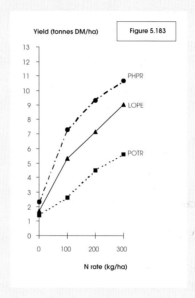

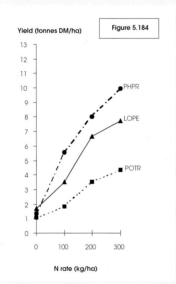

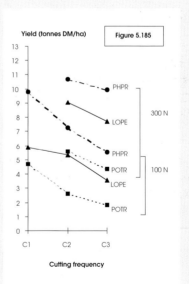

POTR: *P. trivialis;* LOPE: *L. perenne;* PHPR: *P. pratense.* Fig. 5.182: 2 cuts/year (C1); Fig. 5.183: 4 cuts/year (C2); Fig. 5.184: 5–6 cuts/year (C3); Fig. 5.185: 100 and 300 kg N/ha

Source: Peeters and Decamps, 1999

of May, the growth rate of *P. trivialis* is clearly lower than *L. perenne* (Haggar, 1976). Further information on growth rate is shown in Table 5.68. In this experiment, the growth of *P. trivialis* was slower than *Lolium perenne* and it accumulated less standing biomass.

NUTRITIVE VALUE

The digestibility of *P. trivialis* may be lower or higher than *Lolium perenne* (Table 5.69).

The digestibility evolution during the first growth cycle is almost identical to that of *Lolium perenne* (Haggar, 1976). Figure 5.186 shows that the rate of decline of digestibility of *P. trivialis* is faster, but the digestibility values are very close.

However, since *P. trivialis* is markedly less productive than *Lolium perenne*, the DOM yields are clearly inferior for *P. trivialis*. At identical N rates, its N content is higher, especially in summer and in autumn when its growth is slow.

ACCEPTABILITY

Very good intake when grazed before heading. Hay of *P. trivialis* is also very well accepted.

ANIMAL PERFORMANCE

There are no data on performance by animals fed on pure stands of *P. trivialis*. However, judged by the information above, performance would be much lower than on pure stands of *Lolium perenne*.

DISEASES

Susceptible to rusts, it can be attacked by 4 different species. *Puccinia poarum* and *P. poae-nemoralis* seem to be the most frequent but *P. graminis* and *Uromyces dactylidis* are also widespread (*P. poarum* also thrives on *Tussilago farfara*) (Ellis and Ellis, 1997).

This *Poa* is also attacked by *Drechslera poae*.

TABLE 5.68

Development of yields (tonnes DM/ha) of *Poa trivialis* compared with *Lolium perenne* during the first growth cycle of the springs of 1991–93 in Belgium

	Dates					
1991	8 April	15 April	24 April	13 May	27 May	11 June
L. perenne	1.0	1.8	2.8	6.0	8.7	12.9
P. trivialis	0.7	1.3	1.6	4.0	7.3	9.5
1992	15 April	29 April	7 May	19 May	27 May	9 June
L. perenne	0.9	2.8	4.7	7.8	10.1	9.9
P. trivialis	1.0	2.2	3.5	6.6	7.0	7.8
1993	16 April	26 April	4 May	17 May	25 May	1 June
L. perenne	1.7	3.8	5.4	7.7	9.2	9.7
P. trivialis	1.3	3.2	4.9	6.7	7.5	8.3

Note: N fertilization during the first growth cycle in spring: 100 kg/ha
Source: Peeters and Decamps, 1994

TABLE 5.69

Comparison of the digestibility (%) of *Poa trivialis* and *Lolium perenne* according to several authors

Type of data	*P. trivialis*	*L. perenne*	Reference
DOMD (D-value)	71.1	65.4	Haggar (1976)
OMD	76.8	80.0	Frame (1991)
OMD	73.9	78.7	Frame (1989)

USE FOR PURPOSES OTHER THAN FORAGE PRODUCTION

Constituent of wild flora mixtures for neutral (Crofts and Jefferson, 1999) and wet soils.

MAIN QUALITIES

High digestibility, well accepted before heading. Colonizes the empty patches in grassland by establishing among the plants of more productive species like *Lolium perenne* or *Phleum pratense*. Extremely widespread in intensive grasslands and in many cases, its presence certainly increases the yield.

MAIN SHORTCOMINGS

Relatively low production and not very persistent (1–2 years on average). Heads very early in spring and so is difficult to control in grazed swards. Once headed, it is refused by cattle and so it is often necessary to cut the rejected plants to ensure leafy regrowths. Summer and autumn production are very low.

FIGURE 5.186

Evolution of the digestibility of *Poa trivialis* compared with *Lolium perenne* during an uninterrupted growth period in spring (100 kg N/ha) (linear regression equations and determination coefficients shown)

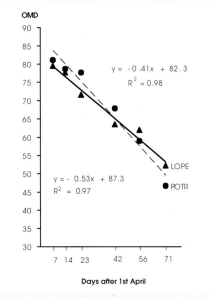

POTR: *P. trivialis;* LOPE: *L. perenne*

Source: Peeters, 1992

TRISETUM FLAVESCENS (L.) BEAUV.

SYNONYM
Avena flavescens L.

ETYMOLOGY
LATIN: *tres* = 3 and *seta* = bristle.
Spikelet with 3 arms.
LATIN: *flavescens* = yellowish.

COMMON NAMES
ENGLISH: Yellow oat-grass, golden oat-grass
FRENCH: Avoine jaunâtre, avoine dorée
GERMAN: Goldhafer

DISTRIBUTION
Native to Europe, western Asia and
Northern Africa.

DESCRIPTION

MORPHOLOGICAL DESCRIPTION
Perennial plant, medium-sized, hairy (a few

FIGURE 5.187
Inflorescence, basis of the plant and part of
the leaf of *Trisetum flavescens*

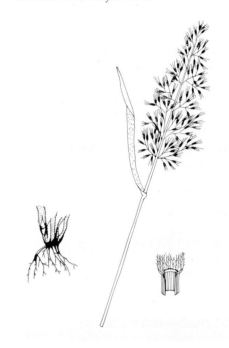

PLATE 5.89
Trisetum flavescens

PLATE 5.90
Trisetum flavescens

exceptions), caespitose, with short rhizomes. Stems erect, 20–80 cm high, not very leafy. Leaf blade rolled when young, rather wide (2–4 mm), with visible veins, more hairy on the upper side than on the lower side which can be hairless. Some forms have no hairs on the blade. Sheath of the lower leaves often very hairy, hairs bottom-oriented. Ligule short (up to 2 mm), dentate. No auricles. Panicle-like inflorescence, spreading, 6–15 cm long, yellowish green to yellowish at seed maturity. Spikelets 2–4-flowered. Lemma with long awn (Figure 5.187) (Plates 5.89 and 5.90).

1000-SEED WEIGHT
0.2–0.3 g (small seeds).

CHROMOSOME NUMBER
2n = 28 (tetraploid).
Another chromosome number has also been observed: 2n = 24.

ECOLOGICAL REQUIREMENTS

ALTITUDE – VEGETATION BELTS
From low altitudes in the hill belt up to the alpine belt, but much more abundant in the mountain and subalpine belts than elsewhere.

SOIL MOISTURE
Optimum on normally drained to slightly dry soils, but large range especially on dry soils. Dominant in irrigated hay meadows in mountain areas. Not adapted to very wet soils (Figure 5.188).

SOIL FERTILITY
Optimum on soils that are nutrient-poor to moderately nutrient-rich (mesotrophic species). Large range of pH from acid to alkaline soils (Figures 5.188 and 5.189).

SOIL TYPE AND TOPOGRAPHY
Indifferent to soil texture.

FIGURE 5.188
Ecological optimum and range for soil pH and humidity of *Trisetum flavescens*

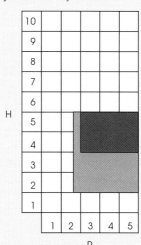

R: 1 = extremely acid and 5 = alkaline
H: 1 = extremely dry and
10 = permanently flooded

Complete key in Chapter 2, section 2.8.

FIGURE 5.189
Ecological optimum and range for nutrient availability and management of *Trisetum flavescens*

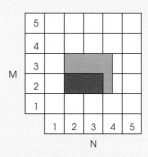

N: 1 = very low and 5 = very high
M: 1 = one defoliation per year and
5 = 5 defoliations per year or more

Complete key in Chapter 2, section 2.8.

CLIMATE REQUIREMENTS
Large climate range, but favoured by warm microclimates.

TOLERANCE OF CUTTING AND GRAZING
Species typical of hay meadows in mountain areas where it is often dominant (Figure 5.189). However, it can also be found at low abundance in lowlands, in dry grazed or cut grasslands. Its ability to colonize gaps in the sward because of early seed production probably explains its presence in these low-density swards. Susceptible to trampling and frequent defoliation.

SOCIABILITY – COMPETITIVENESS – PLANT COMMUNITIES
Moderately competitive. Grasslands dominated by *T. flavescens* usually include a great number of species.

Mostly frequent in *Polygono-Trisetion flavescentis*, dry intensive grasslands, *Arrhenatherion elatioris* and *Mesobromion erecti*.

COMPATIBILITY WITH OTHER FORAGE GRASSES AND LEGUMES
Compatible with other forage species like *Dactylis glomerata, Phleum alpinum, Poa* spp. and *Trifolium* spp. *Trifolium pratense* in particular can form large spontaneous populations in *T. flavescens* grasslands. In Switzerland, *T. flavescens* is used in several mixtures along with *Dactylis glomerata* and *Poa pratensis* for dry areas (Mosimann *et al.*, 1996a).

INDICATOR VALUE
Sward extensively grazed on dry soils or cut for hay.

AGRONOMIC CHARACTERISTICS

SOWING RATES
Pure: 20–30 kg/ha.
Mixed: 3–5 kg/ha in mixtures adapted to altitudes unfavourable to *Lolium perenne* (Lehmann *et al.*, 1992).

VARIETIES
Few forage varieties: Roznovsky (Oseva, CZ), Trisett 51 (Steinach, D).
Ornamental variety: Peter Hall.

FERTILIZATION
Moderate needs.

DRY MATTER YIELDS AND SEASONALITY OF PRODUCTION
The annual yield can reach 15–16 tonnes DM/ha (Table 5.70). In Switzerland, the yield produced in a 3 cuts/year regime was not appreciably different from 4 cuts/year. In the lowlands, *T. flavescens* is always less productive than *Arrhenatherum elatius* and its sward is also more easily invaded by other species (Lehmann *et al.*, 1992).

ACCEPTABILITY
Well accepted as hay.

TOXICITY
Can cause calcinosis (cattle disease characterized by a bone porosity) especially if it is consumed in large amounts (> 50 percent of the feed) and at a young stage.

TABLE 5.70
Annual yields (tonnes DM /ha) of *Trisetum flavescens* for 2 years at different altitudes in Switzerland

	1990			1991	
			Altitude (m)		
Variety	520	560	1 200	520	560
Trisett 51	12.9	15.3	9.2	10.8	13.4
CZ origin	12.4	14.5	8.4	11.1	12.4

Note: 3 cuts/year regime
Source: Lehmann *et al.*, 1992

After heading, the calcinosis risk disappears. The hay does not cause calcinosis but silage will not stop the disease from developing if the fresh forage has shown risk (Köhler *et al.*, 1978).

DISEASES

Two rusts have been observed: *Puccinia graminis* and *P. recondita* spp. *triseti.* A smut, *Ustilago hypodytes,* has also been noted (Ellis and Ellis, 1997).

USE FOR PURPOSES OTHER THAN FORAGE PRODUCTION

Has a graceful, 'light' appearance and golden colour after flowering. Thus, it is a suitable grass for ornamental purposes. Since it is not very competitive, it can be used in species-rich wild flora mixtures for acid to calcareous (pH 4.5–7.5) soils (Crofts and Jefferson, 1999).

MAIN QUALITIES

Well suited to hay meadows at high altitudes on moderately fertile soils. In these conditions, it is one of the most productive grasses and it spontaneously forms the basis of these grasslands. Wilts easily and can produce good quality hay. Can also be managed for silage production (3–4 cuts/year). Suitable as a component of mixtures for grazing with other grasses adapted to mountain areas.

MAIN SHORTCOMINGS

Not sufficiently productive and competitive for use in the lowlands. Ability to cause cattle calcinosis is a very unfavourable factor, but most grasslands dominated by this species are intended for hay production and thus do not present a risk.

PART III

Chapter 6

Advantages of wild or secondary grasses and complex multispecies and multivariety mixtures

6.1 ADVANTAGES OF WILD GRASSES MIXTURES

Lolium perenne, L. multiflorum and their hybrids, *Festuca pratensis, F. arundinacea, Phleum pratense* and *Dactylis glomerata* are the six most selected (bred) and most often sown forage grass species of temperate regions. *Festuca rubra* and *Poa pratensis* are two species often sown in lawns. Nevertheless, many other species are often very abundant in spontaneous swards of permanent grasslands and rangelands. In addition, in Europe, spontaneous grasses cover much greater areas than sown grasslands and these species therefore contribute to a large extent to livestock nutrition.

Data synthesized in the second part of the book on the agronomic potential of species that are either not selected or little-selected, reveal that some of these species are at least as productive as *Lolium perenne*. However, their nutritive value and their digestibility, in particular, is always lower. Nevertheless, their nutritive value can be adequate to feed animals with moderate or low nutritional requirements, like mixed cattle, suckler cows, heifers, oxen or ewes. Some species, such as *Arrhenatherum elatius, Poa* spp. and *Trisetum flavescens* are better suited to conservation as hay than *Lolium perenne*, because they dry more quickly. In contrast, their aptitude for conservation as silage is always less good than that of *Lolium perenne* because they are poorer in soluble carbohydrates. Nevertheless, ensiling these species is perfectly possible, though it necessitates special precautions, including the use of preservatives.

In Europe, vast areas have climates that are unsuited to *Lolium perenne* and *L. multiflorum*. This is particularly true in Central and Eastern Europe because of the summer drought, the intensity of winter frost and the existence of late spring frosts. Moreover, many soils are too wet, too acid or too poor in nutrients for the six most selected grasses, and it would be uneconomic to improve them. Nevertheless, other fodder grasses, *Alopecurus pratensis, Festuca rubra* and *Holcus lanatus* for example, could ensure a production yield on such soils that would allow extensive livestock production systems.

Wild grasses could be selected to improve some of their characteristics, especially disease resistance and digestibility. Although these species may be mainly of interest for permanent grasslands, selected cultivars could be introduced by over-seeding into this type of swards to improve forage production.

Some of these secondary species, which plant breeders have largely ignored for grass production, could be used for rehabilitating sites, erosion control, restoring biodiversity, land-use planning and sowing leisure areas. *Alopecurus pratensis, Anthoxanthum odoratum, Arrhenatherum elatius, Avenula* spp., *Briza media, Bromus erectus, Cynosurus cristatus, Deschampsia* spp., *Glyceria* spp., *Holcus* spp., *Koeleria pyramidata, Molinia caerulea, Nardus stricta, Phalaris arundinacea, Phleum alpinum* and *Poa alpina* are all suited to one or more of the above uses. These species need not always necessarily be bred for these types of use, but simply managed for seed production and sales. The seed multiplication would best be done regionally so as to maintain locally adapted ecotypes of these species and hence conserve their genomes.

6.2 GRASS SPECIES MIXTURES

As explained in Chapter 2, during the second half of the twentieth century, the botanical composition of grasslands in Western Europe was progressively simplified following intensification practices that reduced the number of wild species, together with the simplification of sown mixtures (Lazenby, 1981).

Lolium perenne and *L. multiflorum* have thus become the principal species used (Kley, 1995). *Lolium perenne* is sown, often in pure stand, for grazed pastures and *L. multiflorum*, in the same fashion, for cutting meadows. These species are in fact greatly appreciated for their important production potential and their excellent forage quality.

Are there risks connected with sowing pure stands? Have grass monocultures the same problems as arable crop monocultures? It is known that diseases can easily develop in crop monocultures. In addition, a species can be sensitive to one kind of weather incident while another one can be resistant. A mixture of species can thus be very resistant to different climatic events. The main advantages of mixed swards are discussed below.

Mixing species (*Lolium* spp.) that are very demanding for nutrients with more frugal ones – *Phleum pratense*, *Poa pratensis* or *Holcus lanatus*, for instance – requires less inputs, without great yield reduction, at least until soil fertility declines.

Mixtures of species in the same field, or complementary swards of different species in different fields, permit regular production throughout the year by combining species with high spring growth (*Alopecurus pratensis*, *Bromus* spp., *Holcus lanatus*, *Lolium multiflorum*) with species with good summer growth (*Agrostis* spp., *Dactylis glomerata*, *Festuca rubra*, *Poa pratensis*) and others with sustained autumnal growth (*Bromus* spp., *Dactylis glomerata*, *Lolium multiflorum*). For example, crops of *Lolium multiflorum*, which start their growth very early in spring, allow livestock to start grazing before *L. perenne* or

permanent pastures are ready to be grazed. The presence of *Holcus lanatus* and *Alopecurus pratensis* in a sward allows a silage cut to be taken very early in spring, before *Lolium perenne* has attained enough production. Mixing *Poa pratensis* with *Lolium perenne* allows more summer production during periods of drought, because the growth of *L. perenne* is very slow under dry conditions. *Agrostis capillaris*, *Dactylis glomerata*, *Elymus repens*, *Festuca arundinacea*, *F. pratensis* and *F. rubra* also replace *Lolium perenne* in summer in grasslands where these species are grown in mixture.

In cold climates, the use of frost-resistant grasses increases sward longevity and ensures better spring growth. In cold conditions, the main species for grazing and cutting mixtures are *Festuca pratensis*, *Phleum pratense* and *Poa pratensis*.

Festuca arundinacea, *F. pratensis*, *Glyceria fluitans* and *Phalaris arundinacea* resist winter flooding. In particular, the presence of *Festuca pratensis* is valuable in the bottoms of flood-prone valleys. Pure-sown swards of *Lolium perenne* are very susceptible to flooding.

In regions where there is a long period of snow cover in winter, snow mould can cause significant damage in *Lolium* spp. (Raynal *et al.*, 1989; Ellis and Ellis, 1997). In such conditions, the use of mixtures containing *Festuca pratensis* or *Phleum pratense* is particularly indicated.

In mixed grasslands, the combination of species suited to grazing (*Lolium perenne*) and others suited to cutting (*Phleum pratense*) or to dual purpose use (*Dactylis glomerata*, *Festuca pratensis*, *Poa pratensis*) allows a greater flexibility of management and a better overall persistence of the sward.

6.3 GRASS VARIETY MIXTURES

The simultaneous use of several varieties of a species in the same mixture reduces the risks of severe attacks of diseases because, within the genome of one grass species, there are great differences in susceptibility to fungi or bacterial pathogens (Vanbellinghen, Moreau and Maraite, 1991). The combination of varieties for disease resistance is important for *Dactylis glomerata, Lolium multiflorum, L. perenne* and *Poa pratensis.*

Using several varieties can also be interesting in the case of mixed-use grasslands (grazed and cut). Early or intermediate varieties of *Lolium perenne* adapted to early cutting for silage in spring can be sown with late varieties well adapted to grazing. With *Dactylis glomerata,* one can combine varieties with erect foliage adapted to cutting and varieties with more prostrate foliage suited to grazing.

6.4 GRASS AND DICOTYLEDON MIXTURES

Beyond mixtures of species and varieties of grasses, it may be useful to indicate that mixtures of grasses and dicotyledons, including legumes, can also bring a series of advantages. Dicotyledons are generally richer than grasses in bivalent minerals, calcium and magnesium; they are also often richer in minor elements, such as copper, zinc or manganese. Dicotyledons thus bring these important elements for animal feeding in the forage. The presence of species like *Trifolium repens* or dandelion (*Taraxacum* spp.) in grazed swards increases the amount of forage ingested by livestock (Thomson, 1984; Schneeberger, 1987). The fixation of nitrogen by legumes is obviously a considerable advantage.

6.5 PERMANENT AND TEMPORARY GRASSLANDS

Simple or monospecific mixtures cannot persist for long as such. New species can appear spontaneously and rapidly in these mixtures and the botanical composition of the sward becomes progressively more complex. Very simple swards can only exist in temporary grasslands of very short duration. The discussion about complex swards comes back in part to discussing the advantages of permanent grasslands compared to temporary

grasslands. In Great Britain, Hopkins *et al.* (1990) showed in a multilocation experiment over four years, that permanent grasslands with a complex grass flora produced as much as pure-sown *Lolium perenne* swards grown on the same sites and managed in an identical manner, if the lower yield from *L. perenne* in its establishment year is taken into account. Moreover, even only considering the three first years of full production of *Lolium perenne,* the yields of the two types of sward were nearly identical up to a fertilization of 200–300 kg N/ha. These results have upset much of the perceived wisdom about the presumed superiority of grasslands sown with selected varieties of *Lolium perenne* over swards composed of certain wild ecotypes of diverse grass species receiving adequate fertilization.

6.6 GRASS SEEDS AND GRASSLAND SOWINGS

6.6.1 Seeds and sowings

Most grasses have small or very small seeds (Table 6.1). The seeds are clustered in three categories according to their size: small, average or large, because the weight differences are considerable. For instance, the seeds of *Bromus* spp. can be almost 50 times bigger than the small seeds of *Agrostis* spp. However, even the so-called 'average-sized' seeds, like those of *Lolium perenne,* are much smaller than the seeds of arable crops like cereals, oilseed rape (*Brassica napus* ssp. *oleifera*) or pea (*Pisum sativum*).

The germination of seeds requires good contact between the soil and the seeds. Therefore, the soil must be well cultivated and consolidated to ensure moisture availability from the soil. 'Hollow' or 'puffy' seed beds lead to failures in seed germination. The sowing depth must be very shallow, 1–4 cm according to the seed size, but in most cases 1–2 cm are sufficient. After sowing, the soil must be carefully rolled to ensure soil-seed contact. Germination is often better in the tracks of tractor wheels, showing that the soil

TABLE 6.1
1000-seed weight and number of seeds per kg of selected forage grasses

Species	1000-seed weight		category**	Number of seeds per kg ('000)	
	minimum (g)	maximum (g)		minimum	maximum
Bromus erectus	4.0	5.3	l	190	250
Festuca gigantea	3.0	5.0	l	200	330
Hordeum secalinum	4.0	5.0	l	200	250
Bromus inermis	3.0	4.0	l	250	330
Arrhenatherum elatius	2.5	3.5	l	290	400
Bromus hordeaceus	2.5	3.0	l	330	400
Lolium multiflorum					
ssp. *westerwoldicum*	* D 2.5	3.0	a	330	400
	* T 3.7	5.1	l	200	270
Lolium multiflorum					
ssp. *multiflorum*	* D 2.0	2.5	a	400	500
	* T 3.0	4.6	a-l	220	330
Lolium perenne	* D 1.3	2.7	a	370	770
	* T 2.0	4.0	l	250	500
Avenula pubescens	2.0	2.6	a	385	500
Elymus repens	2.0	2.5	a	400	500
Festuca arundinacea	1.8	2.5	a	400	560
Festuca pratensis	1.7	2.1	a	480	590
Lolium x hybridum	1.7	2.0	a	500	590
Dactylis glomerata	0.8	1.4	a	710	1 250
Glyceria fluitans	1.0	1.2	a	830	1 000
Glyceria maxima	1.0	1.2	a	830	1 000
Festuca rubra	0.9	1.1	a	910	1 110
Festuca ovina	0.8	1.0	a	1 000	1 250
Alopecurus pratensis	0.60	0.99	a	1 010	1 670
Cynosurus cristatus	0.55	0.80	s	1 250	1 820
Phalaris arundinacea	0.60	0.80	s	1 250	1 670
Molinia caerulea	0.50	0.70	s	1 430	2 000
Phleum pratense	0.30	0.70	s	1 430	3 330
Briza media	0.40	0.64	s	1 560	2 500
Anthoxanthum odoratum	0.40	0.63	s	1 590	2 500
Nardus stricta	0.40	0.50	s	2 000	2 500
Poa pratensis	0.30	0.50	s	2 000	3 330
Deschampsia flexuosa	0.30	0.40	s	2 500	3 330
Holcus lanatus	0.30	0.40	s	2 500	3 330
Deschampsia caespitosa	0.20	0.30	s	3 330	5 000
Koeleria pyramidata	0.20	0.30	s	3 330	5 000
Poa annua	0.20	0.30	s	3 330	5 000
Poa trivialis	0.20	0.30	s	3 330	5 000
Trisetum flavescens	0.20	0.30	s	3 330	5 000
Agrostis gigantea	0.08	0.11	s	9 090	12 500
Agrostis canina	0.08	0.10	s	10 000	12 500
Agrostis capillaris	0.06	0.10	s	10 000	16 670
Agrostis stolonifera	0.06	0.09	s	11 110	16 670

* D: diploid, T: tetraploid
** l: large seeds, a: average-sized seeds, s: small seeds

has been insufficiently consolidated outside the tracks.

Seeds can be drilled or broadcast, the latter method being the faster. The soil is more rapidly covered when the seeds are broadcast and weeds are usually better controlled. However, the sowing rate must be increased. For example, a broadcast sowing of *Lolium perenne* seeds needs 30 kg/ha while 15–20 kg/ha are sufficient when it is drilled. The difference in production between the two types of establishment is negligible. Drills are usually 10–15 cm apart.

There are two main sowing periods: at the beginning of the spring or at the end of the summer. The spring sowing (beginning of April in Western Europe) is more suitable to get a good initial establishment. The ground warms up fast in April and soil moisture is normally sufficient. Weeds can easily be controlled, either by a herbicide treatment, or by topping the sward. However, the production is less during the following months than that from an established sward and farmers may not always readily accept this yield loss. Two alternatives are possible: sowing in spring under a cover crop, e.g. a cereal crop, or sowing direct at the end of summer.

When undersowing, the seed mixture should be sown just after the sowing of the spring cereal. The germination of the forage grasses occurs in April and their growth is then very slow but they survive in the understorey of the cereal. The cereal crop is normally harvested in August-September and the sward then grows rapidly and can be grazed in the autumn. This traditional system has several disadvantages: it is difficult to control weeds in the cereal crop if the grassland mixture includes legumes; it is difficult to harvest the cereal because the cereal stems stay wet due to the presence of the forage in the understorey; there may be strong competition to the grassland plants by the cereal, which can cause a significant mortality rate; and there may be lower yield of the cereal than when pure-sown. Therefore, it is preferable to sow the grassland mixture in spring under a cover crop – a spring cereal (oats or barley) sown at low density – and to harvest the cover crop at the end of June for ensiling. To improve the protein content of the arable silage, a legume such as pea (*Pisum sativum* ssp. *arvense*) or vetch (*Vicia* spp.) can be sown with the cereal. This approach ensures better establishment of the grassland than when a cereal crop is harvested for grain.

Sowing at the end of summer should be done preferably before mid-August and no later than mid-September, provided soil moisture is sufficient. Later sowings can be successful if the autumn is mild, but the probability of failure is high. The sward cover develops slowly during the autumn and the winter. If insufficiently developed, the plants may suffer from winter cold and even be killed. Weed control by herbicide or by cutting or grazing may be necessary, but weed invasion is usually less than with spring sowing.

6.6.2 Forage mixtures
In oceanic climates, *Lolium perenne* is undoubtedly the best grass for grazing swards and one of the best for cutting. This species alone represents a big part of the seed trade in forage production in Western Europe. However, when the climate or the soil are less favourable, other species are suitable for both moderately intensive or extensive management systems involving animals such as suckler cows, dry cows or growing heifers, which do not always need top quality forage.

In the first half of the twentieth century, the mixtures were usually very complex and could include more than ten species. As systems became more intensive, there was a marked trend to simplify seed mixtures (Lazenby, 1981). Nowadays, they may only include one species, *Lolium perenne*. However, two to five varieties are generally offered in commercial mixtures. The mixtures recommended in Table 6.2 are typical of many situations in northwestern and Central Europe. They are often more diversified to take into account the diversity of possible ecological situations and future managements.

The mixtures recommended in Switzerland are often much more complex. For information, consult Mosimann *et al.* (2000).

In mainly cut temporary grasslands, the choice of species must be mainly based on the duration of the grass crop (Table 6.3). The persistence can be very different under grazing conditions. A good sward of *Lolium perenne* for instance, can persist for many years in wet and mild climates if it is correctly grazed. The same is true for *Festuca arundinacea* in warmer and drier environments.

TABLE 6.2

Examples of mixtures for diverse uses

Short-lived grasslands, mainly cut

- 0–1 year, for cutting, for sowing in intercropping
M1: *Lolium multiflorum* ssp.*westerwoldicum,* 100% pure-sown or 70% mixed with 30% *Trifolium incarnatum* or 30% *Trifolium alexandrinum*

- 2 years, for cutting
M2: *Lolium multiflorum* ssp. *multiflorum,* 100% pure-sown or 70% mixed with 30% *Trifolium pratense* (moderately acidic to neutral soil)

- 3 years, mainly for cutting
M3: 100% *Lolium perenne,* early varieties in the lowlands or intermediate varieties at altitudes over 500 m; suitable for dairy cows

M4: 70% *Lolium perenne,* intermediate varieties + 30% *Phleum pratense* for areas with hard winters

M5: 100% *Phleum pratense* for areas with very hard winters (e.g. Scandinavia, altitude over 1 000 m); often used for moderately intensive systems

M6: 45% *Lolium perenne,* early or intermediate varieties + 25% *Festuca pratensis* + 30% *Phleum pratense*; all-purpose mixture for areas with hard winters; used for animals with moderate nutritional requirements, e.g. suckler cows

M7: 45% *Lolium perenne,* early or intermediate varieties + 25% *Festuca pratensis* + 30% *Dactylis glomerata*; all-purpose mixture for rather dry areas; used for animals with moderate nutritional requirements, e.g. suckler cows

Trifolium pratense (6 kg/ha) and sometimes *T. repens* (*Hollandicum* varieties or Ladino in areas with a suitable climate) (3–4 kg/ha) can also be added in mixtures M3 to M7. The addition of *Trifolium repens* is mainly advantageous if the mixture is occasionally grazed

M8: 40% *Dactylis glomerata* + 60% *Medicago sativa* for deep, basic soils
    Some people recommend sowing more *Dactylis glomerata* (60%) than *Medicago sativa* (40%). *Arrhenatherum elatius* can be used in place of *D. glomerata*

M9: 50% *Festuca arundinacea* + 50% *Medicago sativa* for warm climates, dry in summer, on deep and base-rich soils

M10: 70% *Bromus* spp. +/- 30% *Medicago sativa* for warm climates, on deep and base-rich soil

M11: 15% *Festuca pratensis* + 85% *Onobrychis viciifolia* for calcareous, shallow soil

- 4–5 years, mainly for cutting
M12: 100% *Festuca arundinacea* for early cuts

M13: 100% *Dactylis glomerata*
     Choose varieties resistant to diseases

These two species (M12 and M13) are mainly suitable in areas with a large water deficit in summer

Short-lived grasslands, grazed or mixed (grazing and cutting)
Fertile soil, normally drained, suitable climate
M14: 85% *Lolium perenne* (late varieties or late + intermediate varieties) +/- 15% *Trifolium repens*

Fertile soil, warm climate, rather dry in summer
M15: 60% *Lolium perenne* (late varieties or late + intermediate varieties) + 30% *Poa pratensis* +/- 10% *Trifolium repens*

Long-lived grassland, grazed or mixed (grazing and cutting)
Fertile soil, normally drained, suitable climate
M16: 85% *Lolium perenne* (late varieties or late + intermediate varieties) +/- 15% *Trifolium repens*

Fertile soil, warm climate, rather dry in summer
M17: 65% *Lolium perenne* (late varieties or late + intermediate varieties) + 20% *Poa pratensis* +/- 15% *Trifolium repens*

M18: 40% *Lolium perenne* (late varieties or late + intermediate varieties) + 30% *Dactylis glomerata* +/- 20% *Poa pratensis* +/- 10% *Trifolium repens*. This mixture is very suitable for horses.

M19: 85% *Festuca arundinacea* +/- 15% *Trifolium repens*; strong drought in summer

(Continued)

TABLE 6.2
Examples of mixtures for diverse uses (Cont.)

Fertile soil but cold winter
M20: 50% *Lolium perenne* (late varieties or late + intermediate varieties) + 20% *Phleum pratense* + 20% *Poa pratensis* +/- 10% *Trifolium repens*

Moderately fertile soil, possible drought in summer
M21: 23% *Lolium perenne* (late varieties) + 23% *Festuca rubra* + 12% *Agrostis capillaris* + 23% *Poa pratensis* +/- 7% *Trifolium repens* +/- 12% *Lotus corniculatus*

Moderately fertile, wet soils
M22: 40% *Lolium perenne* (late varieties) + 40% *Holcus lanatus* (variety resistant to rust) +/- 10% *Agrostis stolonifera* +/- 10% *Trifolium repens* (innovative mixture)

Nutrient-poor, dry soil
M23: 40% *Festuca rubra* + 15% *Agrostis capillaris* + 25% *Poa pratensis* + 15% *Trisetum flavescens* +/- 5% *Lotus corniculatus*

Nutrient-poor, wet soil
M24: 80% *Holcus lanatus* (variety resistant to rust) +/- 20% *Lotus uliginosus*

Note: Mixtures that include legumes should receive moderate or no fertilizer nitrogen (< 100 kg N/ha annually) if the potential of the legumes is to be fully realized

6.6.3 Species-rich grassland restoration

Mixtures for 'flower meadows' or species-rich grasslands should contain only grasses that cover the ground well but are not aggressive. In addition, these grasses should be adapted to the ecological characteristics of the site where they are sown. Seed of these grasses should also be commercially available.

The best adapted species are *Agrostis capillaris*,

TABLE 6.3
Persistence of some widespread species in mainly cut grasslands

Species	Years
Festuca arundinacea, Dactylis glomerata	> 5
Phleum pratense, Lolium perenne, L. x hybridum, Festuca pratensis	3–4
Bromus carthaticus	2–3
Lolium multiflorum ssp. *multiflorum*	2
Lolium multiflorum ssp. *westerwoldicum*	1

TABLE 6.4
Examples of mixtures for species-rich grassland restoration

Simple mixture, average ecological conditions	17% *Agrostis capillaris*, 33% *Festuca rubra* ssp. *commutata*, 20% *Poa pratensis* and 30% dicots
Complex mixture, average ecological conditions	10% *Agrostis capillaris*, 10% *Anthoxanthum odoratum*, 10% *Briza media*, 20% *Festuca rubra* ssp. *commutata*, 10% *Poa pratensis*, 10% *Trisetum flavescens* and 30% dicots
Dry conditions, calcareous soils	20% *Bromus erectus*, 10% *Festuca ovina*, 10% *F. rubra* ssp. *commutata*, 5% *Koeleria pyramidata*, 10% *Poa pratensis*, 15% *Trisetum flavescens* and 30% dicots
Wet conditions	10% *Agrostis stolonifera*, 25% *Alopecurus pratensis*, 25% *Festuca rubra commutata*, 10% *Poa trivialis* and 30% dicots

Festuca rubra and *Poa pratensis*, which can form the basis of most mixtures. *Anthoxanthum odoratum*, *Briza media* and *Trisetum flavescens* are also very suitable; they are not very aggressive and they have aesthetic inflorescences, but their seeds are fairly expensive and not always easy to find on the market. They can be used to complement the three other species (Table 6.4).

On calcareous, dry soils, *Bromus erectus* constitutes the basis of spontaneous swards and can be combined with other grass species. Crested hair-grass (*Koeleria pyramidata*) is also a species typical of dry habitats. It can be sown at lighter rates than the brome. *Festuca ovina, F. rubra* ssp. *commutata, Poa pratensis* and *Trisetum flavescens* also tolerate these difficult conditions for plant growth and can therefore be incorporated into mixtures adapted to calcareous, dry soils (Table 6.4).

Alopecurus pratensis can be mixed with meso-eutrophic dicotyledons in moist situations and soils rich in organic matter. *Festuca rubra, Poa trivialis* and *Agrostis stolonifera* can accompany it in these conditions (Table 6.4).

The use of mixtures containing very aggressive species like *Dactylis glomerata, Lolium* spp., *Phleum pratense* and even *Arrhenatherum elatius* should be avoided.

To restore species-rich grasslands under farming conditions, the cost of seed should naturally be kept within acceptable limits for the farmer. In this case, the mixtures do not contain more than five percent of dicotyledon seeds. However, this low proportion of dicotyledon seeds can be sufficient to ensure a significant establishment of these species in the sward in the long term.

6.6.4 Amenity, soil conservation and restoration

The finest turfs mainly contain fine-leaved fescues (*Festuca rubra* and *F. ovina*) and *Agrostis*. Turf that is heavily trodden, like sports grounds, should contain a dominance of *Lolium perenne* (Table 6.5).

'Turf' varieties of *Lolium perenne* are particularly indicated for sports grounds and for quickly stabilizing the soil on steep slopes in rainy climates. It is the most widely used species in amenity areas. It establishes very easily and very fast. It is remarkably resistant to treading. The foliage is dark green. Cultivars selected for lawns have rather fine leaves and form a dense turf. Their growth in height is less than that of forage varieties (Société française des gazons, 1990).

Agrostis capillaris, A. castellana and *A. stolonifera* are the three species of bents used for amenity areas. *Agrostis capillaris* and highland bent (*A. castellana*, variety Highland bent) are the two main lawn species. *Agrostis stolonifera* replaces these two species on humid soils. *Agrostis stolonifera* is also much used in pure stand for golf greens. These three species form fine turf, which is very dense and can be mown very close. Their resistance to treading is moderate and their establishment is slow. Nevertheless, where there is more than 15 percent of *Agrostis* in a mixture, these species tend in the long term to dominate regularly mown swards; 25 percent of *Agrostis* in a mixture is the absolute maximum (Société française des gazons, 1990).

Three types of *Festuca rubra* are used in lawns: creeping fescue (*F. rubra* ssp. *rubra*), half-creeping fescue (*F. rubra* ssp. *tricophylla*) and chewing's fescue (*F. rubra* ssp. *commutata*). The name *Festuca rubra* ssp. *trichophylla* is sometimes used to describe a group of two closely related species: *F. rubra* ssp. *pruinosa* and *F. rubra* ssp. *littoralis*. All these *Festuca* are quite slow to establish but form a fine and dense turf. Their growth rate is moderate and they can be closely mown. *Festuca rubra* ssp. *rubra* is particularly recommended when soil has to be stabilized, for example on steep slopes, though it does not stand treading very well. *Festuca rubra* ssp. *tricophylla* is more resistant to treading than the other two types of *F. rubra*. *Festuca rubra* ssp. *commutata* is used on flat ground

TABLE 6.5

Examples of mixtures for ornamental lawns and sports fields

Mixtures for ornamental lawns

Fine lawns	15% *Agrostis capillaris*, 30% *Festuca ovina*, 30% *F. rubra* ssp. *commutata*, 25% *Poa pratensis* (normal to slightly dry soils)
	25% *Agrostis capillaris*, 15% *A. stolonifera*, 10% *F. rubra* ssp. *commutata*, 50% *Poa pratensis* (slightly wet soils)
Ornamental lawns for intensive use	100% *Lolium perenne*
	or
	60% *Lolium perenne*, 40% *Poa pratensis*
	or
	15% *Agrostis capillaris*, 25% *Festuca rubra* ssp. *rubra*, 40% *Lolium perenne*, 20% *Poa pratensis*
	or
	60% *Festuca arundinacea*, 40% *Lolium perenne* (dry conditions)
Ornamental lawns in shade	15% *Agrostis capillaris*, 20% *Festuca ovina* ssp. *duriuscula*, 15% *F. rubra* ssp. *commutata*, 10% *F. rubra* ssp. *trichophylla*, 20% *Poa nemoralis*, 20% *P. pratensis*

Mixtures for sports fields

Football fields	20% *Festuca rubra* ssp. *trichophylla*, 35% *Lolium perenne*, 45% *Poa pratensis* (nice lawn)
	50% *Festuca arundinacea*, 10% *F. rubra* ssp. *trichophylla*, 40% *Lolium perenne* (rustic lawn)
	or
	30% *Festuca arundinacea*, 10% *F. ovina* ssp. *duriuscula*, 30% *Lolium perenne*, 30% *Poa pratensis* (dry area)
Rugby fields	60% *Festuca arundinacea*, 30% *Lolium perenne*, 10% *Poa pratensis*
Hockey fields	15% *Festuca rubra* ssp. *trichophylla*, 30% *Lolium perenne*, 55% *Poa pratensis*
Racecourses	35% *Festuca rubra* ssp. *rubra*, 30% *F. rubra* ssp. *trichophylla*, 35% *Lolium perenne* (oceanic climate)
	or
	15% *Festuca rubra* ssp. *commutata*, 20% *Lolium perenne*, 65% *Poa pratensis* (continental climate)
	or
	10% *Festuca ovina* ssp. *duriuscula*, 10% *F. rubra* ssp. *trichophylla*, 60% *F. arundinacea*, 20% *Lolium perenne* (warm climate)
Golf course tees	40% *Lolium perenne*, 60% *Poa pratensis*
	or
	40% *Lolium perenne*, 40% *Poa pratensis*, 20% *Festuca rubra* ssp. *commutata* or *trichophylla*
Fairways	20% *Festuca ovina* ssp. *duriuscula*, 30% *F. rubra* ssp. *rubra*, 20% *F. rubra* ssp. *trichophylla*, 30% *Lolium perenne*
	or
	30% *Festuca ovina duriuscula*, 35% *F. rubra* ssp. *rubra*, 35% *F. rubra* ssp. *trichophylla*
	or
	20% *Festuca ovina* ssp. *duriuscula*, 30% *F. rubra* ssp. *rubra*, 20% *F. rubra* ssp. *trichophylla*, 30% *F. rubra* ssp. *commutata*
	or
	30% *Festuca rubra* ssp. *rubra*, 20% *F. rubra* ssp. *trichophylla*, 20% *F. rubra* ssp. *commutata*, 30% *Lolium perenne*
	or
	25% *Festuca rubra* ssp. *commutata*, 35% *Lolium perenne*, 40% *Poa pratensis*
	or
	30% *Festuca rubra* ssp. *rubra*, 30% *F. rubra* ssp. *commutata*, 40% *Poa pratensis*
Greens	100% *Agrostis stolonifera*
	or
	10% *Agrostis capillaris*, 45% *Festuca rubra* ssp. *trichophylla*, 45% *F. rubra* ssp. *commutata*
	or
	10% *Agrostis capillaris*, 45% *Festuca ovina* ssp. *duriuscula*, 45% *F. rubra* ssp. *commutata*
Rough	Wild grasses
	or
	species-rich mixtures

Source: Société française des gazons, 1990; Vavassori, 2000

and in 'average' ecological conditions. Its resistance to treading is poor to middling. In addition, it is less drought-resistant than the other two types (Société française des gazons, 1990).

Festuca ovina is regularly used in lawns because its foliage is fine and its growth slow. In particular, it replaces or accompanies *Festuca rubra* on dry soils and at altitude. Generally, it performs better than *Festuca rubra* under difficult conditions of soil and climate. Hard fescue (*Festuca ovina* ssp. *duriuscula*) is better than the type species (*F. ovina*) in lawns because it stays greener in summer, but its seeds are unfortunately more expensive. *Festuca ovina* resists treading very badly and is slow to establish.

The taxonomy of *Festuca rubra* and *F. ovina* is very complex and it is still poorly known to which species or subspecies the turf varieties belong. Detailed studies using the most up-to-date techniques of genetic and morphological classification ought to allow these taxa to be more clearly distinguished and characterized in the future (Société française des gazons, 1990).

Poa pratensis can be used in various situations, from fine lawns to sports grounds, though it is slow to establish. It forms a dense turf, with a moderate growth rate (Société française des gazons, 1990). Its resistance to uprooting makes it particularly suitable for sports grounds, notably rugby pitches. This characteristic is due to the presence of rhizomes, which distinguish the species from *Lolium perenne*, for example. Its resistance to diseases and drought is very variable between varieties.

The 'lawn' varieties of *Festuca arundinacea* resist drought well, but they are only moderately resistant to treading and do not tolerate close mowing very well. They are also quite slow to establish. The winter appearance of the 'European' types is not very aesthetic, but they have a good summer appearance, whereas for the 'Mediterranean' types the opposite is true. In fact 'Mediterranean' types

are dormant in summer so their turf does not stay very green in this period of the year (Société française des gazons, 1990).

Seeds of turf varieties are sown at far higher densities than fodder seeds in order to obtain very rapid establishment and a high turf density. The cost of seeds is also a less important factor in the development of amenity areas than in farming. The seed rate varies according to the composition of the mixture, but an average rate is generally between 250–300 kg/ha, which is about ten times more than in agricultural conditions. However, seeding of *Agrostis stolonifera* used in golf greens does not exceed 20–40 kg/ha mainly because it has very small seeds.

The technique of turf in rolls, which appeared at the start of the twentieth century in the USA, developed strongly in Europe from the 1960s. This technique uses the same species, varieties and mixtures as those that are used for direct sowing, but supplies clients with ready-to-lay turf. It is sufficient to unroll the sheets of turf on a previously prepared soil to obtain instant completion.

Table 6.5 illustrates several mixtures for diverse uses.

In addition to the species given for turf, three other species can be used for seeding ski pistes; namely *Dactylis glomerata*, *Cynosurus cristatus* and *Bromus inermis* (Table 6.6).

New species – *Bellardiochloa violacea*, *Poa alpina* and *Phleum alpinum* – could also be used for sowing ski pistes because they are particularly well adapted to the ecological conditions of the mountains (Krautzer, 1996 and 1998; Krautzer *et al.*, 2001). However, their seeds are still difficult to find on the market.

Mixtures used for stabilizing roadside banks (Table 6.7) include *Lolium perenne* to ensure a rapid soil cover, and other longer-lived species, such as the *Festuca* spp., *Poa pratensis* and the *Agrostis* spp. Despite a fairly slow establishment, *Festuca rubra* and *Poa pratensis* stabilize the soil well in the long term thanks to their rhizomes. *Agrostis*

TABLE 6.6
Examples of mixtures for ski pistes

Altitude	Dry conditions	Wet conditions
Above 1 800 m	*Dactylis glomerata, Festuca ovina, Festuca ovina ssp.duriuscula, Poa pratensis*	*Agrostis capillaris, A. stolonifera, Festuca rubra ssp. commutata, Phleum pratense*
Below 1 800 m	*Bromus inermis, Dactylis glomerata, Festuca arundinacea, F. ovina, F. ovina ssp. duriuscula, Poa pratensis*	*Agrostis capillaris, A. stolonifera, Festuca arundinacea, F. pratensis, F. rubra ssp. commutata, Lolium perenne, Phleum pratense*

Source: Dinger in Société française des gazons, 1990

TABLE 6.7
Examples of mixtures for roadside and motorway verges (maritime climate)

Soil types	Species
Clay soils	*25% Festuca ovina, 45% F. rubra ssp.rubra and F. rubra ssp.trichophylla, 5% Lolium perenne, 25% Poa pratensis*
Sandy soils	*10% Agrostis capillaris, 40% Festuca ovina, 35% F. rubra ssp. rubra and F. rubra ssp. trichophylla, 15% Lolium perenne*

Source: Société française des gazons, 1990

TABLE 6.8
Example of a mixture for soil conservation and restoration

10% Agrostis capillaris, 30% Dactylis glomerata, 15% Festuca ovina, 15% F. rubra ssp. rubra and F. rubra ssp. trichophylla, 15% Poa pratensis, 10% Lotus corniculatus, 5% Medicago lupulina

capillaris and *Festuca ovina* are particularly suited to dry sites. Dicotyledons can be added to these mixtures to create a diversity of species, which makes the turf more aesthetic ('flowery') and more attractive to wildlife. Mixtures recommended for restoring species-rich grasslands (section 6.6.3) can also be used in this situation.

The change of use or cleaning up of industrial sites often brings about very unusual soil conditions. This is the case namely with mine tips, stone or sand quarries and soils contaminated with heavy metals. The restoration of eroded soils and the stabilization of rocky slopes after civil engineering works also produce very difficult soil conditions for plant growth. In fact, these soils are often low in soil organic matter and in nutrients, free draining and sometimes unstable. Maintenance after sowing is often reduced to a strict minimum. It is therefore necessary to use undemanding species in the mixtures (Muncy, 1985; Poser and Jochimsen, 1989) and possibly inoculate the seeds with mycorrhizae in order to improve water and phosphorus nutrition of the plants. Drought-resistant legumes can be incorporated in these mixtures to enrich the habitat in N (Table 6.8).

In some situations, it is not possible to use a classic seed drill and tractor combination to sow the mixtures. The steepness of the slope or the presence of stones or rocks make such classical techniques impossible, for example on ski pistes, roadsides, steep banks and disused industrial sites. Under these conditions, the technique of hydroseeding allows the spraying of seeds in liquid suspension onto the surface to be sown. The seed rate used is 100 kg/ha for average situations and up to 200–250 kg/ha in difficult situations and on ski pistes. The suspension often contains soil-stabilizing products (cellulose, resin, hydrogels) as well as spray binders, which allow a homogeneous mixture of water, seed and fixatives that facilitates the contact of the seed with the soil. Hydroseeding is often associated with jute or coco webs to ensure the stability of sloping soil (Société française des gazons, 1990).

6.6.5 Set-asides, field margins, intercropping and forest clearings

Set-asides are used under the Common Agricultural Policy of the European Union as a way of controlling the volume of production of cereals, oilseeds and protein crops. These set-asides can also provide a habitat for wildlife, notably game. Mammals like roe deer (*Capreolus capreolus*), *Lepus capensis*, *Oryctolagus cuniculus*, and birds like *Perdix perdix, Alectoris* spp., *Phasianus colchicus, Chrysolophus* spp., *Coturnix coturnix*, woodcock (*Scolopax rusticola*), *Alauda arvensis, Motacilla flava* and kestrel (*Falco tinnunculus*) can find a habitat there that fulfils several functions: food supply, shelter from predators and hikers, and a nesting site for birds (Peeters and Decamps, 1998).

The feed can be composed mainly of leaves for *Capreolus capreolus* and *Lepus capensis* for example, but also of arthropods (spiders, aphids) for young *Perdix perdix* and *Alectoris* spp., of earthworms for *Scolopax rusticola*, or of small mammals for *Falco tinnunculus*. It is important that the vegetation is not too dense, because otherwise the animals simply cannot reach the food.

Ease of movement in the vegetation is also an essential factor for assuring shelter. The vegetation needs to be 'permeable' enough so that animals can move about easily among the vegetation or nest in it. This is important for birds like *Perdix perdix, Alectoris* spp., *Coturnix coturnix, Phasianus colchicus* or *Chrysolophus* spp. In addition, the vegetation should be tall enough so that the animals feel hidden and invisible to predators. Tall grasses like *Arrhenatherum elatius, Dactylis glomerata* and *Phleum pratense* can form a permeable cover, if they are sown at a low seed rate (Peeters and Decamps, 1998).

These various functions can be fulfilled simultaneously by one kind of vegetation, but some types do not suit one or other of these functions. In fact it is difficult to find an ideal multifunctional vegetation. The one that presents the most advantages is *Medicago sativa* sown pure at low density. It forms tall vegetation, which is not very dense but is rich in feed resources in the form of protein-rich leaves and in arthropods. In general, grasses provide less nourishment for fauna than legumes, because their leaves are less digestible, less rich in protein and they attract

TABLE 6.9

Examples of mixtures for set-asides

Function	Species
Food production	35% *Festuca rubra*, 55% *Lolium perenne*, 10% *Trifolium repens*
Shelter against predators and nesting site	40% *Arrhenatherum elatius*, 40% *Dactylis glomerata*, 20% *Phleum pratense*
Shelter and food production	40% *Dactylis glomerata*, 60% *Medicago sativa*, (*Arrhenatherum elatius, Phleum pratense*) or 60% *Dactylis glomerata*, 25% *Trifolium pratense*, 15% *T. repens*, (*Arrhenatherum elatius, Phleum pratense*)

Source: Peeters and Decamps, 1998

far fewer aphids. Nevertheless, grass-legume mixtures or even mixtures of grasses can be of value for wildlife (Table 6.9) (Peeters and Decamps, 1998).

A mixture of low-growing grasses with *Trifolium repens* constitutes a feed source for *Capreolus capreolus*, *Lepus capensis* and *Oryctolagus cuniculus*. Ideally, it should be sited on the edge of woods since this allows reduction of game damage to neighbouring crops (Peeters and Decamps, 1998).

The *Dactylis glomerata-Medicago sativa* mixture has characteristics close to a *Medicago sativa* monoculture if sown thinly. It produces shelter, food and perhaps a breeding site for birds (Peeters and Decamps, 1998).

A mixture of *Dactylis glomerata* with *Trifolium repens* and *T. pratense* is dense and less easy for wildlife to penetrate. It produces a great deal of feed, but little of it is available to animals. Thus it is advisable to cut serpentine alleys in the vegetation with a rotary mower to increase the border areas where the animals have access to feed. It is also in these border areas that they can build their nests. Two tall, low-tillering grasses, *Arrhenatherum elatius* and *Phleum pratense*, can be mixed with *Dactylis glomerata* in this mixture and the preceding one (Table 6.9) (Peeters and Decamps, 1998).

Simple mixtures can be sown for long-duration set-aside establishment if the only objective is to cover the soil. *Lolium perenne* can be used alone, but it is preferable to mix it with *Trifolium pratense* and *T. repens*, which improve the soil fertility and humus content. A considerable amount of nitrogen (100–200 kg N/ha) is then available for the following crop after the destruction of the set-aside (Clotuche and Peeters, 2000).

The Common Agricultural Policy of the European Union subsidizes the creation of field margins with the aim of improving the environment and reducing agricultural production. These field margins can fulfil several functions: protection of surface water, extensive production of forage and habitat for wildlife (Table 6.10). They are only several metres wide but can be several hundred metres long. They may be cut or grazed late and the cut herbage may be removed.

Field margins may be mainly sited on the edges of watercourses (humid conditions) or on a slope following the contour (moderate or dry conditions). In these two cases, the grass strip stops or slows down runoff and holds back the suspended particles and elements dissolved in the water. In particular, the ions of nitrate and phosphate, as well as pesticides are stopped by the grass strip and

TABLE 6.10

Examples of mixtures for field margins

Function	Species
Run-off control, phosphate and nitrate absorption for the protection of surface water	20% *Agrostis stolonifera*, 60% *Dactylis glomerata*, 20% *Phleum pratense* (wet conditions) or 20% *Agrostis stolonifera*, 40% *Dactylis glomerata*, 40% *Festuca arundinacea* (wet conditions) or 30% *Arrhenatherum elatius*, 40% *Dactylis glomerata*, 30% *Festuca rubra* ssp. *rubra* (average or dry conditions)
Forage production in late cut regimes	30% *Dactylis glomerata*, 20% *Lolium perenne*, 25% *Phleum pratense*, 25% *Medicago sativa*
Habitat for wildlife	25% *Agrostis stolonifera*, 50% *Alopecurus pratensis*, 25% *Poa trivialis* and dicots (wet conditions) or 20% *Agrostis capillaris*, 40% *Festuca rubra* ssp. *commutata*, 40% *Poa pratensis*, (*Anthoxanthum odoratum*, *Briza media*, *Trisetum flavescens*) and dicots (average conditions)

TABLE 6.11
Examples of mixtures for intercropping

100% (or 70%) *Lolium multiflorum* ssp. *westerwoldicum,* (30% *Trifolium incarnatum*)
or
100% (or 70%) *Lolium multiflorum* ssp. *multiflorum,* (30% *Trifolium incarnatum*)
or
100% (or 70%) *Lolium perenne,* (30% *Trifolium incarnatum*)

---

the nutrients can be absorbed by plants. A width of 4 m suffices to stop the bulk of these materials. Mixtures should contain vigorous species, e.g. *Dactylis glomerata, Festuca arundinacea* and *Phleum pratense,* capable of absorbing N and P. These species are persistent and adapted to humid soils. *Arrhenatherum elatius* should replace *Phleum pratense* and *Festuca arundinacea* in field margins established on slopes. All these grasses are not very dense at their base because they are cut infrequently and at late maturity. In order to reduce runoff to a minimum, it is advantageous to add a low-growing grass with good vegetative coverage, e.g. *Agrostis stolonifera* alongside watercourses and *Festuca rubra* ssp. *rubra* on slopes (Table 6.10).

Extensive forage production can be ensured by persistent and productive species like *Dactylis glomerata* and *Phleum pratense.* *Lolium perenne* can be added to the mixture to increase production in the two years following sowing. It will not persist longer under a late cutting regime. *Medicago sativa* allows increased production, improvement of fodder quality and provides a source of food for pollinating insects, notably bees. It also improves the aesthetic interest of the countryside during its flowering period.

Field margins also constitute a habitat for wild fauna and flora. Alongside streams, mixtures should contain grasses adapted to humid soils that are often nutrient-rich: *Agrostis stolonifera, Alopecurus pratensis* and *Poa trivialis* fulfil this requirement, although seed of *Poa trivialis* is not always readily available. It can then be replaced by *Festuca*

rubra ssp. *commutata* or *Poa pratensis.* On normally drained or dry terrain, the mixtures are closely comparable to standard mixtures recommended for restoring species-rich grasslands (section 6.6.3), which contain *Agrostis capillaris, Festuca rubra* and *Poa pratensis.* Dicotyledons may also be included.

Because of their remarkable speed of establishment, *Lolium perenne* and *L. multiflorum* can be used for intercropping in crop rotations (Table 6.11). *Lolium multiflorum* establishes even more rapidly than *L. perenne.* It is therefore preferred for very short growing periods. The species are generally sown in summer or at the start of autumn, after the harvest of a cereal and thereafter ploughed in towards the month of November, or in spring before a crop of maize or beet for example. They can also be sown early in spring to replace a frost-damaged cereal crop. These intercrops can be used as green manure, catch crops or as forage. If they are to be used as green manure or forage, *Trifolium incarnatum,* a legume with rapid establishment and growth, can be included.

Game pasture can be installed in forest clearings to provide fodder for deer. The soils of these clearings are often poor in nutrients and acids and maintenance of these pastures is often minimal. It is therefore necessary to use mixtures of hardy grasses that are undemanding, persistent and of acceptable feeding value. For example, one could propose: 20 percent *Agrostis capillaris,* 40 percent *Festuca rubra* ssp. *commutata,* 30 percent *Poa pratensis* and 10 percent *Trifolium repens.*

Chapter 7

Some prospects for wild and little-selected species

It is hoped that this book will stimulate research on little-known though widely distributed grasses that have been neglected in comparison with the six dominant forage species previously discussed. The synthesis done in this book suggests some lines of future research. Of course, this list is not comprehensive, but rather aims to give some ideas, especially to young researchers.

Concerning grass breeding, the following objectives should take priority:

- Begin a European selection of varieties of *Holcus lanatus*, notably to improve the resistance of the species to diseases and cold, and to improve its palatability and digestibility. This species has great potential for intensive or semi-intensive grazed systems.
- Continue selection of *Poa pratensis*, especially to improve establishment vigour, disease-resistance and digestibility.
- Begin selection of *Festuca rubra* for agricultural rather than amenity purposes. The improvement of digestibility would allow marketing of fodder varieties, adapted to fairly poor soils and extensive meat-producing systems.
- Begin selection of *Elymus repens* to improve its digestibility. The species could be used more widely in cutting grasslands in cold areas (Scandinavia, continental climates, at high altitudes).
- Continue selection of *Arrhenatherum elatius, Alopecurus pratensis,* and *Phalaris arundinacea* to lower their contents of bitter or toxic substances. *Arrhenatherum elatius* is certainly the most interesting species of the three. It could be used in long-duration cutting grasslands. *Alopecurus pratensis*

should be selected to improve its digestibility; it is suited above all to rich, moist soils and to mixed grazing-cutting systems. Selection of *Phalaris arundinacea* could be done in Europe and Asia taking advantage of the American experience in developing varieties. This species is adapted to cutting systems on very humid soils.

As far as species little used up to now are concerned, three species should take priority: *Arrhenatherum elatius, Festuca rubra* and *Holcus lanatus*.

These selection activities could only be implemented through a partnership between public institutions and private companies. Principally, they should aim to cover the needs of the countries of Central and Eastern Europe, where *Lolium perenne* cannot suit all situations.

A better knowledge of the ecophysiology of secondary grasses is certainly a priority research area. In fact, little is known of their phyllocrone, the rate of leaf senescence, the maximum number of leaves per tiller, the allocation of the photosynthesis resources, the life span of their roots and the ratio between the aerial and radicular parts, to name some potential lines of investigation. This knowledge could then be integrated into growth models.

The study of interspecies competition is a theoretical research subject that could have numerous practical applications. It would be important to determine the factors explaining the success or failure of species in binary mixtures, for instance. The relationships between ecophysiological characteristics and competitive advantages could thus be studied and thereafter modelled. Such studies would serve to determine the optimum mixtures of species for long-term or permanent swards.

It would also be important to elucidate the factors that explain the differences in digestibility between the species, and notably to study the relationships between the anatomical structure of leaves and stems and the digestibility of the forage. Some grasses are less rich in soluble sugars than the *Lolium* spp. It is therefore necessary to develop techniques that ensure their successful ensiling, e.g. the use of silage additives.

Several grasses of high production potential are sensitive to early grazing, unlike *Lolium perenne*. Research should therefore define the grazing systems that would ensure their persistence.

Fodder production should be looked at in a more systematic way in the future. Rather than trying to achieve the highest yield and quality in all situations, forage should be produced that corresponds to the real needs of animals. Therefore, within a farm several types of grasslands could be defined and different functions allocated to them, for example, spring pastures, summer pastures, early hay or silage meadows, late hay meadows. The botanical composition of swards ought thereafter be judged in relation to these different functions.

It is also necessary to be able to respond to the specific needs of new uses of grassland; for example, pastures managed by newcomers to the countryside, in rural areas or on the periphery of towns, for horses, cattle of hardy breeds, deer or bison. Atypical or novel swards are also needed for the rehabilitation of degraded sites or for other 'unusual' uses described in Chapter 2. Wild species or varieties of secondary species could certainly be more suitable than *Lolium perenne* in many situations.

The role of grasslands as carbon sinks ought to be more thoroughly studied to allow maximization of carbon fixed in the soil and thus a contribution to the reduction of the greenhouse effect. It is likely that species rich in lignin (*inter alia Molinia caerulea*) induce a greater accumulation of humus than other species.

The percolation of nitrate is very low under cutting grasslands and rangelands, but in a grazed pasture with heavy nitrogen fertilization, the risks of leaching are greater, especially with a fertilization of over 200 kg/ha N (Simon *et al.*, 1997). Species with a high capacity for nitrogen absorption could limit nitrate leaching.

The use of grass strips as anti-pollution barriers along watercourses is a means of conserving water quality. It would be useful to determine the most suitable species, namely those that maximize absorption of nutrients while achieving persistence.

These ideas are of course not exhaustive and more research topics will be necessary. It can also be concluded that much of the research carried out in the past for the major species is needed for the wild species.

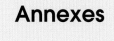

Annexes

Annex 1

Identification key of forage grasses at a vegetative stage

There are many grass identification keys for the whole of Europe (Tutin *et al.,* 1980) or for more restricted geographical areas (e.g. Lambinon *et al.,* 1992; Landwehr, 1977; Lauber and Wagner, 1998; Hubbard, 1978; Rose, 1989; Stace, 1997; Van der Meijden, 1996). However, these keys are mainly based on the morphological characters of the inflorescence, and the ecologist and the agronomist must often identify species at the vegetative stage, either because grasses have not yet headed, or because they have been defoliated by grazing animals or cutting machines. It is thus essential for them to use an identification key that is based on morphological characters of the vegetative organs, e.g. blade, sheath, auricle, ligule, base of stems, presence of rhizomes. The dichotomic key presented below complies with this condition.

Some species appear on several occasions in this key because they present a variability in some of their characters, e.g. leaf flat or in needle shape, rolled or folded when young, hairy or hairless. The key thus allows identification of these species at every particular vegetative case.

The key, which complements the species profiles in Chapter 5, is designed in the following way:

- Needle-like leaf   section A
- Flat leaf, folded when young   section B
  - Auricles   section B1
  - No auricles   section B2
- Flat leaf, rolled when young   section C
  - Auricles   section C1
  - No auricles   section C2

The key starts in paragraph number 1. For each paragraph, there are two options. Each option refers to a species name or to a subsequent paragraph number of the key.

IDENTIFICATION KEY

1.
- Leaves needle-like, at least those from the base (section A)   **2.**
- Leaves flat   **5.**

SECTION A
NEEDLE-LIKE LEAF
2.
- Tuft compressed, leaf blade very hard and perpendicular to the sheath,
  long white ligule   *Nardus stricta*
- Tuft non-compressed, leaf blade not perpendicular to the sheath   **3.**
3.
- Ligule very short, almost not visible   **4.**
- Ligule more or less long, always well visible, leaf blade dark green   *Deschampsia flexuosa*
4.
- Leaf blade often greyish or bluish, stiff, with a thorny extremity,
  rolled on itself, less than 0.6 mm diameter   *Festuca ovina*
- Leaf blade dark green, flexible, a little thorny, rolled on itself   *Festuca rubra*
5.
- Leaf folded when young (section B)   **6.**
- Leaf rolled when young (section C)   **21.**

SECTION B
FLAT LEAF, FOLDED WHEN YOUNG

SECTION B1
AURICLES
6.
- Auricles (section B1)
  Lower side of the blade very shiny, veins well visible on the upper side,
  sheath of the lower leaves red to purplish-red, ligule short, membranous,
  auricles often well visible but always rather short   *Lolium perenne*
- No auricles (section B2)   **7.**

SECTION B2
NO AURICLES
7.
- Transversal nerves visible by transparency on the blade, presence
  of 'ski tracks' on both sides of the central nerve   **8.**
- No transversal nerves visible on the blade   **9.**
8.
- Ligule long and acute, often split, blade veins well visible, blade
  with a slightly hooded tip, narrow (less than 1 cm)   *Glyceria fluitans*
- Ligule average, truncate and pointed in the middle, blade veins
  not well visible, blade rather wide (often more than 1 cm)   *Glyceria maxima*

9.
- No visible ribs on the blade   **10.**
- Ribs well visible on the blade   **16.**

10.
- 'Ski tracks' on both sides of the central nerve of the blade   **11.**
- No 'ski tracks' on both sides of the central nerve of the blade   **15.**

11.
- Leaf distinctly hairy   *Avenula pubescens*
- Leaf hairless or with very short hairs   **12.**

12.
- Ligule short and truncate, translucent, blade rather stiff with a hooded tip   *Poa pratensis*
- Ligule long, especially on the upper leaves, or short and pointed   **13.**

13.
- Sheath clearly hooded on the edge   **14.**
- Sheath not hooded on the edge (though sometimes a little on the
  lower leaves), ligule triangular, pointed   *Avenula pratensis*

14.
- Ligule long or short but always pointed, translucent   *Poa trivialis*
- Ligule long, whitish, not pointed, blade yellowish-green, often
  accordion-pleated in the top third   *Poa annua*

15.
- Edge of the blade hairy   *Bromus erectus*
- Edge of the blade hairless, tillers very compressed at the base   *Dactylis glomerata*

16.
- Ligule long (more than 3 mm), acute   **17.**
- Ligule shorter (less than 3 mm), sometimes almost not visible, non acute   **18.**

17.
- Ligule very long (more than 5 mm), ribs of the blade very marked,
  presence of white lines well visible by transparency between these ribs   *Deschampsia caespitosa*
- Ligule shorter (less than 5 mm), ribs of the blade less marked,
  no white lines visible by transparency between these ribs   *Agrostis canina*

18.
- Leaf hairless, blade not narrowed at its base   **19.**
- Leaf hairy, blade narrowed at its base   *Koeleria pyramidata*

19.
- Blade flat on all leaves, lower side of the blade a little or very shiny, ligule
  short but robust   **20.**
- Blade of lower leaves very narrow, in needle shape, ligule almost
  not visible or absent. Sheath of lower leaves often a little reddish   *Festuca rubra*

20.
- Sheath of the lower leaves often yellowish-brown, lower side of the blade
  a little shiny   *Cynosurus cristatus*
- Sheath of the lower leaves red or purplish-red, lower side of the blade very
  shiny   *Lolium perenne*

SECTION C
FLAT LEAF, ROLLED WHEN YOUNG
- Auricles (section C1)   22.
- No auricles (section C2)   27.

SECTION C1
AURICLES
22.
- Auricles hairy or replaced by hairs   23.
- Auricles hairless   24.
23.
- Ligule very short but vigorous, auricles well developed and hairy on
  the edges, rest of the leaf hairless, ribs well developed on the upper
  side of the blade, lower side of the blade shiny, sheath of the lower leaves
  often red or purplish red   *Festuca arundinacea*
- Ligule elongated, sometimes dark purple, auricles replaced by hairs or
  hairy, some hairs on the blade, lower side of the blade not shiny, ribs
  of the upper side of the blade not visible. Smell and taste of coumarin   *Anthoxanthum odoratum*
24.
- Auricles very thin from the base. Sheath of the lower leaves
  often very hairy.
  Auricles usually long and always narrow, ligule short, blade more
  or less hairy but always rough (rub the blade on your lips from the
  top to the base of the blade), sheath more or less hairy, sheath of
  the lower leaves often very hairy, whitish rhizome   *Elymus repens*
- Auricles large at the base, narrowing to the top   25.
25.
- Sheath hairy, auricles short. Blade usually hairy on the two sides,
  but more hairy on the upper side. Ligule short but a little larger than
  that of *Elymus repens*   *Hordeum secalinum*
- Sheath hairless and red or purplish red on the leaves of the base,
  auricles well developed, lower side of the blade shiny, ribs
  well marked on the blade   26.
26.
- Ligule very short, clearly shorter than the uncoloured base of the blade,
  less than 16 ribs on the upper side of the blade   *Festuca pratensis*
- Ligule short but almost as high as the uncoloured base of the blade,
  more than 16 ribs on the upper side of the blade   *Lolium multiflorum*

SECTION C2
NO AURICLES
27.
- Leaf hairy   28.
- Leaf hairless   37.
28.
- Blade narrowed at its base, greenish to greyish, with strong ribs   *Koeleria pyramidata*
- Blade not narrowed at its base   29.

29.
- Sheath of the lower leaves clearly hairy   **30.**
- Sheath not very hairy or with hairiness concentrated on the node   **34.**

30.
- Sheath of the lower leaves striated by purplish red vertical lines. Plant
  entirely hairy, with soft hairs. Two kinds of hairs on the blade: long
  on the top of the ribs, short in the hollows (examine a transversal cut
  of the blade with a lens).
  Ligule long and white   *Holcus lanatus*
- Sheath of the lower leaves not striated by purplish red vertical lines   **31.**

31.
- Ligule reduced to a hair crown. Some hairs on the upper side
  of the blade.
  Blade narrowing at the base like a guttering   *Molinia caerulea*
- Ligule membranous whitish or absent   **32.**

32.
- No ligule, auricles reduced to some hairs. Smell and taste
  of coumarin   *Anthoxanthum odoratum*
- Ligule well visible   **33.**

33.
- Ribs well visible on the upper side of the blade (always more than 10
  ribs, often about 14). Upper side of the blade as hairy as the lower side.
  Sheath with bottom-oriented hairs, making an angle of about 45° with
  the stem   *Trisetum flavescens*
- Ribs not well visible on the upper side of the blade. Sheath hairs not
  bottom-oriented   *Bromus hordeaceus*

34.
- Hairiness concentrated on the node and decreasing from the node
  to the top of the sheath   *Holcus mollis*
- Hairiness loose, not concentrated on the node   **35.**

35.
- Blade with a hairy edge   *Bromus erectus*
- Blade with a hairless edge   **36.**

36.
- Blade 2–4 mm wide, shorther than 15 cm. Upper side of the blade as hairy
  as the lower side. Sheath with bottom-oriented hairs, making an angle of
  about 45° with the stem   *Trisetum flavescens*
- Blade 4–10 mm wide, usually longer than 15 cm. Often two hairs at the
  top of the blade even if the rest of the blade is hairless. Blade often with a
  small constriction at about the top third and showing light lines visible by
  transparency. Ligule white, roughly dentate, rather long   *Arrhenatherum elatius*

37.
- No visible ribs on the blade   **38.**
- Ribs well visible on the blade   **42.**

38.
- Ligule short, truncate   **39.**
- Ligule long or elongated   **40.**

39.
- Blade edge hairy. Ligule short   *Bromus erectus*
- Blade edge hairless, finely dentate. Top of the blade finishes with a yellow
  point. Ligule very short   *Briza media*

40.
- Root yellow. Ligule white, roughly dentate, rather long. Light lines
  visible by transparency in the blade. Blade often with a small
  constriction at about the top third. Stem rather thin   *Arrhenatherum elatius*
- Root white   **41.**

41.
- Ligule white, long, dentate. Ribs of the blade not well marked.
  Stem robust, often bulging at the base and forming a small bulb   *Phleum pratense*
- Ligule long, membranous, greyish white. Ribs of the blade not
  well marked, less marked than for *Phleum pratense*. Stem robust   *Phalaris arundinacea*

42.
- Ligule short (usually less than 2 mm), not pointed   **43.**
- Ligule pointed or elongated (usually longer than 2 mm)   **49.**

43.
- Blade of the inferior leaves narrow, needle-like. Ligule very short   *Festuca rubra*
- Blade flat   **44.**

44.
- Rhizomatous plant   **45.**
- Non-rhizomatous plant   **47.**

45.
- Ribs of the blade well marked, lower side of the blade often shiny, blade
  dark green. Ligule short, oblique at the base (look at the back side of the
  junction blade-sheath). Plant robust with an erect stem   *Alopecurus pratensis*
- Ribs of the blade not well marked, lower side of the blade not shiny.
  Ligule short or long, not oblique at the base. Stem creeping or erect   **46.**

46.
- Ligule truncate and asymmetrical   *Agrostis capillaris*
- Ligule long and pointed, not asymmetrical   *Agrostis gigantea*

47.
- Blade edge hairy   *Bromus erectus*
- Blade edge hairless or almost   **48.**

48.
- Sheath of the lower leaves often yellowish-brown, lower side of the
  blade a little shiny   *Cynosurus cristatus*
- Sheath of the lower leaves whitish, lower side of the blade not shiny   *Trisetum flavescens*

49.
- Ligule very pointed, blade narrow (often 1–2 mm width), pale green   *Agrostis canina*
- Ligule elongated but not very pointed, blade wider   **50.**

50.
- Ribs of the blade very marked, with a pointed top (lens!), blade
  dark green   *Alopecurus geniculatus*
- Ribs of the blade less marked, with a rounded top (lens!), blade
  green   *Agrostis stolonifera*

Annex 2

Lists of species in alphabetical order in Latin, English, French and German

Scientific name	English name	French name	German name
Agrostis canina L.	Velvet bent	Agrostis des chiens	Sumf straussgras
Agrostis capillaris L.	Common bent	Agrostis commun	Gemeines straussgras
Agrostis gigantea Roth	Black bent	Agrostis géant	Riesen straussgras
Agrostis stolonifera L.	Creeping bent	Agrostis stolonifère	Fioringras
Alopecurus geniculatus L.	Marsh foxtail	Vulpin genouillé	Knick-fuchsschwanz
Alopecurus pratensis L.	Meadow foxtail	Vulpin des prés	Wiesen-fuchsschwanz
Anthoxanthum odoratum L.	Sweet vernal-grass	Flouve odorante	Gemeines ruchgras
Arrhenatherum elatius (L.) Beauv. ex J. et C. Presl	Tall oat-grass	Fromental	Glatthafer
Avenula pratensis (L.) Dum.	Meadow oat-grass	Avoine des prés	Rauher wiesenhafer
Avenula pubescens (Huds.) Dum.	Hairy oat-grass	Avoine pubescente	Flaumiger wiesenhafer
Briza media L.	Common quaking grass	Amourette commune	Gewöhnliches zittergras
Bromus catharticus Vahl	Rescue grass	Brome purgatif	Ahrengrasähnliche trespe
Bromus erectus Huds.	Upright brome	Brome dressé	Aufrechte trespe
Bromus hordeaceus L.	Soft brome	Brome mou	Weiche trespe
Bromus inermis Leyss.	Smooth brome	Brome inerme	Unbegrannte trespe
Cynosurus cristatus L.	Crested dog's tail	Crételle	Wiesen kammgras
Dactylis glomerata L.	Cocksfoot	Dactyle vulgaire	Gemeines knäuelgras
Deschampsia caespitosa (L.) Beauv.	Tufted hair-grass	Canche cespiteuse	Rasenschmiele
Deschampsia flexuosa (L.) Trin.	Wavy hair-grass	Canche flexueuse	Geschlängelte schmiele
Elymus repens (L.) Gould	Couch	Chiendent commun	Gemeine quecke
Festuca arundinacea Schreb.	Tall fescue	Fétuque élevée	Rohr schwingel
Festuca ovina L.	Sheep's fescue	Fétuque ovine	Schaf schwingel
Festuca pratensis Huds.	Meadow fescue	Fétuque des prés	Wiesen schwingel
Festuca rubra L.	Red fescue	Fétuque rouge	Roter schwingel
Glyceria fluitans (L.) R. Brown	Floating sweet-grass	Glycérie flottante	Flutender schwaden
Glyceria maxima (Hartm.) Holmberg	Reed sweet-grass	Glycérie aquatique	Wasser schwaden
Holcus lanatus L.	Yorkshire fog	Houlque laineuse	Wolliges honiggras
Holcus mollis L.	Creeping soft-grass	Houlque molle	Weiches honiggras
Hordeum secalinum Schreb.	Meadow barley	Orge faux-seigle	Roggen gerste
Koeleria pyramidata (Lam.) Beauv.	Crested hair-grass	Koelérie pyramidale	Grosses schillergras
Lolium x *hybridum* Haussk.	Hybrid ryegrass	Ray-grass hybride	Bastard-Raigras
Lolium multiflorum Lam.	Italian ryegrass	Ray-grass d'Italie	Italienisches Raigras
Lolium perenne L.	Perennial ryegrass	Ray-grass anglais	Englisches Raigras
Molinia caerulea (L.) Moench	Purple moor-grass	Molinie	Pfeifengras
Nardus stricta L.	Mat-grass	Nard	Borstgras
Phalaris arundinacea L.	Reed canary-grass	Baldingère	Rohr glanzgras
Phleum alpinum L.	Alpine cat's-tail	Fléole des Alpes	Alpen timothe
Phleum pratense L.	Timothy	Fléole des prés	Wiesen lieschgras
Poa alpina L.	Alpine meadow grass	Pâturin des Alpes	Alpen rispengras
Poa annua L.	Annual meadow grass	Pâturin annuel	Einjähriges rispengras
Poa pratensis L.	Smooth meadow grass	Pâturin des prés	Wiesen rispengras
Poa trivialis L.	Rough meadow grass	Pâturin commun	Gemeines rispengras
Trisetum flavescens (L.) Beauv.	Yellow oat-grass	Avoine jaunâtre	Goldhafer

English name	Scientific name	French name	German name
Alpine cat's-tail	*Phleum alpinum* L.	Fléole des Alpes	Alpen timothe
Alpine meadow grass	*Poa alpina* L.	Pâturin des Alpes	Alpen rispengras
Annual meadow grass	*Poa annua* L.	Pâturin annuel	Einjärhriges rispengras
Black bent	*Agrostis gigantea* Roth	Agrostis géant	Riesen straussgras
Cocksfoot	*Dactylis glomerata* L.	Dactyle vulgaire	Gemeines knäuelgras
Common bent	*Agrostis capillaris* L.	Agrostis commun	Gemeines straussgras
Common quaking grass	*Briza media* L.	Amourette commune	Gewöhnliches zittergras
Couch	*Elymus repens* (L.) Gould	Chiendent commun	Gemeine quecke
Creeping bent	*Agrostis stolonifera* L.	Agrostis stolonifère	Fioringras
Creeping soft-grass	*Holcus mollis* L.	Houlque molle	Weiches honiggras
Crested dog's tail	*Cynosurus cristatus* L.	Crételle	Wiesen kammgras
Crested hair-grass	*Koeleria pyramidata* (Lam.) Beauv.	Koelérie pyramidale	Grosses schillergras
Floating sweet-grass	*Glyceria fluitans* (L.) R. Brown	Glycérie flottante	Flutender schwaden
Hairy oat-grass	*Avenula pubescens* (Huds.) Dum.	Avoine pubescente	Flaumiger wiesenhafer
Smooth brome	*Bromus inermis* Leyss.	Brome inerme	Unbegrannte trespe
Hybrid ryegrass	*Lolium* x *hybridum* Hausskn.	Ray-grass hybride	Bastard-Raigras
Italian ryegrass	*Lolium multiflorum* Lam.	Ray-grass d'Italie	Italienisches Raigras
Marsh foxtail	*Alopecurus geniculatus* L.	Vulpin genouillé	Knick-fuchsschwanz
Mat-grass	*Nardus stricta* L.	Nard	Borstgras
Meadow barley	*Hordeum secalinum* Schreb.	Orge faux-seigle	Roggen gerste
Meadow fescue	*Festuca pratensis* Huds.	Fétuque des prés	Wiesen schwingel
Meadow foxtail	*Alopecurus pratensis* L.	Vulpin des prés	Wiesen-fuchsschwanz
Meadow oat-grass	*Avenula pratensis* (L.) Dum.	Avoine des prés	Rauher wiesenhafer
Perennial ryegrass	*Lolium perenne* L.	Ray-grass anglais	Englisches Raigras
Purple moor-grass	*Molinia caerulea* (L.) Moench	Molinie	Pfeifengras
Red fescue	*Festuca rubra* L.	Fétuque rouge	Roter schwingel
Reed canary-grass	*Phalaris arundinacea* L.	Baldingère	Rohr glanzgrass
Reed sweet-grass	*Glyceria maxima* (Hartm.) Holmberg	Glycérie aquatique	Wasser schwaden
Rescue grass	*Bromus catharticus* Vahl	Brome purgatif	Ahrengrasähnliche trespe
Rough meadow grass	*Poa trivialis* L.	Pâturin commun	Gemeines rispengras
Sheep's fescue	*Festuca ovina* L.	Fétuque ovine	Schaf schwingel
Smooth meadow grass	*Poa pratensis* L.	Pâturin des prés	Wiesen rispengras
Soft brome	*Bromus hordeaceus* L.	Brome mou	Weiche trespe
Sweet vernal-grass	*Anthoxanthum odoratum* L.	Flouve odorante	Gemeines ruchgras
Tall fescue	*Festuca arundinacea* Schreb.	Fétuque élevée	Rohr schwingel
Tall oat-grass	*Arrhenatherum elatius* (L.) Beauv. ex J. et C. Presl	Fromental	Glatthafer
Timothy	*Phleum pratense* L.	Fléole des prés	Wiesen lieschgras
Tufted hair-grass	*Deschampsia caespitosa* (L.) Beauv.	Canche cespiteuse	Rasen schmiele
Upright brome	*Bromus erectus* Huds.	Brome dressé	Aufrechte trespe
Velvet bent	*Agrostis canina* L.	Agrostis des chiens	Sumf straussgras
Wavy hair-grass	*Deschampsia flexuosa* (L.) Trin.	Canche flexueuse	Geschlängelte schmiele
Yellow oat-grass	*Trisetum flavescens* (L.) Beauv.	Avoine jaunâtre	Goldhafer
Yorkshire fog	*Holcus lanatus* L.	Houlque laineuse	Wolliges honiggras

French name	Scientific name	English name	German name
Agrostis commun	*Agrostis capillaris* L.	Common bent	Gemeines straussgras
Agrostis des chiens	*Agrostis canina* L.	Velvet bent	Sumf straussgras
Agrostis géant	*Agrostis gigantea* Roth	Black bent	Riesen straussgras
Agrostis stolonifère	*Agrostis stolonifera* L.	Creeping bent	Fioringras
Amourette commune	*Briza media* L.	Common quaking grass	Gewöhnliches zittergras
Avoine des prés	*Avenula pratensis* (L.) Dum.	Meadow oat-grass	Rauher wiesenhafer
Avoine jaunâtre	*Trisetum flavescens* (L.) Beauv.	Yellow oat-grass	Goldhafer
Avoine pubescente	*Avenula pubescens* (Huds.) Dum.	Hairy oat-grass	Flaumiger wiesenhafer
Baldingère	*Phalaris arundinacea* L.	Reed canary-grass	Rohr glanzgrass
Brome dressé	*Bromus erectus* Huds.	Upright brome	Aufrechte trespe
Brome inerme	*Bromus inermis* Leyss.	Smooth brome	Unbegrannte trespe
Brome mou	*Bromus hordeaceus* L.	Soft brome	Weiche trespe
Brome purgatif	*Bromus catharticus* Vahl	Rescue grass	Ahrengrasähnliche trespe
Canche cespiteuse	*Deschampsia caespitosa* (L.) Beauv.	Tufted hair-grass	Rasen schmiele
Canche flexueuse	*Deschampsia flexuosa* (L.) Trin.	Wavy hair-grass	Geschlängelte schmiele
Chiendent commun	*Elymus repens* (L.) Gould	Couch	Gemeine quecke
Crételle	*Cynosurus cristatus* L.	Crested dog's tail	Wiesen kammgras
Dactyle vulgaire	*Dactylis glomerata* L.	Cocksfoot	Gemeines knäuelgras
Fétuque des prés	*Festuca pratensis* Huds.	Meadow fescue	Wiesen schwingel
Fétuque ovine	*Festuca ovina* L.	Sheep's fescue	Schaf schwingel
Fétuque élevée	*Festuca arundinacea* Schreb.	Tall fescue	Rohr schwingel
Fétuque rouge	*Festuca rubra* L.	Red fescue	Roter schwingel
Fléole des Alpes	*Phleum alpinum* L.	Alpine cat's-tail	Alpen timothe
Fléole des prés	*Phleum pratense* L.	Timothy	Wiesen lieschgras
Flouve odorante	*Anthoxanthum odoratum* L.	Sweet vernal-grass	Gemeines ruchgras
Fromental	*Arrhenatherum elatius* (L.) Beauv. ex J. et C. Presl	Tall oat-grass	Glatthafer
Glycérie aquatique	*Glyceria maxima* (Hartm.) Holmberg	Reed sweet-grass	Wasser schwaden
Glycérie flottante	*Glyceria fluitans* (L.) R. Brown	Floating sweet-grass	Flutender schwaden
Houlque laineuse	*Holcus lanatus* L.	Yorkshire fog	Wolliges honiggras
Houlque molle	*Holcus mollis* L.	Creeping soft-grass	Weiches honiggras
Koelérie pyramidale	*Koeleria pyramidata* (Lam.) Beauv.	Crested hair-grass	Grosses schillergras
Molinie	*Molinia caerulea* (L.) Moench	Purple moor-grass	Pfeifengras
Nard	*Nardus stricta* L.	Mat-grass	Borstgras
Orge faux-seigle	*Hordeum secalinum* Schreb.	Meadow barley	Roggen gerste
Pâturin annuel	*Poa annua* L.	Annual meadow grass	Einjähriges rispengras
Pâturin commun	*Poa trivialis* L.	Rough meadow grass	Gemeines rispengras
Pâturin des Alpes	*Poa alpina* L.	Alpine meadow grass	Alpen rispengras
Pâturin des prés	*Poa pratensis* L.	Smooth meadow grass	Wiesen rispengras
Ray-grass anglais	*Lolium perenne* L.	Perennial ryegrass	Englisches Raigras
Ray-grass d'Italie	*Lolium multiflorum* Lam.	Italian ryegrass	Italienisches Raigras
Ray-grass hybride	*Lolium* x *hybridum* Hausskn.	Hybrid ryegrass	Bastard-Raigras
Vulpin des prés	*Alopecurus pratensis* L.	Meadow foxtail	Wiesen-fuchsschwanz
Vulpin genouillé	*Alopecurus geniculatus* L.	Marsh foxtail	Knick-fuchsschwanz

German name	Scientific name	English name	French name
Ahrengrasähnliche trespe	*Bromus catharticus* Vahl	Rescue grass	Brome purgatif
Alpen rispengras	*Poa alpina* L.	Alpine meadow grass	Pâturin des Alpes
Alpen timothe	*Phleum alpinum* L.	Alpine cat's-tail	Fléole des Alpes
Aufrechte trespe	*Bromus erectus* Huds.	Upright brome	Brome dressé
Bastard-Raigras	*Lolium* x *hybridum* Hausskn.	Hybrid ryegrass	Ray-grass hybride
Borstgras	*Nardus stricta* L.	Mat-grass	Nard
Einjärhriges rispengras	*Poa annua* L.	Annual meadow grass	Pâturin annuel
Englisches Raigras	*Lolium perenne* L.	Perennial ryegrass	Ray-grass anglais
Fioringras	*Agrostis stolonifera* L.	Creeping bent	Agrostis stolonifère
Flaumiger wiesenhafer	*Avenula pubescens* (Huds.) Dum.	Hairy oat-grass	Avoine pubescente
Flutender schwaden	*Glyceria fluitans* (L.) R. Brown	Floating sweet-grass	Glycérie flottante
Gemeine quecke	*Elymus repens* (L.) Gould	Couch	Chiendent commun
Gemeines knäuelgras	*Dactylis glomerata* L.	Cocksfoot	Dactyle vulgaire
Gemeines rispengras	*Poa trivialis* L.	Rough meadow grass	Pâturin commun
Gemeines ruchgras	*Anthoxanthum odoratum* L.	Sweet vernal-grass	Flouve odorante
Gemeines straussgras	*Agrostis capillaris* L.	Common bent	Agrostis commun
Geschlängelte schmiele	*Deschampsia flexuosa* (L.) Trin.	Wavy hair-grass	Canche flexueuse
Gewöhnliches zittergras	*Briza media* L.	Common quaking grass	Amourette commune
Glatthafer	*Arrhenatherum elatius* (L.) Beauv. ex J. et C. Presl	Tall oat-grass	Fromental
Goldhafer	*Trisetum flavescens* (L.) Beauv.	Yellow oat-grass	Avoine jaunâtre
Grosses schillergras	*Koeleria pyramidata* (Lam.) Beauv.	Crested hair-grass	Koelérie pyramidale
Italienisches Raigras	*Lolium multiflorum* Lam.	Italian ryegrass	Ray-grass d'Italie
Knick-fuchsschwanz	*Alopecurus geniculatus* L.	Marsh foxtail	Vulpin genouillé
Pfeifengras	*Molinia caerulea* (L.) Moench	Purple moor-grass	Molinie
Rasen schmiele	*Deschampsia caespitosa* (L.) Beauv.	Tufted hair-grass	Canche cespiteuse
Rauher wiesenhafer	*Avenula pratensis* (L.) Dum.	Meadow oat-grass	Avoine des prés
Riesen straussgras	*Agrostis gigantea* Roth	Black bent	Agrostis géant
Roggen gerste	*Hordeum secalinum* Schreb.	Meadow barley	Orge faux-seigle
Rohr glanzgras	*Phalaris arundinacea* L.	Reed canary-grass	Baldingère
Rohr schwingel	*Festuca arundinacea* Schreb.	Tall fescue	Fétuque élevée
Roter schwingel	*Festuca rubra* L.	Red fescue	Fétuque rouge
Schaf schwingel	*Festuca ovina* L.	Sheep's fescue	Fétuque ovine
Sumf straussgras	*Agrostis canina* L.	Velvet bent	Agrostis des chiens
Unbegrannte trespe	*Bromus inermis* Leyss.	Smooth brome	Brome inerme
Wasser schwaden	*Glyceria maxima* (Hartm.) Holmberg	Reed sweet-grass	Glycérie aquatique
Weiche trespe	*Bromus hordeaceus* L.	Soft brome	Brome mou
Weiches honiggras	*Holcus mollis* L.	Creeping soft-grass	Houlque molle
Wiesen kammgras	*Cynosurus cristatus* L.	Crested dog's tail	Crételle
Wiesen lieschgras	*Phleum pratense* L.	Timothy	Fléole des prés
Wiesen rispengras	*Poa pratensis* L.	Smooth meadow grass	Pâturin des prés
Wiesen schwingel	*Festuca pratensis* Huds.	Meadow fescue	Fétuque des prés
Wiesen-fuchsschwanz	*Alopecurus pratensis* L.	Meadow foxtail	Vulpin des prés
Wolliges honiggras	*Holcus lanatus* L.	Yorkshire fog	Houlque laineuse

References

References

A'Brook, J.A. & Heard, A.J. 1975. The effect of ryegrass mosaic virus on the yield of perennial ryegrass swards. *Ann. Appl. Biol.*, 80: 163–168.

Allen, V. coord. 1991. *Terminology for grazing lands and grazing animals.* Blacksburg, USA, Pocahontas Press Inc. 38 pp.

Al Mufti, M.M., Sydes, C.L., Furness, S.B., Grime, J.B. & Band, S.R. 1977. A quantitative analysis of shoot phenology and dominance in herbaceous vegetation. *J. Ecol.*, 65: 759–761.

Ammerman, A. & Cavalli-Sforza, L.L. 1971. Measuring the rate of spread of early farming in Europe. *Man*, 6: 674–688.

Anonymous. 1979. *Results of experiments.* Agronomy Department, West of Scotland Agricultural College. Auchincruive, UK.

Anonymous. 1881. *Annuaire des cinq départements normands* [Yearbook of the five Norman departments]. Caen, France, Editions Caen 1. 478 pp.

Armstrong, R.H., Common, T.G. & Smith, H.K. 1986. The voluntary intake and *in vivo* digestibility of herbage harvested from indigenous hill plant communities. *Grass Forage Sci.*, 41: 53–60.

Armstrong, R.H., Grant, S.A., Common, T.G. & Beattie, M.M. 1997. Controlled grazing studies on *Nardus* grasslands: effects of between-tussock sward height and species of grazer on diet selection and intake. *Grass Forage Sci.*, 52: 219–231.

Bacon, C.W. & Siegel, M.R. 1988. Endophyte parasitism of tall fescue. *J. Prod. Agr.*, 1: 45–55.

Baert, J., Verbruggen, I. & Carlier, L. 1994. About the ryegrass endophyte (*Neotyphodium lolii*) in Belgium. *In* L. 't Mannetje & J. Frame, eds. *Grassland and society.* Proceedings of the 15th General Meeting of the European Grassland Federation, The Netherlands, pp. 142–144.

Barnes, R.F., Miller, D.A. & Jerry, Nelson C. 1995. *Forages.* Vol. I. *An introduction to grassland agriculture.* 5th ed. Ames, USA, Iowa State University Press. 516 pp.

Beddows, A.R. 1959. *Dactylis glomerata* L. *J. Ecol.*, 47: 223–239.

Beddows, A.R. 1961. *Holcus lanatus* L. *J. Ecol.*, 49: 421–430.

Beddows, A.R. 1967. *Lolium perenne* L. *J. Ecol.*, 55: 567–587.

Beddows, A.R. 1973. *Lolium multiflorum* Lam. (*L. perenne* L. ssp. *multiflorum* (Lam.) Husnot, *L. italicum* A. Braun). *J. Ecol.*, 61: 587–600.

Benoit, M. 1994. Risques de pollution des eaux sous prairie et sous culture. Influence des pratiques d'apport d'engrais de ferme [Water pollution risk in grassland and arable land. Influence of manure]. *Fourrages*, 140: 407–420.

Betin, M. & Gillet, M. 1983. Etude complémentaire sur le comportement de différentes espèces de bromes en France: *catharticus, sitchensis, carinatus, valdivianus* [Complementary study on the performances of several brome species in France: *catharticus, sitchensis, carinatus, valdivianus*]. *Fourrages*, 96: 81–104.

Betin, M. & Mansat, P. 1979. Valeur agronomique de différentes espèces de bromes en France [Agronomic value of several brome species in France]. *Fourrages*, 78: 51–66.

Bianchi, A.A. & Ciriciofolo, E. 1978. Valutazione della produttivita di ecotipi di specie foraggere in Umbria [Evaluation of the productivity of ecotypes of forage species in Umbria]. *Annali della Facolta di Agraria, Università degli Studi di Perugia*, 32: 143–168.

Bonnier, G. 1897. *Flore complète illustrée en couleurs de France, Suisse et Belgique* [Illustrated complete flora of France, Switzerland and

Belgium]. Libraire générale de l'enseignement, 12(1): 131 pp.

Borin, M., Giupponi, C. & Morari F. 1997. Effects of four cultivation systems for maize on nitrogen leaching. 1. Field experiment. *Eur. J. Agron.*, 6: 101–112.

Borin, **M. & Bigon**, **E.** 2002. Abatement of NO_3N concentration in agricultural waters by narrow buffer strips. *Environ. Pollut.*, 117(1): 165–168.

Borrill, **M.** 1976. Temperate grasses. *Lolium, Festuca, Dactylis, Phleum, Bromus (Gramineae)*. *In* N.W. Simmonds, ed. *Evolution of crop plants*, pp. 137–142. London, Longman Group Ltd.

Borrill, **M. & Carroll**, **C.P.** 1969. A chromosome atlas of the genus *Dactylis* (part 2). *Cytologia*, 34: 6.

Bottoni, P., Keizer, J. & Funari, E. 1996. Leaching indices of some major triazine metabolites. *Chemosphere*, 32(7): 1401–1411.

Bournérias, **M., Arnal**, **G. & Bock**, **C.** 2001. *Guide des groupements végétaux de la région parisienne* [Guide to plant communities of the Paris region]. Paris, Belin. 640 pp.

Braun-Blanquet, J. 1964. *Pflanzensoziologie* [Plant sociology]. 3rd ed. Vienna, Springer. 865 pp.

Brinkforth, B. 1997. *Les couvre-sols pour parcs et jardins* [Soil covers for parks and gardens]. Stuttgart, Germany, Editions Eugen Ulmer. 184 pp.

Broersma, K. 1991. *Meadow foxtail: how beneficial is it as a forage grass species for the central interior of B.C.?* Prince George, Experimental Farm, British Columbia. pp. 13–18.

Broersma, K. & Mir, P.S. 1990. *Comparison of four grasses for nutrient content and digestibility.* Prince George, Experimental Farm, British Columbia. p. 23.

Brüschweiler, **S**. 1999. *Plantes et savoirs des Alpes: l'exemple du Val d'Anniviers* [Plants and knowledge of the Alps: the example of the Anniviers valley]. Sierre, Switzerland, Editions monographic SA. 283 pp.

Buchkina, N.P. & Balashov, E.V. 2001. The influence of grass-clover mixture on soil organic matter and aggregation of a podzolic loamy sand soil. *In* R.M. Rees, B.C. Ball, C.D. Campbell & C.A. Watson, eds, *Sustainable*

management of soil organic matter, pp. 214–219. Wallingford, UK, CAB.

Bush, **L.P. & Burrus**, **P.B.** 1988. Tall fescue forage quality and agronomic performance as affected by the endophyte. *J. Prod. Agr.*, 1: 55–60.

Caputa, J. 1967. *Les plantes fourragères* [Forage plants]. Paris, Payot. 205 pp.

Castelle, A.J., Johnson, A.W. & Conolly, C. 1994. Wetland and stream buffer size requirements. A review. *J. Environ. Qual.*, 23: 878–882.

Castillon, **P.** 2000. Pertes d'azote dans les systèmes de culture à base de maïs ensilé. Causes et remèdes [Nitrogen losses in silage maize based systems. Causes and remedies]. *Fourrages*, 163: 283–291.

Catherall, **P.L.** 1987. Effects of barley yellow dwarf and ryegrass mosaic viruses alone and in combination on the productivity of perennial and Italian ryegrasses. *Plant Pathol.*, 36: 73–78.

Chadwick, M.J. 1960. *Nardus stricta* L. *J. Ecol.*, 48: 255–267.

Chapman, **G.P.** 1998. *The biology of grasses*. Wallingford, UK, CAB. 273 pp.

Charles, A.H., Jones, J.L., Thornton, M.S. & Thomas, T.A. 1978. Comparison of ryegrass with some common unsown grasses sown separately and in mixtures. *In* A.H. Charles & R.J. Haggar, eds. *Changes in sward composition and productivity*. British Grassland Society, Occasional Symposium No. 10, pp. 25–29.

Clarke, J. 1993. Set-aside: a weeder and seeder of volunteer crops. *Asp. Appl. Biol.*, 35: 215–222.

Clotuche, P. 1998. *Influence de l'introduction du gel des terres rotationel sur les apports azotés à la culture subséquente des nitrates et le risque de lessivage des nitrates*. [Influence of the introduction of rotational set-aside on the nitrogen availability for the following crop and the nitrate leaching risk]. Laboratoire d'Ecologie des Prairies, Université catholique de Louvain, Louvain la Neuve, Belgium. (PhD thesis)

Clotuche, P. & Peeters, A. 2000. Nitrogen uptake by Italian ryegrass after destruction of non-fertilized set-aside covers at different times in autumn and winter. *J. Agron. Crop Sci.*, 184: 121–131.

Clotuche, P., Peeters, A. & Van Bol, V. 1998. En Belgique, valorisation fourragère des jachères rotationnelles après la période légale de gel des terres [Forage production after the legal period of rotational set-asides in Belgium]. *Fourrages*, 154: 197–209.

Clotuche, P., Peeters, A., Van Bol, V., Imbrecht, O. & Decamps, C. 1997. L'entretien des jachères rotationnelles. Etude réalisée en Belgique sur couverts semés et adventices [Management of rotational set-aside. Study carried out in Belgium on sown swards and weeds]. *Phytoma*, 498: 31–38.

Common, T.G., Hunter, E.A., Eadie, J., Floate, M.J.S. & Hodgson, J. 1991a. The long-term effects of a range of pasture treatments applied to three semi-natural hill grassland communities. 1. Pasture production and botanical composition. *Grass Forage Sci.*, 46: 239–251.

Common, T.G., Hunter, E.A., Eadie, J., Floate, M.J.S. & Hodgson, J. 1991b. The long-term effects of a range of pasture treatments applied to three semi-natural hill grassland communities. 2. Animal performance. *Grass Forage Sci.*, 46: 253–263.

Common, T.G., Wright, I.A. & Grant, S.A. 1998. The effect of grazing by cattle on animal performance and floristic composition in *Nardus*-dominated swards. *Grass Forage Sci.*, 53: 260–269.

Cook, F.G. 1975. Production loss estimation in *Drechslera* infection of ryegrass. *Ann. Appl. Biol.*, 81: 251–256.

Cooper, A. 1976. The vegetation of carboniferous limestone soils in South Wales. 2. Ecotypic adaptations in response to calcium and magnesium. *J. Ecol.*, 64: 147–155.

Coppel, B. & Etienne, M. 1992. Production prairiale dans les Préalpes: les pelouses à *Bromus erectus* [Grassland production in the Pre-Alps: the dry rangelands of *Bromus erectus*]. *Fourrages*, 131: 271–282.

Couplan, F. 1989. *Le régal végétal: plantes sauvages comestibles* [The vegetal treat: wild edible plants]. Flers, France, Equilibres-Aujourd'hui. 453 pp.

Cowling, D.W. & Lockyer, D.R. 1965. A comparison of the reaction of different grass species to fertilizer nitrogen and to growth in association with white clover. I. Yield of dry matter. *J. Brit. Grass. Soc.*, 20: 197–204.

Crofts, A. & Jefferson, R.G., eds. 1999. *The lowland grassland management handbook*. 2nd ed. Peterborough, UK, English Nature/The Wildlife Trusts.

Curtin, D., Selles, F., Campbell, C.A. & Biederbeck, V.O. 1994. *Canadian prairie agriculture as a source and sink of the greenhouse gases, carbon dioxide and nitrous oxide.* Agriculture & Agri-food Canada Publication No. 379M0082.

Daccord, R. 1988. Digestibilité de quelques dicotylédones ('crutes') [Digestibility of some dicotyledons]. *Rech. Agron. Suisse*, 26(2): 139–151.

Davies, D.A. & Munro, J.M.M. 1974. Potential pasture production in the uplands of Wales. 4. Nitrogen response from sown and natural pastures. *J. Brit. Grass. Soc.*, 29: 149–158.

Davies, H. & Williams, A.E. 1970. The effect of mildew and leaf blotch on yield quality of cv. Lior Italian ryegrass. *Plant Pathol.*, 19: 135–138.

Davies, T.H. & Riley, J. 1992. Reseeding in the Falkland Islands. Herbage production of a range of grass species and cultivars. *Grass Forage Sci.*, 47(1): 62–69.

Davy, A.J. 1980. *Deschampsia caespitosa* (L.) Beauv. (*Aira cespitosa* L., *Deschampsia cespitosa* (L.) Beauv.). *J. Ecol.*, 68: 1075–1096.

Delpech, R. 1960. *Critères de jugement de la valeur agronomique des prairies* [Trial criteria of the agronomic value of grasslands]. Proceedings du Colloque de la Société Botanique de France, pp. 83–97.

Demarquilly, C. coord. 1981. *Prévision de la valeur nutritive des aliments des ruminants* [Prediction of the nutritive value of ruminant foods]. Paris, INRA. 580 pp. + tables.

Demon, M., Schiavon, M., Portal, J.M. & Andreux, F. 1994. Seasonal dynamics of atrazine in three soils under outdoor conditions. *Chemosphere*, 28(3): 453–466.

Dillaha, E.A., Reneau, R.B., Mostaghimi, S. & Lee, D. 1989. Vegetative filter strips for agricultural non-point source pollution control. *Trans. ASAE*, 32: 513–519.

Di Muccio, A., Chirico, M., Dommarco, R., Funari, E., Musmeci, L., Santilio, A., Vergori, F., Zapponi, G., Giulliaro, G. & Sparacino, A.C. 1990. Vertical mobility of soil contaminants: preliminary results on the herbicide atrazine. *Int. J. Environ. Anal. Chem.*, 38: 211–220.

Dixon, J.M. 1991. *Avenula* (Dumort.) Dumort. *J. Ecol.*, 79: 829–865.

Dixon, J.M. 2002. *Briza media* L. *J. Ecol.*, 90: 737–752.

Dousset, S., Mouvet, C. & Schiavon, M. 1995. Leaching of atrazine and some of its metabolites in undisturbed field lysimeters of three soil types. *Chemosphere*, 30(3): 511–524.

Duvigneaud, P. 1974. *La synthèse écologique* [The ecological synthesis]. Paris, Doin. 296 pp.

Ellenberg, H., Weber, H.E., Düll, R., Wirth, V., Werner, W. & Paulissen, D. 1991. *Zeigerwerte von pflanzen in mitteleuropa* [Ecological indicatory value of plants from Central Europe]. Göttingen, Germany, Goltze E.K.G. 247 pp.

Ellis, M.B. & Ellis, J.P. 1997. *Microfungi on land plants: an identification handbook*. Slough, UK, Richmond Publishing. 868 pp.

European Commission. 1989. Catalogue commun des variétés des espèces agricoles. Annexe au Journal officiel des Communautés européennes C 326 A [Common catalogue of varieties of agricultural species]. 16th ed. Luxembourg, Office des Publications officielles des Communautés européennes. 387 pp.

European Commission. 1999. Catalogue commun des variétés des espèces agricoles. Annexe au Journal officiel des Communautés européennes C 321 A [Common catalogue of varieties of agricultural species]. 21st ed. Luxembourg, Office des Publications officielles des Communautés européennes. 451 pp.

Faliu, L. 1981. *Botanique appliquée* [Applied botany]. Toulouse, France, Ecole Nationale Vétérinaire de Toulouse. 351 + 106 pp.

FAO. 1990. *Tropical grasses*, by P.J. Skerman & F. Riveros. Rome. 832 pp.

FAO. 2002. FAOSTAT Agriculture Data – Land Use (available at: http://apps.fao.org/page/collections?subset =agriculture)

Favarger, C. 1962. *Flore et végétation des Alpes. I. Etage alpin* [Flora and vegetation of the Alps. I. Alpine belt]. Neuchatel, Switzerland, Delachaux & Niestlé. 293 pp.

Favarger, C. 1966. *Flore et végétation des Alpes. II. Etage subalpin* [Flora and vegetation of the Alps. II. Subalpine belt]. Neuchatel, Switzerland, Delachaux & Niestlé. 301 pp.

Frame, J. 1982. *Yield and quality response of secondary grasses to fertilizer nitrogen.* Proceedings of the 9th General Meeting of the European Grassland Federation, Reading, UK, pp. 292–294.

Frame, J. 1989. Herbage productivity of a range of grass species under a silage cutting regime with high fertilizer nitrogen application. *Grass Forage Sci.*, 44: 267–276.

Frame, J. 1990. Herbage productivity of a range of grass species in association with white clover. *Grass Forage Sci.*, 45: 57–64.

Frame, J. 1991. Herbage production and quality of a range of secondary grass species of five rates of fertilizer nitrogen application. *Grass Forage Sci.*, 46: 139–151.

Frame, J., Hunt, I.V. & Harkess, R.D. 1971. *The role of timothy* (Phleum pratense L.) *in high yielding conservation leys.* Proceedings of the 4th General Meeting of the European Grassland Federation, Switzerland, pp. 198–202.

Frame, J. & Morrison, M.W. 1991. Herbage productivity of prairie grass, reed canary-grass and phalaris. *Grass Forage Sci.*, 46: 417–425.

Frame, J. & Tiley, G.E.D. 1988. *Herbage productivity of secondary grass species.* Proceedings of the 12th General Meeting of the European Grassland Federation, Dublin, pp. 189–193.

Fribourg, H.A., Hoveland, C.S. & Codron, P. 1991. La fétuque élevée et l'endophyte

Neotyphodium coenophialum. Aperçu de la situation aux Etats-Unis [Tall fescue and the endophyte *Neotyphodium coenophialum*. Overview of the situation in the United States]. *Fourrages*, 126: 209–223.

Fribourg, H.A., McLaren, J.B., Barth, K.M., Bryan, J.M. & Connelly J.T. 1979. Productivity and quality of Bermudagrass and orchardgrass-Ladino clover pastures for beef steers. *Agron. J.*, 71(2): 315–320.

Fribourg, H.A., McLaren, J.B., Chestnut, A.B. & Waller, J.C. 1989. *Recent effects of* Neotyphodium coenophialum *on the performance of beef cattle grazing* Festuca arundinacea. Proceedings of the XVI International Grassland Congress, Nice, France, pp. 705–706.

Funari, E., Barbieri, L., Bottoni, P., Del Carlo, G., Forti, S., Giuliano, G., Marinelli, A., Santini, C. & Zavatti, A. 1998. Comparison of the leaching properties of alachlor, metolachlor, triazines and some of their metabolites in an experimental field. *Chemosphere*, 36(8): 1759–1773.

Garde, L. 1990. *Ressources pastorales en Haute-Provence et modélisation de la relation végétation/troupeau* [Rangeland resources in Haute-Provence and modelling the vegetation/herd relationship]. Aix-Marseille III, France. (Thèse UDES)

Garwood, E.A., Tyson, K.C. & Clement, C.R. 1977. *A comparison of yield and soil conditions during 20 years of grazed grass and arable cropping*. Technical Report 21. Grassland Research Institute, Hurley, UK.

Gibson, D.J. & Newman, J.A. 2001. *Festuca arundinacea* Schreber (*F. elatior* L. ssp. *arundinacea* (Schreber) Hackel). *J. Ecol.*, 89: 304–324.

Gillet, M. 1980. *Les graminées fourragères* [Forage grasses]. Paris, Gauthier-Villars. 306 pp.

Gilliam, J.W. 1994. Riparian wetlands and water quality. *J. Environ. Qual.*, 23: 869–900.

Girre, L. 2001. *Les plantes et les médicaments: l'origine végétale de nos médicaments* [Plants and drugs: the vegetal origin of our medicines]. Neuchatel, Switzerland, Delachaux & Niestlé. 253 pp.

Grant, S.A., Suckling, D.E., Smith, H.K., Torvell, L., Forbes, T.D.A. & Hodgson, J. 1985. Comparative studies of diet selection by sheep and cattle: the hill grasslands. *J. Ecol.*, 73: 987–1004.

Grant, S.A., Torvell, L., Sim, E.M. & Small, J.L. 1996. Controlled grazing studies on *Nardus* grassland: effects of between-tussock sward height and species of grazer on *Nardus* utilization and floristic composition in two fields in Scotland. *J. Appl. Ecol.*, 33: 1053–1064.

Granval, P., Muys, B. & Leconte, D. 2000. Intérêt faunistique de la prairie permanente pâturée [Faunistic interest of grazed permanent grassland]. *Fourrages*, 162: 157–167.

Green, B.H. 1990. Agricultural intensification and the loss of habitat, species and amenity in British grasslands: a review of historical change and assessment of future prospects. *Grass Forage Sci.*, 45: 365–372.

Greenlee, J. 2000. *The encyclopedia of ornamental grasses*. Emmaus, USA, Rodale Press. 186 pp.

Greenwood, D.J., Lemaire, G., Goose, G., Cruz P., Draycott, A. & Neeteson, J.J. 1990. Decline in percentage N of C3 and C4 crops with increasing plant mass. *Ann. Bot.*, 66: 425–436.

Grime, J.P. 1979. *Plant strategies and vegetation processes*. Chicester, UK, John Wiley & Sons. 222 pp.

Grime, J.P., Hodgson, J.G. & Hunt, R. 1988. *Comparative plant ecology: a functional approach to common British species*. London, Unwyn Hyman. 742 pp.

Grounds, R. 2002. *The plantfinder's guide to ornamental grasses*. Portland, USA, Timber Press Inc. 192 pp.

Grubb, L.M. 1968. *Grass species: soils division research report*. Dublin, An Foras Taluntais. 99 pp.

Haggar, R.J. 1976. The seasonal productivity, quality and response to nitrogen of four indigenous grasses compared with *Lolium perenne*. *J. Brit. Grass. Soc.*, 31: 197–207.

Haggar, R.J. & Standell, C.J. 1982. *The effect of mefluidide on yield and quality of eight grasses*. Proceedings of the 1982 British Crop

Protection Conference – Weeds, UK, pp. 395–399.

Hancock, J.F. 1992. *Plant evolution and the origin of crop species*. Inglewood Cliffs, USA, Prentice-Hall Inc. 305 pp.

Hansen, R. & Stahl, F. 1992. *Les plantes et leurs milieux* [Plants and their habitats]. Stuttgart, Germany, Editions Eugen Ulmer. 571 pp.

Hanson, A.A. & Carnahan, H.L. 1956. *Breeding perennial forage grasses*. USDA Technical Bulletin, 1145. 116 pp.

Hartley, W. & Williams, G.L. 1956. *Centres of distribution of cultivated pasture grasses*. Proceedings of the 7th International Grassland Congress, New Zealand, pp. 190–199.

Harvey, D.M.R., Crothers, S.H. & Hayes, P. 1984. Dry matter and quality of herbage harvested from *Holcus lanatus* and *Lolium perenne* grown in monocultures and in mixtures. *Grass Forage Sci.*, 39: 159–165.

Haycock, N.E. & Pinay, G. 1993. Groundwater nitrate dynamics in grass and poplar vegetated riparian buffer strips during the winter. *J. Environ. Qual.*, 21: 273–278.

Haycock, N.E., Burt, T.P., Goulding, K.W.T. & Pinay, G., eds. 1997. *Buffer zones: their processes and potential in water protection*. Harpenden, UK, Quest Environmental. 320 pp.

Hayes, P. 1979. *Competition of sown grasses with volunteer* Gramineae. Annual report on Research and Technical Work of the Department of Agriculture for Northern Ireland, pp. 95–96.

Henderson, J.L., Edwards, R.S. & Hammerton, J.L. 1962. The productivity of five grass species at six levels of compound fertilizer application. *J. Agr. Sci.*, Cambridge, 59: 5–11.

Hopkins, A., ed. 1990. *Grass: its production and utilization*. 3rd ed. Oxford, UK, Blackwell Science Ltd. 440 pp.

Hopkins, A., Gilbey, J., Dibb, C., Bowling, P.J. & Murray, P.J. 1990. Response of permanent and reseeded grassland to fertilizer nitrogen. 1. Herbage production and herbage quality. *Grass Forage Sci.*, 45: 43–55.

Hopkins, A., Martyn, T.M., Johnson, R.H., Sheldrick, R.D. & Lavender, R.H. 1996.

Forage production by two *Lotus* species as influenced by companion grass species. *Grass Forage Sci.*, 51: 343–349.

Hubbard, C.E. 1978. *Grasses*. 2nd ed. Harmondsworth, UK, Penguin Books. 463 pp.

Hubert, D. 1978. *Evaluation du rôle de la végétation des parcours dans le bilan écologique et agro-économique des Causses* [Evaluation of the role of the rangeland vegetation in the ecological and agro-economic balance of the Causses]. USTL Montpellier, 247 pp. (Thèse Ecologie)

Hume, D.E. 1991. Primary growth and quality characteristics of *Bromus wildenowii* and *Lolium multiflorum*. *Grass Forage Sci.*, 46: 313–324.

Hutchinson, C.S. & Seymour, G.B. 1982. *Poa annua* L. *J. Ecol.*, 70: 887–901.

Hutson, M.A. 1979. A general hypothesis of species diversity. *Am. Nat.*, 113: 81–101.

Imaizumi, S., Nishino, T., Miyabe, K., Fujimori, T. & Yamada, M. 1997. Biological control of annual bluegrass (*Poa annua* L.) with a Japanese isolate of *Xanthomonas campestris* pv. *poae* (JT-P482). *Biol. Control*, 8: 7–14.

Ivins, J.D. 1974. *Pastures new: changing roles for grasses and clovers in modern management*. Proceedings of the 13th NIAB Crop Conference, Cambridge, UK: 1.

Jarrige, R., coord. 1979. Aspects biologiques et techniques de la remise en exploitation des hauts pâturages dégradés des Monts Dore [Biological and technical aspects of the restarting of the management of the degraded high altitude pastures of the Monts Dore]. *In* G. Molenat & R. Jarrige, eds. *Utilisation par les ruminants des pâturages d'altitude et parcours méditerranéens* [Utilization of high altitude and Mediterranean rangelands by ruminants], pp. 57–135. Paris, INRA.

Jarrige, R. 1981. Les constitutants glucidiques des fourrages: variations, digestibilité et dosage [Carbohydrate components of forage: variation, digestibility and analysis]. *In* C. Demarquilly, ed. *Prévision de la valeur nutritive des aliments des ruminants* [Prediction of the nutritive value of ruminant food], pp. 13–40. Paris, INRA.

Jarrige, R. coord. 1988. *Alimentation des bovins, ovins et caprins* [Feeding of cattle, sheep and goats]. Paris, INRA. 471 pp.

Jeangros, B. & Schmid, W. 1991. Production et valeur nutritive des prairies permanentes riches en espèces [Production and nutritive value of species-rich permanent grasslands]. *Fourrages*, 126: 131–136.

Jeangros, B., Troxler, J. & Schmid, W. 1991. Prairies de Suisse riches en espèces: description et rendement [Swiss species-rich grasslands: description and yield]. *Rev. Suisse Agr.*, 23, 1: 26–35.

Jones, K., Carroll, C.P. & Borrill, M. 1961. A chromosome atlas of the genus *Dactylis*. *Cytologia*, 26: 333.

Jones, M.B. & Lazenby, A., eds. 1988. The grass crop. The physiological bases of production. London, Chapman and Hall Ltd. 369 pp.

Kastirr, U. 1998. Resistance evaluation of *Lolium perenne* L. to *Rhynchosporium orthosporum* Caldwell. In *Naar een duurzame grasland- en groenvoederuitbating. Onderzoek naar de integratie van landbouwkundige en ecologische doelstellingen bij grasland- en groenvoederwinning*, pp. 30–31 [Towards sustainable grassland and forage production. Research on the integration of agricultural and ecological objectives in grassland and forage production]. Brussels, Ministerie van Middenstand en Landbouw.

Keane, M.G. & Allen, P. 1998. Effects of production system intensity on performance, carcass composition and meat quality of beef cattle. *Livest. Prod. Sci.*, 56: 203–214.

Klapp, E. 1965. *Grünland – vegetation und standort* [Grassland vegetation and environment]. Hamburg, Germany, Parey. 384 pp.

Kley, G. 1995. *Seed production in grass and clover species in Europe*. Proceedings of the 3rd International Herbage Seed Conference, Germany, pp. 12–22.

Kline, P. & Broersma, K. 1983. The yield nitrogen and nitrate content of reed canarygrass, meadow foxtail and timothy fertilized with nitrogen. *Can. J. Plant Sci.*, 63: 943–950.

Köhler, H., Libiseller, R., Schmid, S. & Swoboda, R. 1978. Zur kalzinose der rinder in Osterreich. VII Untersuchungen zur bedeutung der aufwuschsstadien sowie der gewinnung (silage, heu) von goldhafer (*Trisetum flavescens*) für die entstehung der kalzinose [On cattle calcinosis in Austria. VII Analysis of the importance of the wilting stage after a cut (silage, hay) of yellow oat-grass (*Trisetum flavescens*) for the development of calcinosis]. *Zbl. Vet. Med. A.*, 25: 617–633.

Kornas, J. 1983. Man's impact on flora and vegetation in Central Europe. *Geobotany*, 5: 277–286.

Kozlowski, S. & Golinska, B. 1994. Szybkosc odkladania lignin, celulozy I hemiceluloz przez *Alopecurus pratensis* [Lignin, cellulose and hemi-cellulose contents of *Alopecurus pratensis*]. *Zeszyty problemowe Postepow nauk rolniczych* [Journal of Advances in Agricultural Sciences], 412: 121–124.

Krautzer, B. 1996. Recultivation of alpine areas with seed of alpine plants. In G. Parente, J. Frame & S. Orsi, eds. *Grassland and land use systems*, Proceedings of the 16th General Meeting of the European Grassland Federation, Italy, pp. 775–779.

Krautzer, B. 1998. Re-establishment of species-rich alpine meadows with seeds of indigenous plants. In *Breeding for a multifunctional agriculture*. Proceedings of the 21st Meeting of the Fodder Crops and Amenity Grasses, EUCARPIA, Switzerland, pp. 118–122.

Krautzer, B., Bohner, A., Partl, C., Venerus, S. & Parente, G. 2001. New approaches to restoration of alpine ski slopes. In J. Isselstein, G. Spatz & M. Hofmann, eds. *Organic grassland farming*. Proceedings of the International Occasional Symposium of the European Grassland Federation, Germany, pp. 193–196.

Kreil, W., Kaltofen, H. & Wacker, G. 1964. Weitere versuchsergebnisse uber die dungung einer weide mit verscieden hohen N-gaben (1961–1963) [Further research results on the fertilization of a meadow with different N rates (1961–1963)]. *Zeitschrift für Landeskultur*, 5 H 3: 221–244.

Kruijne, A.A. & De Vries, D.M. 1963. *Data concerning important herbage plants*. Instituut van biologische scheikunde onderzoek op landbouwgewassen, Wageningen, The Netherlands. Mededeling [Communication] 225: 44 pp.

Kruijne, A.A. & De Vries, D.M. 1976. *Vegetatieve herkenning van onze grasland planten* [Vegetative identification of our grassland plants]. Wageningen, The Netherlands, Veenman & Zonen. 112 pp.

Kwaad, F.J.P.M. 1994. Cropping systems of fodder maize to reduce erosion of cultivated loess soils. *In* R.J. Rickson, ed. *Conserving soil resources, European perspectives*, pp. 354–368, Wallingford, UK, CAB.

Kwaad, F.J.P.M., van der Zijp, M. & van Dijk, P.M. 1998. Soil conservation and maize cropping systems on sloping loess soils in the Netherlands. *Soil Till. Res.*, 46(1–2): 13–21.

Laissus, R. 1979. La valeur fourragère du vulpin des prés [Forage value of meadow foxtail]. *Fourrages*, 79: 75–87.

Lam, A. 1984. *Drechslera siccans* from ryegrass fields in England and Wales. *Trans. Brit. Myc. Soc.*, 83(2): 305–311.

Lam, A. 1985. Effect of fungal pathogens on digestibility and chemical composition of Italian ryegrass (*Lolium multiflorum*) and tall fescue (*Fesctuca arundinacea*). *Plant Pathol.*, 34: 190–199.

Lambinon, J., De Langhe, J.E., Delvosalle, L. & Duvigneaud, J. 1992. *Nouvelle flore de la Belgique, du Grand-Duché de Luxembourg, du Nord de la France et des régions voisines* [New flora of Belgium, Grand-Duchy of Luxemburg, north of France and neighbouring regions]. 4th ed. Meise, Belgium, Editions du patrimoine du Jardin botanique national de Belgique. 1092 pp.

Lancashire, J.A. & Latch, G.C.M. 1966. Some effects of crown rust (*Puccinia coronata* Corda) on the growth of two ryegrass varieties in New Zealand. *New Zeal. J. Agr. Res.*, 9: 628–640.

Landwehr, J. 1977. *Atlas van de Nederlandse grassen* [Atlas of Dutch grasses]. 2nd ed. Zutphen, The Netherlands, Thieme-Zutphen. 362 pp.

Lauber, K. & Wagner, G. 1998. *Flora Helvetica.* Paris, Belin. 1616 pp.

Laurent, F., Machet, J.M., Pellot, P. & Trochard, R. 1995. Cultures intermédiaires pièges à nitrates: *comparaison des espèces* [Catch crops in intercropping: comparison of species]. *In* F. Laurent ed. *Azote et interculture* [Nitrogen and intercropping], pp. 38–49. Paris, Institut Technique des Céréales et des Fourrages.

Lazenby, A. 1981. British grasslands; past, present and future. *Grass Forage Sci.*, 36: 243–266.

Leafe, E.L. 1988. Introduction: the history of improved grasslands. *In* M.B. Jones & A. Lazenby eds. *The grass crop. The physiological basis of production*, pp. 1–23, London, Chapman and Hall Ltd.

Le Gall, A., Legarto, J. & Pfimlin, A. 1997. Place du maïs et de la prairie dans les systèmes fourragers laitiers. III. Incidence sur l'environnement [Place of maize and grassland in the dairy forage systems. III. Impact on the environment]. *Fourrages*, 150: 147–169.

Lehmann, J., Briner, H.U., Mosimann, E. & Chalet, C. 1992. Diversification des espèces pour prairies temporaires [Species diversification for temporary grasslands]. *Rev. Suisse Agr.*, 24(3): 159–163.

Lingorsky, V. 1994. Note sur l'effet du rythme d'exploitation de diverses graminées prairiales dans les zones de contreforts en Bulgarie [Note on the effect of the defoliation frequency on several forage grasses in hill areas in Bulgaria]. *Fourrages*, 138: 157–164.

Lodge, R.W. 1959. *Cynosurus cristatus* L. *J. Ecol.*, 47: 511–518.

Maljean, J.F. & Peeters, A. 2002. *Integrated farming and biodiversity: impacts and political measures.* High-level Pan-European conference on agriculture and biodiversity, Paris. 28 pp.

Marten, G.C. & Jordan, R.M. 1974. *Significance of palatability differences among* Phalaris arundinacea L., Bromus inermis *Leyss. and* Dactylis glomerata *L. grazed by sheep.* Proceedings of the 12th International Grassland Congress, Moscow, USSR, 3: 305–312.

Marten, G.C., Sheaffer, C.C. & Wyse, D.L. 1987. Forage nutritive value and palatability of perennial weeds. *Agron. J.*, 79: 980–986.

Maurizio, S. 1933. *Histoire de l'alimentation végétale* [History of vegetal feeding]. Paris, Payot. 150 pp.

Moloney, A. 1999. *The quality of meat from beef cattle: is it influenced by diet?* R & H Hall Technical Bulletin Issue 4.

Morris, R.M. & Thomas, J.G. 1972. The seasonal pattern of dry matter production of grasses in the North Pennines. *J. Brit. Grass. Soc.*, 27: 163–172.

Morton, J.D., Bolton, G.R. & Hodgson, J. 1992. The comparative performance of *Holcus lanatus* and *Lolium perenne* under sheep grazing in the Scottish uplands. *Grass Forage Sci.*, 47: 143–152.

Mosimann, E., Lehmann, J., Rosenberg, E. & Bassetti, P. 1996a. Mélanges standards pour la production fourragère [Standard mixtures for forage production]. *Rev. Suisse Agr.*, 28(6): 353–364.

Mosimann, E., Chalet, C., Lehmann, J. & Briner, H.U. 1996b. Essais de variétés de pâturin des prés et de vulpin des prés 1993–1995 [Trials of smooth meadow grass and meadow foxtail varieties (1993–1995)]. *Rev. Suisse Agr.*, 28(2): 77–80.

Mosimann, E., Chalet, C., Lehmann J., Briner, H.U. & Bassetti, P. 1998. Liste 1999–2000 des variétés recommandées de plantes fourragères [List (1999–2000) of recommended varieties of forage plants]. *Rev. Suisse Agr.*, 30(5): 16 pp.

Mosimann, E., Lehmann, J. & Rosenberg, E. 2000. Mélanges standards pour la production fourragère. Révision 2001–2004 [Standard mixtures for forage production. Revision 2001–2004]. *Rev. Suisse Agr.*, 32(5): 1–12.

Mousset, C. 2000. Rassemblement, utilisation et gestion des ressources génétiques de dactyle à l'INRA de Luzignan [Collecting, utilizing and managing the genetic resources of cocksfoot at the INRA station of Lusignan]. *Fourrages*, 162: 121–139.

Muller, J. 1981. Fossil pollen records of extinct angiosperms. *Bot. Rev.*, 47: 1–145.

Muncy, J. 1985. *Reclamation of abandoned mica, feldspar, and kaolin surface mines and associated tallings disposal sites in western North Carolina.* Proceedings of American Society for Surface Mining Operation, West Virginia, pp. 153–157.

Narashimhalu, P., Winter, K.A. & Kunelius, H.T. 1982. *In vivo* utilization of wilted grass silages prepared in the maritime provinces of Canada. *Can. J. Plant Sci.*, 62: 391–397.

Neuteboom, J.H. 1975. Variability of *Elytrigia repens* (L.) Desv. (Syn. *Agropyron repens* (L.) P.B.) on Dutch agricultural soils. Landbouwhogeschool, Wageningen, Mededelingen [Communications] 75–77.

Neuteboom, J.H. 1980. Variability of couch (*Elytrigia repens* (L.) Desv.) in grasslands and arable fields in two localities in the Netherlands. *Acta Bot. Neerl.*, 29(5–6): 407–417.

Neuteboom, J.H. 1981. Effects of different mowing regimes on the growth and development of four clones of couch (*Elytrigia repens* (L.) Desv., syn *Agropyron repens* (L.) Beauv.) in monoculture and mixtures with perennial ryegrass (*Lolium perenne* L.). Landbouwhogeschool, Wageningen, The Netherlands. Mededeling [Communication] 15.

Ninane, V., Goffart, J.P., Meeus-Verdinne, K., Destain, J.P., Guiot, J., François, E., Ducat, N. & Bock, L. 1995. Enfouissement de matières organiques: conséquences agronomiques et environnementales [Ploughing in of organic matter: agronomical and environmental consequences]. *In* M. Geypens & J.P. Honnay, eds. *Matières organiques dans le sol: conséquences agronomiques et environnementales* [Soil organic matter: agronomical and environmental consequences], pp. 67–128. Brussels, IRSIA.

Oberdorfer, E. 1977. *Suddeutsche pflanzengesellschaften*. Teil I [Plant communities of South Germany. Volume I]. Stuttgart, Germany, Gustav Fischer Verlag. 311 pp.

Oberdorfer, E. 1978. *Suddeutsche pflanzengesellschaften*. Teil II [Plant communities of South Germany. Volume II]. Stuttgart, Germany, Gustav Fischer Verlag. 355 pp.

Oberdorfer, E. 1983. *Suddeutsche pflanzengesellschaften*. Teil III [Plant communities of South Germany. Volume III]. Stuttgart, Germany, Gustav Fischer Verlag. 455 pp.

Oberdorfer, E. 2001. *Pflanzensoziologische exkursionsflora für Deutschland und angrenzende gebiete* [Plant sociology excursion flora for Germany and neighbouring regions]. Stuttgart, Germany, Verlag Eugen Ulmer. 1051 pp.

O'Rourke, C.J. 1967. Grass rusts. *Plant Pathol.*, 36: 455–461.

O'Rourke, C.J. 1976. *Diseases of grasses and forage legumes in Ireland*. Dublin, An Foras Taluntais. 401 pp.

Ovington, J.D. & Scurfield, G. 1956. *Holcus mollis* L. *J. Ecol.*, 44: 272–280.

Ozenda, P. 1985. *La végétation de la chaîne alpine dans l'espace montagnard européen*. [The vegetation of the alpine range in the European mountain area]. Paris, Masson. 331 pp.

Ozenda, P. 1994. *Végétation du continent européen* [Vegetation of the European continent]. Neuchatel, Switzerland, Delachaux & Niestlé. 271 pp.

Palmer, J.H. & Sagar, G.R. 1963. *Agropyron repens* (L.) Beauv. (*Triticum repens* L.; *Elytrigia repens* (L.) Nevski). *J. Ecol.*, 51: 783–794.

Parente, G. 1996. Grassland and land use systems. *In* G. Parente, J. Frame & S. Orsi, eds. *Grassland and land use systems*. Proceedings of the 16th General Meeting of the European Grassland Federation, Italy, pp. 23–34.

Peeters, A. 1989. *Techniques d'exploitation, végétation et qualité alimentaire de l'herbe: étude de leurs relations triangulaires dans les systèmes herbagers* [Management, vegetation and feeding quality of grasslands: study of their triangular relationships in forage systems]. Université catholique de Louvain, Louvain la Neuve, Belgium. (PhD thesis)

Peeters, A. 1992. *Potential and quality of secondary grasses as the basis of sustainable grassland management*. Proceedings of the 14th Meeting of the European Grassland Federation, Finland, pp. 522–524.

Peeters, A. & Decamps, C. 1994. *Rendement et qualité de graminées secondaires comparées au ray-grass anglais et à la fléole (1991–1993)* [Yield and quality of secondary grasses compared to perennial ryegrass and timothy (1991–1993)].

Rapport d'essais. Laboratoire d'Ecologie des Prairies, Université catholique de Louvain, Louvain la Neuve, Belgium. 20 pp.

Peeters, A. & Decamps, C. 1998. *Choix et gestion de couverts herbacés dans les jachères et les tournières faunistiques* [Choice and management of herbaceous covers in set-asides and field margins for wildlife]. Proceedings of the XXIIIrd Congress of the International Union of Game Biologists. *Gibier Faune Sauvage – Game and Wildlife*, 15(1): 117–129.

Peeters, A. & Decamps, C. 1999. Rendement et qualité de graminées secondaires comparées au ray-grass anglais et à la fléole, selon 3 régimes de coupes et 4 niveaux de fertilisation azotée (1996–1998) [Yield and quality of secondary grasses compared with perennial ryegrass and timothy, in 3 cutting regimes and 4 nitrogen fertilization rates (1996–1998)]. Rapport d'essais. Laboratoire d'Ecologie des Prairies, Université catholique de Louvain, Louvain la Neuve, Belgium. 30 pp.

Peeters, A., Decamps, C. & Janssens, F. 1999. Rendement et proportion de graminées secondaires en mélange avec le ray-grass anglais ou la fléole et exploités selon 3 régimes de coupes et à 4 niveaux de fertilisation azotée (1996–1998) [Yield and proportion of secondary grasses mixed with perennial ryegrass or timothy, in 3 cutting regimes and 4 nitrogen fertilization rates (1996–1998)]. Rapport d'essais. Laboratoire d'Ecologie des Prairies, Université catholique de Louvain, Louvain la Neuve, Belgium. 28 pp.

Peeters, A. & Kopec, C. 1996. Production and productivity of cutting grasslands in temperate climates of Europe. *In* G. Parente, J. Frame & S. Orsi, eds. *Grassland and land use systems*, pp. 59–73, Proceedings of the 16th European Grassland Federation, Italy.

Peeters, A., Moens, A., Hendrickx, C. & Lambert, J. 1991. Intérêt fourrager du chiendent *Elymus repens*. Données de l'Ardenne belge [Forage value of couch grass *Elymus repens*. Data from Belgian Ardenne]. *Fourrages*, 126: 173 186.

Pfitzenmeyer, C.D.C. 1962. *Arrhenatherum elatius* (L.) J. & C. Presl (*A. avenaceum* Beauv.). *J. Ecol.*, 50: 235–245.

Poser, A. & Jochimsen, M. 1989. Vegetationskundliche analyse einer landespfegerish begrunten [Vegetation analysis of a managed plot]. *Bergehalte Gesellshaft für Okologie*, 18: 93–99.

Potter, L.R. 1987. Effect of crown rust on regrowth, competitive ability and nutritional quality of perennial and Italian ryegrasses. *Plant Pathol.*, 36: 455–461.

Pousset, J. 2000. *Engrais verts et fertilité des sols* [Green manure and soil fertility]. Paris, Editions Agridécisions. 287 pp.

Pringle, W.L. & Kline, P.M. 1982. Effects of date of cutting on yield and quality of meadow foxtail. Annual Meeting of the Canadian Society of Agronomy, *Can. J. Plant Sci.*, 63: 347.

Pulli, S. 1976. Cellulase digestion technique compared with the *in vitro* digestibility of Forages. *J. Sci. Agr. Soc. Fin.*, 48: 187–194.

Pulli, S. 1983. Productivity and forage quality of couch grass (*Agropyron repens*). British Grassland Society, Occasional Symposium, No. 14: 263–267.

Ramade, F. 1992. *Eléments d'écologie. Ecologie appliquée* [Ecology elements. Applied ecology]. Paris, Ediscience International. 578 pp.

Rameau, J.C., Mansion, D., Dumé, G., Timbal, J., Lecointe, A., Dupont, P. & Keller, R. 1989. *Flore forestière française. 1. Plaines et Collines* [French forest flora. 1. Lowlands and hills]. Paris, Institut pour le développement forestier. 1785 pp.

Rameau, J.C., Mansion, D., Dumé, G., Lecointe, A., Timbal, J., Dupont, P. & Keller, R. 1993. *Flore forestière française. 2. Montagnes* [French forest flora. 2. Uplands]. Paris, Institut pour le développement forestier. 2421 pp.

Raynal, G. 1988. Du nouveau sur la quenouille du dactyle [News on cocksfoot choke]. *Phytoma*, 403: 39–41.

Raynal, G., Gondran, J., Bournoville, R. & Courtillot, M. 1989. *Ennemis et maladies des prairies* [Pests and diseases of grasslands]. Paris, INRA. 249 pp.

Reagan, J.O., Carpenter, J.A., Bauer, F.T. & Lowrey, R.S. 1977. Packaging and palatability characteristics of grass and grass-grain fed beef. *Anim. Sci.*, 46: 716–721.

Roberts, D.A., Sherwood, R.T., Fezer, K.D. & Ramamurthi, C.S. 1955. Diseases of forage crops in New York. *Plant Dis. Rep.*, 39: 316–317.

Rode, L.M. 1986. Inhibitory effect of meadow foxtail (*Alopecurus pratensis*) on the growth of steers. *Can. J. Anim. Sci.*, 66: 303–305.

Rode, L.M. & Pringle, W.L. 1986. Growth, digestibility and voluntary intake by yearling steers grazing timothy (*Phleum pratense*) or meadow foxtail (*Alopecurus pratensis*) pastures. *Can. J. Anim. Sci.*, 66: 463–472.

Rodwell, J.S., ed. 1992. *British plant communities.* Vol. 3. *Grasslands and montane communities.* Cambridge, UK, Cambridge University Press. 540 pp.

Rose, F. 1989. *Colour identification guide to the grasses, sedges, rushes and ferns of the British Isles and Northwestern Europe.* London, Viking, Penguin Group. 240 pp.

Salo, M.L., Nykanen, A. & Ormunen, R. 1975. Composition, pepsin-HCl solubility and *in vitro* digestibility of forages at different growth stages. *J. Sci. Agr. Soc. Fin.*, 47: 480–490.

Santilocchi, R. 1989. *Scenario Umbro-Marchigiano. Distribuzione della produzione dei pascoli in ambienti marginali italiani* [Umbro-Marchigiano scenario. Distribution of pasture production in an Italian marginal environment], pp. 128–132. Palermo, Italy, CNR-IPRA.

Schneeberger, H. 1987. La réalisation de performances élevées à partir du fourrage de base [Reaching high performances from roughage]. *Rev. Suisse Agr.*, 19(2): 63–69.

Scholefield, D., Lockyer, D.R., Whitehead, D.C. & Tyson, K.C. 1991. A model to predict transformations and losses of nitrogen in UK pastures grazed by beef cattle. *Plant Soil*, 132: 165–177.

Schröder, F. & Adolf, G. 1993. Der einfluss des schnittzeitpunktes im ersten aufwuchs und unterschiedlich langer nachwuschsdauer im zweiten aufwuchs auf bestandsentwicklung, ertrag und futterqualität der grasart *Alopecurus* *pratensis* L. in abhängigkeit vom standort [The influence of cutting height during the first growth cycle and different periods of regrowth during the second growth cycle on the persistency, yield and feeding quality of *Alopecurus pratensis* L. in relation with the environment]. *Zeitschrift das Wirtschaftseigene Futter*, 43(1): 83–86.

Schroeder, J.W., Cramer, D.A., Bowling, R.A. & Cook, C.W. 1980. Palatability, shelflife and chemical differences between forage- and grass-grain fed beef. *Anim. Sci.*, 50: 852–859.

Schultes, R.E., Hofmann, A. & Rätsch, C. 2000. *Les plantes des dieux. Pouvoirs magiques des plantes psychédéliques* [Gods' plants. Magic powers of psychedelic plants]. Paris, Les Editions du lézard. 208 pp.

Scurfield, G. 1954. *Deschampsia flexuosa* (L.) Trin. *J. Ecol.*, 42: 225–233.

Shantz, H.L. 1954. The place of grasslands in the earth's cover of vegetation. *Ecology*, 35: 142–145.

Sheaffer, C.C. & Marten, G.C. 1995. Reed canarygrass. *In* R.F. Barnes, D.A. Miller & C.J. Nelson, eds. *Forages*. I. *An introduction to grassland agriculture*, pp. 335–344. Ames, USA, Iowa State University Press.

Sheldrick, R.D., Lavender, R.H. & Martyn, T.M. 1990. Dry matter yield and response to nitrogen of an *Agrostis stolonifera*-dominant sward. *Grass Forage Sci.*, 45: 203–213.

Simmons, R.C., Gold, A.J. & Groffman, P.M. 1992. Nitrate dynamics in riparian forests: groundwater studies. *J. Environ. Qual.*, 21: 659–665.

Simon, J.C. & Le Corre, L. 1988. Lessivage d'azote en monoculture de maïs, en sol granitique du Finistère [Nitrogen leaching in maize monoculture, on granitic soil in Finistere (Britanny)]. *Fourrages*, 114: 193–207.

Simon, J.C., Le Corre, L. & Coppenet, M. 1983. Essai comparatif de dix graminées fourragères dont six cultivars de bromes dans le Finistère [Comparative trial of ten forage grasses including six brome cultivars in the Finistère]. *Fourrages*, 93: 85–108.

Simon, J.C., Peyraud, J.L., Decau, M.L.,

Delaby, L., Vertès, F. & Delagarde, R. 1997. Gestion de l'azote dans les systèmes prairiaux permanents ou de longue durée [Nitrogen management in permanent or long duration grassland systems]. *In* G. Lemaire & B. Nicolardot, eds. *Maîtrise de l'azote dans les agrosystèmes* [Nitrogen control in the agrosystems], pp. 201–216. Reims, France, Les Colloques No. 83.

Skipp, R.A. & Hampton, J.G. 1996. Fungal and bacterial diseases of pasture plants in New Zealand. *In* S. Chakraborty, K.T. Leath, R.A. Skipp, G.A. Pederson, R.A. Bray, G.C.M. Latch & F.W. Nutter Jr, eds. *Pasture and forage crop pathology*, pp. 213–236. Madison, USA, ASA, CSSA, SSSA.

Sleper, D.A. & Buckner, R.C. 1995. The fescues. *In* R.F. Barnes, D.A. Miller & C.J. Nelson, eds. *Forages. An introduction to grassland agriculture*, pp. 345–356. Ames, USA, Iowa State University Press.

Sletten, A. 1989. Detection methods for *Xanthomonas campestris* pv. *graminis* on forage grasses. *EPPO Bull.* , 19: 69–72.

Smith, A. & Allcock, P.J. 1985. The influence of species diversity on sward yield and quality. *J. Appl. Ecol.*, 22: 185–198.

Smith B.D. 1995. *The emergence of agriculture.* New York, USA, W.H. Freeman, Scientific American Library. 231 pp.

Smith, J.D. 1970. Resistance of timothy cultivars to *Heterosporium phlei*, *Drechslera phlei* and frost injury. *Can. Plant Dis. Surv.*, 50: 95–97.

Smith, J.D., Jackson, N. & Woolhouse, A.R. 1989. *Fungal diseases of amenity turf grasses.* London, E & F.N. Spon. 401 pp.

Société française des gazons. 1990. *L'encyclopédie des gazons* [The lawn encyclopedia]. Boulogne, France, S.E.P.S. 360 pp.

Sougnez, N. & Limbourg, P. 1963. Les herbages de la Famenne et de la Fagne [The grasslands of Famenne and Fagne]. *Bull. Inst. Agron. Stat. Rech.*, 3: 360–413.

Spedding, C.R.W. & Diekmahns, E.C. 1972. *Grasses and legumes in British agriculture.* Farnham Royal, UK, CAB. 509 pp.

Stace, C. 1997. *New flora of the British Isles* 2nd ed.

Cambridge, UK, Cambridge University Press. 1130 pp.

Stapledon, R.G. & Davies, W. 1948. *Ley farming*. London, Faber and Faber.

Stebbins, G.L. & Zohary, D. 1959. *Cytogenetic and evolutionary studies in the genus* Dactylis. I. *Morphology, distribution, and interrelationships of the diploid subspecies*. University of California, Publication of Botany, Berkeley 31 (1).

Steen, R. & Larsson, K. 1986. Carbohydrates in roots and rhizomes of perennial grasses. *New Phytol.*, 104: 339–346.

Straëbler, M. 1994. Quelles espèces semer au printemps sur jachère annuelle? [Which species to sow in spring on annual set-aside?]. *Cultivar*, 357: 61.

Stuedemann, J.A. & Hoveland, C.S. 1988. Fescue endophyte: history and impact on animal agriculture. *J. Prod. Agr.*, 1: 39–44.

Sydes, C. 1984. A comparative study of leaf demography in limestone grassland. *J. Ecol.*, 72(1): 331–345.

Talamucci, P. 1975. Potentialities of tall fescue (*Festuca arundinacea* Schreb.) in Tuscany (central Italy). In *Pasture and forage production in seasonally arid climates*. Proceedings of the 6th General Meeting of the European Grassland Federation, Spain, pp. 109–120.

Taylor, D.R. & Aarssen, L.W. 1988. An interpretation of phenotypic plasticity in *Agropyron repens* (*Gramineae*). *Am. J. Bot.*, 75(3): 401–413.

Taylor, K., Rowland, A.P. & Jones, H.E. 2001. *Molinia caerulea* (L.) Moench. *J. Ecol.*, 89: 126–144.

Thélier-Huché, L., Simon, J.C., Le Corre, L. & Salette, J. 1994. Valorisation, sur prairies et maïs, de la fertilisation organique et minérale. Etude sur le long terme [Use of organic and mineral fertilization on grassland and maize. Long term study]. *Fourrages*, 138: 145–155.

Thomas, J.E. 1991. Diseases of established grassland. In *Strategies for weed, disease and pest control in grassland*. Proceedings of British Grassland Society, Conference, Gloucester, UK. Chap. 3, pp. 1–12.

Thomas, J.G. & Morris, R.M. 1973. Seasonal patterns of digestible organic matter and protein production from grasses in the North Pennines. *J. Brit. Grass. Soc.*, 28: 31–40.

Thomsen, I.K. & Christensen, B.T. 1999. Nitrogen conserving potential of successive ryegrass catch crops in continuous spring barley. *Soil Use Man.*, 15: 195–200.

Thomson, D.J. 1984. The nutritive value of white clover. In D.J. Thomson, ed. *Forage legumes*, pp. 78–92. British Grassland Society, Occasional Symposium No. 16.

Tingle, J.N. & Elliot, C.R. 1975. Forage yield and quality of cultivated perennial grasses harvested at the early heading stage. *Can. J. Plant Sci.*, 55: 271–278.

Tingle, J.N. & van Adrichem, M.C.J. 1974. Meadow foxtail lengthens the spring growing season. *Can. Agri.*, 19(2): 26.

't Mannetje, L. 1994. Towards sustainable grassland management in The Netherlands. In L. 't Mannetje & J. Frame, eds. *Grassland and society*. Proceedings of the 15th General Meeting of the European Grassland Federation, The Netherlands, pp. 3–18.

Toussaint, B. & Lambert, J. 1973. Etude comparative des réactions variétales de 5 graminées de prairie temporaire en fonction de conditions écologiques bien précises [Comparative study of the variety performances of 5 ley grasses in relation to well defined ecological conditions]. *Rev. Agr.*, 26(3): 575–593.

Troxler, J., Jans, F. & Floch, C. 1990. Utilisation et entretien des zones marginales sèches par la pâture des ovins et des vaches allaitantes. II. Influence sur la végétation [Use and maintenance of marginal dry areas by sheep and suckler cow grazing. II. Influence on vegetation]. *Rev. Suisse Agr.*, 22: 231–238.

Tutin, T.G., Heywood, V.H., Burges, N.A., Moore, D.M., Valentine, D.H., Walters, S.M. & Webb, D.A. 1980. *Flora Europaea*. Vol. 5. Cambridge, UK, Cambridge University Press. 452 pp.

van Adrichem, M.C.J. & Tingle, J.N. 1975. Effects of nitrogen and phosphorus on the yield and chemical composition of meadow foxtail. *Can. J. Plant Sci.*, 55: 949–954.

Vanbellinghen, C., Moreau V. & Maraite, H. 1991. Comportement variétal du ray-grass anglais (*Lolium perenne* L.) à la rouille et aux helminthosporioses en Haute et Moyenne Belgique [Variety sensitivity of perennial ryegrass (*Lolium perenne* L.) to rust and *Helminthosporium* diseases in High and Mid-Belgium]. *Parasitica*, 55,(2–3): 105–118.

Van der Meijden, R. 1996. *Heukels' flora van Nederland* [Heukels' flora of the Netherlands]. Groningen, The Netherlands, Wolters-Noordhoff. 678 pp.

Van Dijk, P.M., van der Zijp, M. & Kwaad, F.J.P.M. 1996. Soil erodibility parameters under various cropping systems of maize. *Hydrol. Process*, 10(8): 1061–1067.

Van Veen, J.A. & Paul, E.A. 1981. Organic carbon dynamics in grassland soils. I. Background information and mathematical simulation. *Can. J. Soil Sci.*, 61: 185–201.

Vavassori, A. 2000. *Pelouses et gazons* [Lawns and swards]. Paris, De Vecchi. 48 pp.

Vertès, F. & Decau, M.L. 1992. Suivi d'azote minéral dans les sols: risque de lessivage de nitrate selon le couvert végétal [Follow up of soil mineral nitrogen: nitrate leaching risk according the vegetal cover]. *Fourrages*, 129: 11–28.

Viaux, P., Bodet, J.M. & Le Gall, A. 1999. Complémentarité herbe-culture dans les rotations [Grass-crop complementarity in the rotations]. *Fourrages*, 160: 345–358.

Vivier, M. 1972. Potentialité fourragère et fertilisation azotée des prairies. L'exemple du Calvados [Forage potential and nitrogen fertilization of grasslands. The example of Calvados]. *Fourrages*, 50: 43–81.

Voroney, R.P., van Veen, J.A. & Paul, E.A. 1981. Organic dynamics in grassland soils. 2. Model validation and simulation of the long-term effects of cultivation and rainfall erosion. *Can. J. Soil Sci.*, 61: 211–224.

Vought, L.B.M., Pinay, G., Fuglsang, A. & Ruffinoni, C. 1995. Structure and function of buffer strips from a water quality perspective in agricultural landscapes. *Landsc. Urban Plan.*, 31 (1–3): 323–331.

Watkin, B.R. & Robinson, G.S. 1974. *Dry matter production of 'Massey Basyn' Yorkshire fog* (Holcus lanatus). Proceedings of the New Zealand Grassland Association, 35: 278–283.

Watson, L. & Dallwitz, M.J. 1994. *The grass genera of the world*. Wallingford, UK, CAB. 1081 pp.

Watt, T.A. 1978. The biology of *Holcus lanatus* L. (Yorkshire fog) and its significance in grassland. *Herb. Abs.*, 48: 195–204.

Watt, T.A. 1987. A comparison of two cultivars of *Holcus lanatus* with *Lolium perenne*, under cutting. *Grass Forage Sci.*, 42: 43–48.

Watt, T.A. & Haggar, R.J. 1980. The effect of defoliation upon yield, flowering and vegetative spread of *Holcus lanatus* growing with and without *Lolium perenne*. *Grass Forage Sci.*, 35: 227–234.

Young, A.W. & Kauffman, R.G. 1978. Evaluation of beef from steers fed grain, corn silage or haylage-corn silage diets. *Anim. Sci.*, 46: 41.

Zinkernagel, G. 1995. *La beauté des graminées* [The beauty of grasses]. Stuttgart, Germany, Editions Eugen Ulmer. 95 pp.

Zohary, D. 1986. The origin and early spread of agriculture in the old world. *In* C. Barigozzi, ed. *The origin and domestication of cultivated plants*. Amsterdam, Elsevier.

Index

Index

Sales and Marketing Group, Information Division, FAO
Viale delle Terme di Caracalla, 00100 Rome, Italy
Tel.: (+39) 06 57051 – Fax: (+39) 06 57053360
E-mail: publications-sales@fao.org
www.fao.org/catalog/giphome.htm

أماكن بيع مطبوعات المنظمة
当地何处可以购买粮农组织出版物
WHERE TO PURCHASE FAO PUBLICATIONS LOCALLY
POINTS DE VENTE DES PUBLICATIONS DE LA FAO
PUNTOS DE VENTA DE PUBLICACIONES DE LA FAO

06/04

• ANGOLA
Empresa Nacional do Disco e de
Publicações, ENDIPU-U.E.E.
Rua Cirilo da Conceição Silva, N° 7
C.P. N° 1314-C, Luanda

• ARGENTINA
Librería Hemisferio Sur
Pasteur 743, 1028 Buenos Aires
Correo electrónico: adolfop@hemisferi
osur.com.ar
World Publications S.A.
Av. Córdoba 1877, 1120 Buenos Aires
Tel./Fax: (+54) 11 48158156

• AUSTRALIA
Hunter Publications (Tek Imaging
Pty. Ltd)
PO Box 404, Abbotsford, Vic. 3067
Tel.: (+61) 3 9417 5361
Fax: (+61) 3 9419 7154
E-mail: admin@tekimaging.com.au

• AUSTRIA
Uno Verlag
Am Hofgarten, 10
D-53113 Bonn
Tel.: (+49) 228 94 90 20
Fax: (+49) 228 94 90 222
E-mail: info@uno-verlag.de
Web site: www.uno-verlag.de

• BELGIQUE
M.J. De Lannoy
202, avenue du Roi, B-1060 Bruxelles
CCP: 000-0808993-13
Courriel: jean.de.lannoy@infoboard.be

• BOLIVIA
Los Amigos del Libro
Av. Heroínas 311, Casilla 450
Cochabamba;
Mercado 1315, La Paz
Correo electrónico:
gutten@amigol.bo.net

• BOTSWANA
Botsalo Books (Pty) Ltd
PO Box 1532, Gaborone
Tel.: (+267) 312576
Fax: (+ 267) 372608
E-mail: botsalo@botsnet.bw

• BRAZIL
Fundação Getúlio Vargas
Praia do Botafogo 190, C.P. 9052
Rio de Janeiro
Correio eletrônico: livraria@fgv.br

• CANADA
Renouf Publishing
5369 chemin Canotek Road, Unit 1
Ottawa, Ontario K1J 9J3
Tel.: (+1) 613 745 2665
Fax: (+1) 613 745 7660
E-mail: order.dept@renoufbooks.com
Web site: www.renoufbooks.com

• CHILE
Librería - Marta Caballero
c/o FAO, Oficina Regional para América
Latina y el Caribe (RLC)
Avda. Dag Hammarskjold, 3241
Vitacura, Santiago
Tel.: (+56) 2 33 72 314
Correo electrónico:
german.rojas@field.fao.org
Correo electrónico:
caballerocastillo@hotmail.com

• CHINA
China National Publications
Import & Export Corporation
16 Gongti East Road, Beijing 100020
Tel.: (+86) 10 6506 3070
Fax: (+86) 10 6506 3101
E-mail: serials@cnpiec.com.cn

• COLOMBIA
INFOENLACE LTDA
Cra. 15 No. 86A–31
Santafé de Bogotá
Tel.: (+57) 1 6009474-6009480
Fax: (+57) 1 6180195
Correo electrónico:
servicliente@infoenlace.com.co
Sitio Web: www.infoenlace.com.co

• CONGO
Office national des librairies
populaires
B.P. 577, Brazzaville

• COSTA RICA
Librería Lehmann S.A.
Av. Central, Apartado 10011
1000 San José
Correo electrónico:
llehmann@solracsa.co.cr

• CÔTE D'IVOIRE
CEDA
04 B.P. 541, Abidjan 04
Tél.: (+225) 22 20 55
Télécopie: (+225) 21 72 62

• CUBA
Ediciones Cubanas
Empresa de Comercio Exterior
de Publicaciones
Obispo 461, Apartado 605, La Habana

• CZECH REPUBLIC
Myris Trade Ltd
V Stinhlach 1311/3, PO Box 2
142 01 Prague 4
Tel.: (+420) 2 34035200
Fax: (+420) 2 34035207
E-mail: myris@myris.cz
Web site: www.myris.cz

• DENMARK
Gad Import Booksellers
c/o Gad Direct
31-33 Fiolstraede
DK-1171 Copenhagen K
Tel.: (+45) 3313 7233
Fax: (+45) 3254 2368
E-mail: info@gaddirect.dk

• ECUADOR
Libri Mundi, Librería Internacional
Juan León Mera 851
Apartado Postal 3029, Quito
Correo electrónico:
librimu1@librimundi.com.ec
Web site: www.librimundi.com
Universidad Agraria del Ecuador
Centro de Información Agraria
Av. 23 de julio, Apartado 09-01-1248
Guayaquil
Librería Española
Murgeón 364 y Ulloa, Quito

• EGYPT
MERIC The Middle East Readers'
Information Centre
2 Baghat Aly Street, Appt. 24
El Masry Tower D
Cairo/Zamalek
Tel.: (+20) 2 3413824/34038818
Fax: (+20) 2 3419355
E-mail: info@mericonline.com

• ESPAÑA
Librería Agrícola
Fernando VI 2, 28004 Madrid
Librería de la Generalitat
de Catalunya
Rambla dels Estudis 118 (Palau Moja)
08002 Barcelona
Tel.: (+34) 93 302 6462
Fax: (+34) 93 302 1299
Mundi Prensa Libros S.A.
Castelló 37, 28001 Madrid
Tel.: +34 91 436 37 00
Fax: +34 91 575 39 98
Sitio Web: www.mundiprensa.com
Correo electrónico:
libreria@mundiprensa.es
Mundi Prensa - Barcelona
Consejo de Ciento 391
08009 Barcelona
Tel.: (+34) 93 488 34 92
Fax: (+34) 93 487 76 59

• FINLAND
Akateeminen Kirjakauppa
PL 23, 00381 Helsinki
(Myymälä/Shop: Keskuskatu 1
00100 Helsinki)
Tel.: (+358) 9 121 4385
Fax: (+358) 9 121 4450
E-mail: akatilaus@akateeminen.com
Web site: www.akateeminen.com/
suurasiakkaat/palvelut.htm

• FRANCE
Lavoisier Tec & Doc
14, rue de Provigny
94236 Cachan Cedex
Courriel: livres@lavoisier.fr
Site Web: www.lavoisier.fr
Librairie du commerce international
10, avenue d'Iéna
75783 Paris Cedex 16
Courriel: librarie@cfce.fr
Site Web: www.planetexport.fr

• GERMANY
Alexander Horn Internationale
Buchhandlung
Friedrichstrasse 34
D-65185 Wiesbaden
Tel.: (+49) 611 9923540/9923541
Fax: (+49) 611 9923543
E-mail: alexhorn1@aol.com
TRIOPS - Tropical Scientific Books
S. Toeche-Mittler
Versandbuchhandlung GmbH
Hindenburstr. 33
D-64295 Darmstadt
Tel.: (+49) 6151 336 65
Fax: (+49) 6151 314 048
E-mail for orders: orders@net-library.de
E-mail for info.: info@net-library.de /
triops@triops.de
Web site: www.net-library.de /
www.triops.de
Uno Verlag
Am Hofgarten, 10
D-53113 Bonn
Tel.: (+49) 228 94 90 20
Fax: (+49) 228 94 90 222
E-mail: info@uno-verlag.de
Web site: www.uno-verlag.de

• GHANA
Readwide Bookshop Ltd
PO Box 0600 Osu, Accra
Tel.: (+233) 21 22 1387
Fax: (+233) 21 66 3347
E-mail: readwide@africaonline.cpm.gh

• GREECE
Librairie Kauffmann SA
28, rue Stadiou, 10564 Athens
Tel.: (+30) 1 3236817
Fax: (+30) 1 3230320
E-mail: ord@otenet.gr

• GUYANA
Guyana National Trading
Corporation Ltd
45-47 Water Street, PO Box 308
Georgetown

• HONDURAS
Escuela Agrícola Panamericana
Librería RTAC
El Zamorano, Apartado 93, Tegucigalpa
Correo electrónico:
libreriazam@zamorano.edu.hn

• HUNGARY
Librotrade Kft.
PO Box 126, H-1656 Budapest
Tel.: (+36) 1 256 1672
Fax: (+36) 1 256 8727

• INDIA
Allied Publisher Ltd
751 Mount Road
Chennai 600 002
Tel.: (+91) 44 8523938/8523984
Fax: (+91) 44 8520649
E-mail: allied.mds@smb.sprintrpg.
ems.vsnl.net.in

EWP Affiliated East-West
Press PVT, Ltd
G-I/16, Ansari Road, Darya Gany
New Delhi 110 002
Tel.: (+91) 11 3264 180
Fax: (+91) 11 3260 358
E-mail: affiliat@nda.vsnl.net.in
Monitor Information Services
203, Moghal Marc Ratan Complex
Narayanguda
Hyderabad – 500029, Andhra Pradesh
Tel.: (+91) 40 55787065
Fax: (+91) 40 27552290
E-mail: helpdesk@monitorinfo.com
Web site: www.monitorinfo.com
M/S ResearchCo Book Centre
25-B/2, New Rohtak Road (near Liberty
Cinema), New Delhi 110 005
Tel.: (+91) 11 551 50445
Fax: (+91) 11 287 16134
E-mail: researchco@dishnetdsl.net
Oxford Book and Stationery Co.
Scindia House
New Delhi 110001
Tel.: (+91) 11 3315310
Fax: (+91) 11 3713275
E-mail: oxford@vsnl.com
Periodical Expert Book Agency
G-56, 2nd Floor, Laxmi Nagar
Vikas Marg, Delhi 110092
Tel.: (+91) 11 2215045/2150534
Fax: (+91) 11 2418599
E-mail: pebe@vsnl.net.in
Bookwell
Head Office:
2/72, Nirankari Colony, New Delhi 10009
Tel.: (+91) 11 725 1283
Fax: (+91) 11 328 13 15
Sales Office:
24/4800, Ansari Road
Darya Ganj, New Delhi - 110002
Tel.: (+91) 11 326 8786
E-mail: bkwell@nde.vsnl.net.in

• INDONESIA
P.F. Book
Jl. Setia Budhi No. 274, Bandung 40143
Tel.: (+62) 22 201 1149
Fax: (+62) 22 201 2840
E-mail: pfbook@bandung.wasantara
.net.id

• IRAN
The FAO Bureau, International
and Regional Specialized
Organizations Affairs
Ministry of Agriculture of the Islamic
Republic of Iran
Keshavarz Bld, M.O.A., 17th oor
Teheran

• ITALY
FAO Bookshop
Viale delle Terme di Caracalla
00100 Roma
Tel.: (+39) 06 57053597
Fax: (+39) 06 57053360
E-mail: publications-sales@fao.org
Il Mare Libreria Internationale
Via di Ripetta 239
00186 Roma
Tel.: (+39) 06 3612155
Fax: (+39) 06 3612091
E-mail: ilmare@ilmare.com
Web site: www.ilmare.com
Libreria Commissionaria Sansoni
S.p.A. - Licosa
Via Duca di Calabria 1/1
50125 Firenze
Tel.: (+39) 55 64831
Fax: (+39) 55 64 2 57
E-mail: licosa@ftbcc.it
Libreria Scientifica "AEIOU"
dott. Lucio de Biasio
Via Coronelli 6, 20146 Milano
Tel.: (+39) 02 48954552
Fax: (+39) 02 48954548
E-mail: in@aeioulib.com or
commerciale@aeioulib.com

Sales and Marketing Group, Information Division, FAO
Viale delle Terme di Caracalla, 00100 Rome, Italy
Tel.: (+39) 06 57051 – Fax: (+39) 06 57053360
E-mail: publications-sales@fao.org
www.fao.org/catalog/giphome.htm

أماكن بيع مطبوعات المنظمـة
当地何处可以购买粮农组织出版物
**WHERE TO PURCHASE FAO PUBLICATIONS LOCALLY
POINTS DE VENTE DES PUBLICATIONS DE LA FAO
PUNTOS DE VENTA DE PUBLICACIONES DE LA FAO**

06/04

• **JAPAN**
**Far Eastern Booksellers
(Kyokuto Shoten Ltd)**
12 Kanda-Jimbocho 2 chome
Chiyoda-ku - PO Box 72
Tokyo 101-91
Tel.: (+81) 3 3265 7531
Fax: (+81) 3 3265 4656
Maruzen Company Ltd
5-7-1 Heiwajima, Ohta-Ku
Tokyo 143-0006
Tel.: (+81) 3 3763 2259
Fax: (+81) 3 3763 2830
E-mail: o_miyakawa@maruzen.co.jp

• **KENYA**
Text Book Centre Ltd
Kijabe Street
PO Box 47540, Nairobi
Tel.: +254 2 330 342
Fax: +254 2 22 57 79
Legacy Books
Mezzanine 1, Loita House, Loita Street
Nairobi, PO Box 68077
Tel.: (+254) 2 303853
Fax: (+254) 2 330854
E-mail: info@legacybookshop.com

• **LUXEMBOURG**
M.J. De Lannoy
202, avenue du Roi
B-1060, Bruxelles (Belgique)
Courriel: jean.de.lannoy@infoboard.be

• **MADAGASCAR**
**Centre d'Information et de
Documentation Scientifique et
Technique**
Ministère de la recherche appliquée
au développement
B.P. 6224, Tsimbazaza, Antananarivo

• **MALAYSIA**
MDC Publishers Printers Sdn Bhd
MDC Building
2717 & 2718, Jalan Parmata Empat
Taman Permata, Ulu Kelang
53300 Kuala Lumpur
Tel.: (+60) 3 41086600
Fax: (+60) 3 41081506
E-mail: inquiries@mdcbd.com.my
Web site: www.mdcppd.com.my

• **MAROC**
La Librairie Internationale
70, rue T'ssoule
B.P. 302 (RP), Rabat
Tél.: (+212) 37 75 0183
Fax: (+212) 37 75 8661

• **MÉXICO**
**Librería, Universidad Autónoma de
Chapingo**
56230 Chapingo
Libros y Editoriales S.A.
Av. Progreso Nº 202-1º Piso A
Apartado Postal 18922
Col. Escandón, 11800 México D.F.
Correo electrónico: lyesa99@mail.com/
ventas@lyesa.com
Mundi Prensa Mexico, S.A.
Río Pánuco, 141 Col. Cuauhtémoc
C.P. 06500, México, DF
Tel.: (+52) 5 533 56 58
Fax: (+52) 5 514 67 99
Correo electrónico:
resavbp@data.net.mx

• **NETHERLANDS**
Roodveldt Import b.v.
Brouwersgracht 288
1013 HG Amsterdam
Tel.: (+31) 20 622 80 35
Fax: (+31) 20 625 54 93
E-mail: roodboek@euronet.nl
Swets & Zeitlinger b.v.
PO Box 830, 2160 Lisse
Heereweg 347 B, 2161 CA Lisse
E-mail: infono@swets.nl
Web site: www.swets.nl

• **NEW ZEALAND**
Legislation Direct
c/o Securacopy, PO Box 12 418
1st oor, 242 Thorndon Quay, Wellington
Tel.: (+64) 4 496 56 94
Fax: (+64) 4 496 56 98

E-mail: Jeanette@legislationdirect.co.nz
Web site: www.gplegislation.co.nz

• **NICARAGUA**
Librería HISPAMER
Costado Este Univ. Centroamericana
Apartado Postal A-221, Managua
Correo electrónico:
hispamer@munditel.com.ni

• **NIGERIA**
University Bookshop (Nigeria) Ltd
University of Ibadan, Ibadan

• **PAKISTAN**
Mirza Book Agency
65 Shahrah-e-Quaid-e-Azam
PO Box 729, Lahore 3

• **PARAGUAY**
**Librería Intercontinental
Editora e Impresora S.R.L.**
Caballero 270 c/Mcal Estigarribia
Asunción

• **PERU**
**Librería de la Biblioteca Agrícola
Nacional - Universidad Nacional
Agraria**
Av. La Universidad s/n
La Molina, Lima
Tel.: (+51) 1 3493910
Fax: (+51) 1 3493910
Correo electrónico: an@lamolina.edu.pe
Web site: http://tumi.lamolina.edu.pe/
ban.htm

• **PHILIPPINES**
**International Booksource Center,
Inc.**
1127-A Antipolo St, Barangay Valenzuela
Makati City
Tel.: (+63) 2 8966501/8966505/8966507
Fax: (+63) 2 8966497
E-mail: ibcdina@pacific.net.ph

• **POLAND**
Ars Polona S.A.
ul. Obronców 25
03-933 Warsaw
Tel.: (+48) 22 8261201
Fax: (+48) 22 8266240
E-mail: arspolona@arspolona.com.pl
Web site: http://www.arspolona.com.pl

• **PORTUGAL**
**Livraria Portugal, Dias e Andrade
Ltda.**
Rua do Carmo, 70-74
Apartado 2681, 1200 Lisboa Codex
Correio electrónico: liv.portugal@mail.
telepac.pt

• **REPÚBLICA DOMINICANA**
CEDAF - Centro para el Desarrollo
Agropecuario y Forestal, Inc.
Calle José Amado Soler, 50 - Urban.
Paraíso
Apartado Postal, 567-2, Santo Domingo
Tel.: (+001) 809 5440616/5440634/
5655603
Fax: (+001) 809 5444727/5676989
Correo electrónico: fda@Codetel.net.do
Web site: www.cedaf.org.do

• **RUSSIAN FEDERATION**
tsdatelstovo VES MIR
9a, Kolpachniy pereulok
101831 Moscow
Tel.: (+7) 095 9236839/9238568
Fax: (+7) 095 9254269
E-mail: orders@vesmirbooks.ru
Web site: www.vesmirbooks.ru

• **SERBIA AND MONTENEGRO**
Jugoslovenska Knjiga DD
Terazije 27
POB 36, 11000 Beograd
Tel.: (+381) 11 3340 025
Fax: (+381) 11 3231 079
E-mail: juknjiga@eunet.yu or
babicmius@yahoo.com

• **SINGAPORE**
Select Books Pte Ltd
Tanglin Shopping Centre
19 Tanglin Road, #03-15,
Singapore 247909
Tel.: (+65) 732 1515
Fax: (+65) 736 0855
E-mail: info@selectbooks.com.sg
Web site: www.selectbooks.com.sg

• **SLOVAK REPUBLIC**
**Institute of Scientific and Technical
Information for Agriculture**
Samova 9, 950 10 Nitra
Tel.: (+421) 87 522 185
Fax: (+421) 87 525 275
E-mail: uvtip@nr.sanet.sk

• **SOUTH AFRICA**
Preasidium Books (Pty) Ltd
810 - 4th Street, Wynberg 2090
Tel.: (+27) 11 88 75994
Fax: (+27) 11 88 78138
E-mail: pbooks@global.co.za

• **SUISSE**
UN Bookshop
Palais des Nations
CH-1211 Genève 1
Site Web: www.un.org
Adeco - Editions Van Diermen
Chemin du Lacuez, 41
CH-1807 Blonay
Tel.: (+41) (0) 21 943 2673
Fax: (+41) (0) 21 943 3605
E-mail: mvandier@ip-worldcom.ch
Münstergass Buchhandlung
Docudisp, PO Box 584
CH-3000 Berne 8
Tel.: (+41) 31 310 2321
Fax: (+41) 31 310 2324
E-mail: docudisp@muenstergass.ch
Web site: www.docudisp.ch

• **SURINAME**
Vaco n.v. in Suriname
Domineestraat 26, PO Box 1841
Paramaribo

• **SWEDEN**
Swets Blackwell AB
PO Box 1305, S-171 25 Solna
Tel.: (+46) 8 705 9750
Fax: (+46) 8 27 00 71
E-mail: awahlquist@se.swetsblackw
ell.com
Web site: www.swetsblackwell.com/se/
Bokdistributören
c/o Longus Books Import
PO Box 610, S-151 27 Södertälje
Tel.: (+46) 8 55 09 49 70
Fax: (+46) 8 55 01 76 10; E-mail:
lis.ledin@hk.akademibokhandeln.se

• **THAILAND**
Suksapan Panit
Mansion 9, Rajdamnern Avenue,
Bangkok

• **TOGO**
Librairie du Bon Pasteur
B.P. 1164, Lomé

• **TRINIDAD AND TOBAGO**
Systematics Studies Limited
St Augustine Shopping Centre
Eastern Main Road, St Augustine
Tel.: (+001) 868 645 8466
Fax: (+001) 868 645 8467
E-mail: tobe@trinidad.net

• **TURKEY**
DUNYA ACTUEL A.S.
"Globus" Dunya Basinevi,
100. Yil Mahallesi
34440 Bagcilar, Istanbul
Tel.: (+90) 212 629 0808
Fax: (+90) 212 629 4689
E-mail: aktuel.info@dunya.comr
Web site: www.dunyagazetesi.com.tr/

• **UNITED ARAB EMIRATES**
Al Rawdha Bookshop
PO Box 5027, Sharjah
Tel.: (+971) 6 538 7933
Fax: (+971) 6 538 4473
E-mail: alrawdha@hotmail.com

• **UNITED KINGDOM**
The Stationery Office
51 Nine Elms Lane
London SW8 5DR
Tel.: (+44) (0) 870 600 5522 (orders)
        (+44) (0) 207 873 8372 (enquiries)
Fax: (+44) (0) 870 600 5533 (orders)
        (+44) (0) 207 873 8247 (enquiries)
E-mail: ipa.enquiries@theso.co.uk
Web site: www.clickso.com
**and through The Stationery Office
Bookshops**
E-mail: postmaster@theso.co.uk
Web site: www.the-stationery-office.co.uk
Intermediate Technology Bookshop
103-105 Southampton Row
London WC1B 4HH
Tel.: (+44) 207 436 9761
Fax: (+44) 207 436 2013
E-mail: orders@itpubs.org.uk
Web site: www.developmentbookshop.com

• **UNITED STATES**
Publications:
BERNAN Associates (ex UNIPUB)
4611/F Assembly Drive
Lanham, MD 20706-4391
Toll-free: (+1) 800 274 4447
Fax: (+1) 800 865 3450
E-mail: query@bernan.com
Web site: www.bernan.com
United Nations Publications
Two UN Plaza, Room DC2-853
New York, NY 10017
Tel.: (+1) 212 963 8302/800 253 9646
Fax: (+1) 212 963 3489
E-mail: publications@un.org
Web site: www.unog.ch
UN Bookshop (direct sales)
The United Nations Bookshop
General Assembly Building Room 32
New York, NY 10017
Tel.: (+1) 212 963 7680
Fax: (+1) 212 963 4910
E-mail: bookshop@un.org
Web site: www.un.org
Periodicals:
Ebsco Subscription Services
PO Box 1943
Birmingham, AL 35201-1943
Tel.: (+1) 205 991 6600
Fax: (+1) 205 991 1449
The Faxon Company Inc.
15 Southwest Park
Westwood, MA 02090
Tel.: (+1) 617 329 3350
Telex: 95 1980
Cable: FW Faxon Wood

• **URUGUAY**
Librería Agropecuaria S.R.L.
Buenos Aires 335, Casilla 1755
Montevideo C.P. 11000

• **VENEZUELA**
Tecni-Ciencia Libros
CCCT Nivel C-2
Caracas
Tel.: (+58) 2 959 4747
Fax: (+58) 2 959 5636
Correo electrónico: tclibros@attglobal.net
Fudeco, Librería
Avenida Libertador-Este
Ed. Fudeco, Apartado 254
Barquisimeto C.P. 3002, Ed. Lara
Tel.: (+58) 51 538 022
Fax: (+58) 51 544 394
Librería FAGRO
Universidad Central de Venezuela (UCV)
Maracay

• **YUGOSLAVIA**
See Serbia and Montenegro

• **ZIMBABWE**
Prestige Books
The Book Café
Fife Avenue Shops
Harare
Tel.: (+263) 4 336298/336301
Fax: (+263) 4 335105
E-mail: books@prestigebooks.co.zw

**FOOD AND AGRICULTURE ORGANIZATION
OF THE UNITED NATIONS**
Viale delle Terme di Caracalla
00100 Rome, Italy
Tel.: (+39) 06 57051
Fax: (+39) 06 57053152
Internet: www.fao.org

BLACKWELL PUBLISHING LTD
9600 Garsington Road, Oxford OX4 2DQ, UK
Tel.: +44 (0) 1865 776868
BLACKWELL PUBLISHING PROFESSIONAL
2121 State Avenue, Ames, Iowa, 50014, USA
Tel.:+1 515 292 0140
BLACKWELL PUBLISHING ASIA
550 Swanston Street, Carlton, Victoria 3053, Australia
Tel.: +61 (0) 3 8359 1011
Internet: www.blackwellpublishing.com